Sustainable Development Goals Series

The **Sustainable Development Goals Series** is Springer Nature's inaugural cross-imprint book series that addresses and supports the United Nations' seventeen Sustainable Development Goals. The series fosters comprehensive research focused on these global targets and endeavours to address some of society's greatest grand challenges. The SDGs are inherently multidisciplinary, and they bring people working across different fields together and working towards a common goal. In this spirit, the Sustainable Development Goals series is the first at Springer Nature to publish books under both the Springer and Palgrave Macmillan imprints, bringing the strengths of our imprints together.

The Sustainable Development Goals Series is organized into eighteen subseries: one subseries based around each of the seventeen respective Sustainable Development Goals, and an eighteenth subseries, "Connecting the Goals", which serves as a home for volumes addressing multiple goals or studying the SDGs as a whole. Each subseries is guided by an expert Subseries Advisor with years or decades of experience studying and addressing core components of their respective Goal.

The SDG Series has a remit as broad as the SDGs themselves, and contributions are welcome from scientists, academics, policymakers, and researchers working in fields related to any of the seventeen goals. If you are interested in contributing a monograph or curated volume to the series, please contact the Publishers: Zachary Romano [Springer; zachary.romano@springer.com] and Rachael Ballard [Palgrave Macmillan; rachael.ballard@palgrave.com].

John E. Bradshaw

Can Potatoes Feed the World?

Springer

John E. Bradshaw
Edinburgh, UK

ISSN 2523-3084 ISSN 2523-3092 (electronic)
Sustainable Development Goals Series
ISBN 978-3-031-92889-5 ISBN 978-3-031-92890-1 (eBook)
https://doi.org/10.1007/978-3-031-92890-1

This Springer imprint is published by the registered company Springer Nature Switzerland AG
The registered company address is: Gewerbestrasse 11, 6330 Cham, Switzerland

Contribution of the potato to global food security comes from both smallholders (bottom photograph, China) and vast commercial, mechanized farms (top photograph, England)

Preface

This book and its title arose from talks that I have given to gardening clubs and horticultural societies in Scotland since I retired in 2009, after a 34-year career as a plant breeder and geneticist at what is now the James Hutton Institute. My audiences knew how to grow potatoes, and some present, unlike me, would have won prizes for their displays at vegetable shows. I wanted them to sit up and look beyond their vegetable gardens and allotments, to consider the potato as a global food crop that could help to feed the world, both now and in the future. The previous year had been the United Nations (UN) International Year of the Potato 2008 (IYP 2008). It was in recognition of the contribution of the potato to the United Nations Millennium Development Goals of providing food security and eradicating poverty. As 2008 progressed, the importance of the potato in food security became more widely appreciated, as exemplified by two examples. On August 2, 2008, the front cover of *New Scientist* had a picture of potatoes under the heading "Super Crop: How the humble potato could feed the world," and a cover story "Let them eat spuds." Then, on October 27, 2008, the front page of the *International Herald Tribune*, the global edition of the *New York Times*, carried a picture of potatoes being harvested in India. The heading was "Global food crisis lifts up the potato." The article began: "With governments having trouble feeding the growing number of hungry poor and grain prices fluctuating wildly, food scientists are proposing a novel solution for the global food crisis: Let them eat potatoes." The articles also reminded readers of the danger of relying solely on the potato for food security: the late blight epidemics of 1845, 1846, 1848, and 1849 in Ireland resulted in crop failures and famine with profound societal consequences, both at the time and to the present day.

Seven years after IYP2008, in 2015, the United Nations "2030 Agenda for Sustainable Development" was adopted and started on January 1, 2016. The agenda has 17 goals, the second of which (SDG2) is to end hunger, achieve food security and improved nutrition, and promote sustainable agriculture. By 2030, the aim of the agenda is to "ensure access by all people, in particular the poor and people in vulnerable situations, including infants, to safe, nutritious and sufficient food all year round." The 2023 Sustainable Development Goals Report gave figures for 2022 of 735 million facing chronic hunger and 2.4 billion as moderately or severely food insecure as a result of inadequate food. Furthermore, the projection for 2030 was not zero hunger, but 600 million people still facing hunger. Hence, a greater sense of urgency is required, and also a need to look beyond 2030 to 2050.

By 2050, the United Nations (UN) predicts a world population of 9.7 billion compared with 8 billion in 2022. Increases in food and energy production and the supply of fresh water will be needed to sustain this population, while reducing greenhouse gas emissions to limit global warming. The average global temperature is likely to be 2 to 3 degrees Celsius above that in pre-industrial times, unless there is a greater sense of urgency following the UN Climate Change Conference (COP26) in Glasgow in 2021, which still wanted to limit the rise to 1.5 degrees. There is also increasing concern about the loss of biodiversity on Earth from human activity, including farming, as seen in the outcomes of the UN Biodiversity Conference (COP15) in Montreal in 2022. More encouragingly, on December 1, 2023, 134 countries at COP28 in Dubai endorsed the landmark sustainable agriculture, resilient food systems and climate action declaration which put food systems transformation on the global climate agenda. "While food systems are vital for meeting societal needs and enabling adaptation to climate impacts, they are also responsible for as much as a third of global greenhouse gas emissions."

A warmer climate and loss of biodiversity will make life more difficult for humankind, but can potatoes at least help with food security? It is a scientific and technological question set in a political, economic, and societal context. It has arisen because potatoes have made the journey from wild species to global food crop. The need is to further increase potato production and improve the nutritional value of potatoes, in particular to alleviate micronutrient deficiencies (hidden hunger), while reducing the risk of serious crop failures. A vast amount of relevant and detailed information for potato professionals can be found in a number of recently published books, mainly between 450 and 650 pages in length. Furthermore, Volume 67 of the journal *Potato Research* for the year 2024 ran to 2020 pages. In contrast, this book is relatively short and uses the question "Can potatoes feed the world?" to provide a different perspective for a broader audience on the biology and history of the potato crop and its potential for providing food security. Today our understanding of biology requires a knowledge of biochemistry and molecular biology, as well as mathematics, physics and chemistry. However, I have not provided a glossary of terms as I assume that the reader will quickly find any additional information they require through the Internet.

In writing this book over the period December 2022 to February 2025, I have drawn on my experiences over a 20-year period as a potato breeder and geneticist at the former Scottish Crop Research Institute in Dundee, but the views expressed are mine alone. Throughout this time, I benefited from discussions with colleagues and the wider potato community, particularly members of EAPR (The European Association for Potato Research), EUCARPIA (The European Association for Research on Plant Breeding), SSCR (The Scottish Society for Crop Research), and PBGA (The Pre Basic Growers Association). I also benefited from participation in teaching, working groups, and editorial work. In writing another book for Springer, I have once again received much help and encouragement from Kenneth Teng, as well as help from Arun Siva Shanmugam and Rekha Kumar during the production process. I hope that the book will be of value to academic courses on potatoes, especially adapting the crop to new farming and food systems through breeding; to researchers and practitioners on food security; and to policy makers and development planners who are seeking options to address the challenges of food security; indeed, anyone interested in the role of potatoes in meeting the challenge of feeding a growing world population during a period of climate change.

Edinburgh, UK
4 February 2025

John E. Bradshaw

Contents

About the Author

John E. Bradshaw spent his 34-year career (1975–2009) as a plant breeder and geneticist at what is now the James Hutton Institute in Dundee, Scotland, having studied genetics and applied genetics in England at the Universities of Cambridge (BA), Birmingham (MSc), and East Anglia (PhD). He worked on agricultural crops including 20 years on potatoes, doing research on the applications of genetics to plant breeding as well as breeding finished cultivars. He has written extensively on plant breeding and genetics, including two books, one on *Plant Breeding: Past, Present and Future* (Springer Nature, 2016) and the other on *Potato Breeding: Theory and Practice* (Springer Nature, 2021). He has also co-edited books on *Potato Genetics* (CAB International, 1994), *Potato Biology and Biotechnology* (Elsevier, 2007), and *Approaches for Potato Crop Improvement and Stress Management* (Springer Nature, 2024), and edited one on *Root and Tuber Crops* (Springer, 2010). He has been an Honorary Fellow of the Indian Potato Association since 2008. In 2010, he received the British Potato Industry Award in recognition of his contribution to potato research and knowledge transfer. Throughout his career, he was a member of EUCARPIA, the European Association for Research on Plant Breeding, and having served as Chairperson of the Section Potatoes from 2000 to 2009 was made an honorary member in May 2012. He is a member of the Genetical Society of Great Britain and served on the Editorial Board of their journal *Heredity* as Plant Breeding Subject Editor from 2010 to 2017. He is a member of EAPR, the European Association for Potato Research, and served as Co-ordinating Editor of their journal *Potato Research* from December 2017 to December 2021.

1 Let Them Eat Potatoes

"Can potatoes feed the world?" became a serious question in 2008 when a global food crisis coincided with the United Nations International Year of the Potato. There was a realization that poor countries could achieve greater food security by growing more potatoes to supplement or even replace imported grains. It remains a serious question today, when more potatoes are grown in Asia (53%) and Africa (8%) than in Europe (27%) and North America (7%), and potatoes are the third most important food crop after wheat and rice, given that maize has other uses. A longer and more accurate form of the question is, "Can potatoes make a greater contribution to global food security during a period of world population growth, lack of new land for food production, and climate change?" It is a scientific and technological question set in a political, economic and societal context. The answer requires an understanding of the history and biology of potato cultivation, as well as farming and food systems. A positive outcome needs action by many players, from the Food and Agriculture Organization of the United Nations to individual farmers, and including governments, international organizations and multinational companies.

1.1 Paris, Monday, 27 October 2008

On Monday, 27 October 2008, I was about to leave Paris following a family christening. There on a table in the hotel lobby was a copy of the International Herald Tribune, the global edition of the *New York Times* [1]. On the front page, much to my surprise, was a picture of potatoes being harvested in India. The heading was "Global food crisis lifts up the potato". The article began, "With governments having trouble feeding the growing number of hungry poor and grain prices fluctuating wildly, food scientists are proposing a novel solution for the global food crisis: Let them eat potatoes".

There is an important difference between potatoes and grains as staple foods. Potatoes are a bulky source of carbohydrates (starch) because about 80% of their content is water. Hence, they are consumed close to where they are grown to meet local needs. Grains, in contrast, can be shipped and traded as commodities far from where they are grown and have long been the currency of food aid. The period from the beginning of 2007 to the middle of 2008 saw the

J. E. Bradshaw, *Can Potatoes Feed the World?*, Sustainable Development Goals Series,
https://doi.org/10.1007/978-3-031-92890-1_1

international prices of wheat, maize and soybeans double and rice triple, all as a result of food scarcity [2]. Global production of major food staples was lower than consumption: supply could not meet demand. The human population was continuing to grow; there was an increasing use of crops for fuel rather than food, particularly maize in the United States, and an increasing use of cereals and soybeans for animal feed to meet demand for meat and dairy products in China. Extreme weather events such as high temperatures, droughts, storms and floods did not help and were expected to get worse with global warming [2]. One consequence of the food shortages was riots in 2008 in over 30 countries, with Egypt and Haiti often given as examples [2, 3]. Even in Britain, on 8 June 2008, the Sunday Telegraph carried a headline "The spectre of ration book Britain. Retailers and politicians are seeking ways to keep the global food crisis from hitting the UK" [3]. By 27 October 2008, however, there was a realization that poor countries could achieve food security by growing more potatoes to supplement or even replace imported grains. At last, the world's press understood why 2008 was the International Year of the Potato (IYP). I should add that *New Scientist* had appreciated this earlier in the year [4]. On 2 August 2008, their front cover had a picture of potatoes under the heading "Super crop: How the humble potato could feed the world" and a cover story "Let them eat spuds".

1.2 International Year of the Potato

The initiative for an International Year of the Potato (IYP) had come in 2005 from Peru and the Group of Latin America and Caribbean countries. IYP 2008 was officially launched at the United Nations (UN) headquarters in New York on 18 October 2007 by the Director-General of FAO (Food and Agriculture Organization of the United Nations). It was in recognition of the contribution of the potato to the United Nations Millennium Development Goals of providing food security and eradicating poverty [5].

By 2008, there were more potatoes being produced in Asia, Africa and Latin America (Group 1, Fig. 1.1) than in Europe, North America, Australia and New Zealand (Group 2). Commonly used names for Groups 1 and 2 are developing and industrialized countries, respectively, but these are not without controversy. Since 1991, world potato production has been rising because increases in Asia more than compensated for decreases in Europe (Fig. 1.1). In 1961, 82% of production was in Europe, whereas in 2021, 53% was in Asia [6]. The story of the potato could no longer be told simply from a European perspective. Potatoes were now being grown as a winter crop in the lowland subtropics and tropics of Asia as well as a summer crop in the lowland temperate regions of Europe and North America, having first been cultivated and domesticated as a summer crop in the high Andes of South America.

In Fig. 1.2, it can be seen that most of the increase in Asia came from China and India, with these countries accounting for 76% of Asian production in 2021. These two countries are also the most populous on Earth, each with 1.4 billion people in 2022, the year when the global population hit 8 billion in November [7]. However, the population of China, but not India, is now expected to decline by as much as 50% in the years to 2100 as a result of a falling birth rate, the consequence of rapid social and economic changes [7]. The decrease in potato production in Europe was more complicated than shown in Fig. 1.2. In 1961, the USSR, Poland and Germany were still the big producers in Europe (72% of production). The subsequent decline was large in Poland and Germany but less so in the former USSR (84.31–53.66 million tonnes in 2021), where most of the production (80%) in 2021 was in Russia, Ukraine and Belarus [6].

The consequences of these changes can be seen in Table 1.1, which shows the 15 countries that on average produced more than 5 million tonnes of tubers per year over the period 2016–2020, when there was not much variation from year to year. Together they accounted for nearly 75% of the world's potato production of

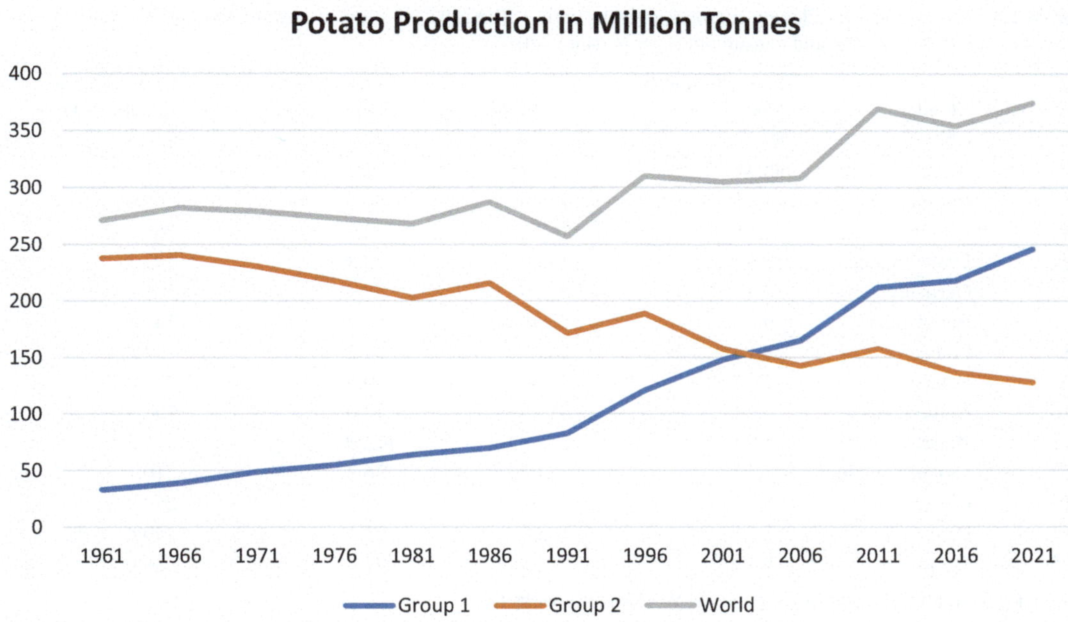

Fig. 1.1 Potato production in million tonnes fresh-weight for Group 1 [Asia (52.6%), Africa (7.6%), Latin America (5.6%)] and Group 2 [Europe (27.0%), North America (6.7%), Australia and New Zealand (0.5%)] countries where percentages in brackets are for 2021. Statistics from http://faostat.fao.org

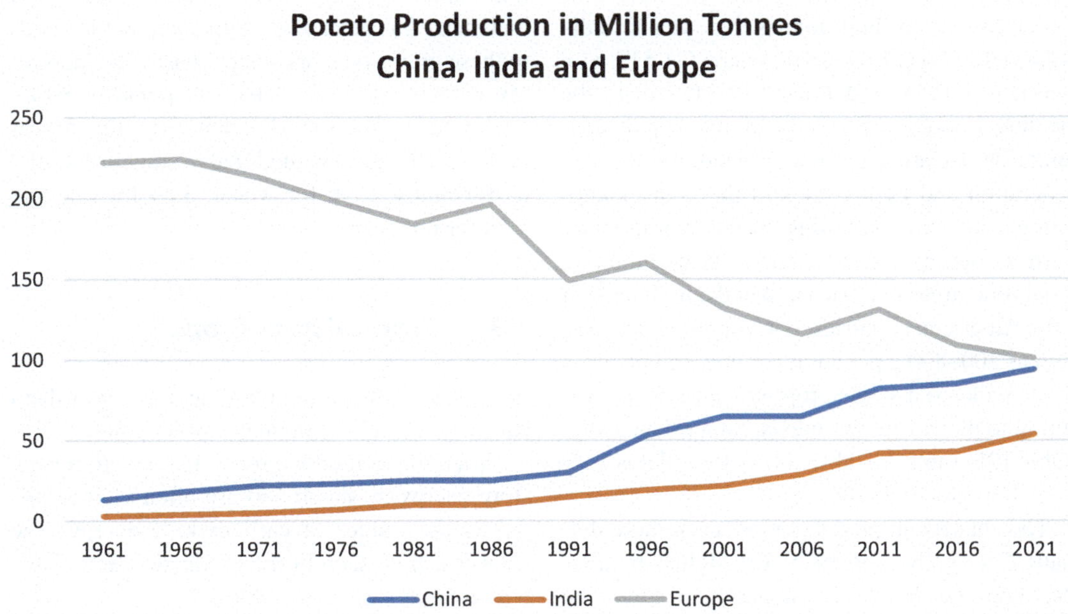

Fig. 1.2 Potato production in million tonnes fresh weight for China, India and Europe. Statistics from http://faostat.fao.org

Table 1.1 Top 15 potato-producing countries over the period 2016–2020 (million tonnes fresh weight per year), yields over the period 2016–2020, and consumption per person in 2018

Position	Country	Production million tonnes	Yield (t/ha)	Consumption (kg/person/year)	Consumption (g/person/day)
1	China	89.19	19.38	43.9	120.3
2	India	48.41	22.71	26.0	71.2
3	Russia	21.64	16.70	94.9	260.0
4	Ukraine	21.51	16.32	132.0	361.6
5	United States	19.92	49.83	52.3	143.3
6	Germany	10.74	41.69	58.9	161.4
7	Bangladesh	9.74	20.44	51.7	141.6
8	France	8.12	40.87	49.2	134.8
9	Poland	7.94	27.56	99.5	272.6
10	Netherlands	6.79	41.85	72.0	197.3
11	Belarus	5.92	21.89	172.0	471.2
12	United Kingdom	5.50	38.62	76.7	210.1
13	Canada	5.32	38.52	77.7	212.9
14	Peru	5.24	16.08	91.1	249.6
15	Egypt	5.18	27.32	31.4	86.0

Statistics from http://faostat.fao.org and the Worldometer website

364 million tonnes of tubers per year (http://faostat.fao.org). The top four producers were now China (89.19 million tonnes per year), India (48.41), Russia (21.64) and Ukraine (21.51), with Germany (10.74) and Poland (7.94) still in the top ten. However, in terms of per capita consumption, Belarus, Ukraine, Poland, the Russian Federation and Peru were still the great potato-eating countries, according to the Worldometer website figures for 2018 [8]. Furthermore, in some rural areas of Africa and in the highlands of Latin American countries, consumption can be as high as 300–800 g per adult per day, and potatoes are considered a staple crop with a major contribution to dietary requirements, rather than a vegetable [9]. Table 1.1 also shows that yields vary from 16.08 to 49.83 t/ha.

The changes in production are even more dramatic if we go back to 1925. T.P. McIntosh, in his 1927 book *The Potato* [10], quotes figures from the *International Yearbook of Agricultural Statistics* for 1925–1926, which show that 81% of the world production of 187.68 million tonnes for 1925 came from seven countries: USSR (44.28 million tonnes), Germany (41.72), Poland (29.11), France (15.20), the United States (8.87), Czechoslovakia (7.50) and the United Kingdom (5.47).

Today the potato is the world's third most important food crop after wheat and rice, given that maize has other uses, with comparable yields to these grains on a dry weight basis (i.e., assuming cereals are 15% water and potatoes 80%). Yields have not varied much over the 5-year period 2016–2020 with global averages of 4.90 t/ha for maize, 2.96 for wheat, 3.89 for rice and 4.28 for potatoes.

1.3 Tropical Root Crops

Why "Let them eat potatoes" and not "Let them eat cassava" or "Let them eat sweet potatoes"?

After all, their edible roots, like potato tubers, store energy as starch and are often viewed primarily as sources of carbohydrate energy. The answer can be seen in the production and distribution figures given in Table 1.2.

All three crops were domesticated from wild relatives in either Central or South America. Whereas potato cultivation spread to Europe and then to the rest of the world, cassava and sweet potato cultivation spread to Africa and then to Asia. Potatoes can now be grown as both temperate and tropical crops, whereas cassava and sweet

Table 1.2 Production and yield averaged over the period 2016–2020 (million tonnes fresh weight per year and tonnes fresh weight per hectare)

Crop	Region	Production Mt/year	Yield t/ha
Potatoes	World	364.48	21.41
Solanum tuberosum	Europe	107.13	23.00
	North America	25.24	46.89
	Africa	25.24	14.38
	Asia	184.44	20.55
Cassava	World	302.64	10.72
Manihot esculenta	South America	25.44	14.39
	Africa	186.97	8.52
	Asia	87.88	21.27
Sweet potatoes	World	90.73	12.17
Ipomoea batatas	Africa	27.33	6.65
	Asia	58.85	20.14

Statistics from http://faostat.fao.org [6]

potatoes remain primarily tropical ones. Today most potatoes are grown in Asia (51%) and Europe (29%), followed by North America (7%) and Africa (7%). The top five producers are China (89.19 million tonnes per year), India (48.41), the Russian Federation (21.64), Ukraine (21.51) and the United States (19.92). Most cassava is grown in Africa (62%) and Asia (29%), followed by South America (8%). The top four producers are Nigeria (59.95 million tonnes per year), the Democratic Republic of the Congo (39.22), Thailand (30.22) and Ghana (20.95). Finally, most sweet potatoes are grown in Asia (65%) and Africa (30%), with 57% of world production in China (51.49 million tonnes per year), where half of the crop goes for animal feed. Global yields of potatoes are higher than cassava and sweet potatoes, even after allowing for the lower dry-matter content of potatoes. Furthermore, for all three crops, yields are lower in Africa than in Asia.

More information about root and tuber crops can be found in the "*Handbook of Plant Breeding 7*", published by Springer in 2010 [11]. Here, we can conclude that in terms of complementing grain crops for food security, potatoes will be the crop of choice in many countries and possibly globally if African countries can increase both production and yields. However, Africa is a vast continent where farming systems vary. Large commercial farms with irrigation are common in Egypt and South Africa, where yields for 2016–2020 averaged 27.32 and 36.55 t/ha, respectively. In contrast, the tropical highland zones are intensely cultivated under rainfed conditions by smallholder farmers with less than 5 ha of land, of which potatoes usually occupy less than 1 ha and yields are much lower [12]. Hence, increasing potato production and yields need to be considered for different farming systems in North Africa and Sub-Saharan Africa (East, South, West and Central Africa).

1.4 Nigeria

Africa's most populous country is Nigeria, with 224 million people. However, this number is expected to rise to 377 million by 2050, making it globally the third largest country by population after India (1670 million) and China (1313 million) and just ahead of the United States (375 million) [7]. The country's biggest problem is providing enough food for its growing population, despite 70% of the population still being engaged in farming. Currently it spends $22 billion a year on food imports, notably wheat and fish, yet 19 million people were reported to be experiencing food insecurity in 2022. Can potatoes be of help, or is production going to remain low at the 2016–2020 average of 1.21 million tonnes per year, which nevertheless is the sixth highest in Africa. The Plateau State accounts for 95% of production, mainly on smallholder farms of less than 3 ha, with up to 0.5 devoted to potatoes. The state is 8°N with an altitude of 1200 masl, average annual rainfall of 1400 mm with a rainy season from April to October, and an average temperature of 13–22 °C all year round. Despite adequate growing conditions, yield per hectare is extremely low at 3.80 t/ha, compared with a global figure of 21.41 t/ha (2016–2020). There appears to be scope for increased production, including yield, if a number of problems can be solved: the high costs of inputs; lack of credit facilities and clean planting material; disease, pest and soil fertility problems; inadequate storage facilities; and finally, attacks on farms. In October 2023, it was therefore encouraging to see Nigeria's Federal Ministry of Agriculture and Food Security (FMAFS) launch "The National Potato Strategy for Nigeria 2023–2028", with an emphasis on delivering sustainable food and nutrition security in a commercially oriented potato sector that encourages trade and income generation (see the website of The International Potato Centre (CIP): https://cipotato.org/blog/unlocking-prosperity-national-potato-strategy-nigeria/). Hence the potato should be able to make a greater contribution to food security, as well as providing a cash crop, both in Nigeria and elsewhere in Africa [13, 14].

1.5 Political, Economic and Societal Upheavals in Major Potato-Producing Countries

Although "Can potatoes feed the world?" is a scientific and technological question, it is set in a political, economic and societal context, which can have a big influence on the outcome. During the last hundred years, there have been major political, economic and societal upheavals in some of the major potato-producing countries, particularly World War II (1939–1945) and its aftermath. Indeed, it was a consequence of the war that the United Nations and its Food and Agriculture Organization were founded in October 1945.

Europe witnessed the formation of the Union of Soviet Socialist Republics (USSR or Soviet Union for short), with 15 members in 1922, and its collapse in 1991. The period after the collapse saw the eastward expansion of the European Union (EU) to 28 members in 2013 and then back to 27 when the United Kingdom left in 2020. The EU was created in 1993 but dates back to the post-war European Coal and Steel Community formed in 1952 by Belgium, France, Italy, Luxembourg, the Netherlands and West Germany. India gained independence from the United Kingdom on 15 August 1947 and is now the world's largest democracy. The People's Republic of China came into existence on 1 October 1949 as a Marxist-Leninist single-party state controlled by the Chinese Communist Party. Finally, one cannot ignore the post-war effects of American (United States) foreign policy as initiated in 1949 with Harry S. Truman's inaugural address after he won a full 4-year term in the 1948 election [15]. As well as supporting the United Nations, Europe's recovery from war, and the security of the North Atlantic area through NATO, the policy aimed to assist underdeveloped countries to prosper and fight Communism through technology-driven economic growth. The effects on agriculture are usually seen in terms of grain production, but later in this book we will see some of the consequences for potatoes.

1.5.1 USSR and Russia

Joseph Stalin's rule of the USSR from 1924 to 1953 is remembered, amongst other things for Industrialization, unproductive collective farming and suppression of the new science of genetics and its application to animal and crop improvement. The early 1930s (particularly 1933) saw famine in Ukraine with an estimated four million deaths. In her book *Red Famine*, published in 2017, Anne Applebaum argues that the famine was Stalin's war on Ukraine [16]. As I started to write this book in December 2022, once again there was war in now-independent Ukraine following the Russian invasion on 24 February 2022. However, as Vladimir Putin's Russia (from 2000 to present day) had rapidly become a net exporter of wheat through increased production [17], as had Ukraine, the war threatened to block 9% of global wheat exports, thus driving wheat prices to unprecedented heights [18].

1.5.2 China

Mao Zedong's China (1949–1976) is also remembered for famine as a result of his "Great Leap Forward". He thought that he could catapult China past its western competitors by simultaneously transforming agriculture and industry through a planned economy involving collectivization and so create a modern communist society of plenty for all. We now know that the result was rather different. Possibly as many as 45 million people died of starvation between 1958 and 1962 [19]. After the death of Chairman Mao, there was a period of uncertainty over the direction the country would take, particularly after "The Massacre" in Tiananmen Square on 4 June 1989 (referred to as the June Fourth Incident in Chinese) as the People's Liberation Army crushed student-led demonstrations [20]. However, by the time of the collapse of the Soviet Union in 1991, China was on the road to the world superpower that it is today, a Communist dictatorship with political control of its people and economy [20]. Since then, the Chinese state has been encouraging potato production, and today potato production is seen as a way of ensuring food security as well as individual health and well-being [21]. According to Professor Jin of the Chinese Academy of Agricultural Sciences, the Ministry of Agriculture (MOA) issued documents in 2006 and 2016 on accelerating the development of the Chinese potato industry. The China Potato Research System and the MOA's Key Laboratories dedicated to potato research were built in 2008 and 2011, respectively [22].

1.5.3 India

In contrast, India is usually associated with the "Green Revolution" in wheat production (1963–1974), which involved collaboration between Norman Borlaug (1914–2009) and Monkombu Sambasivan Swaminathan (1925–2023) and resulted in India being self-sufficient in all cereals. Borlaug was the wheat breeder who won the Nobel Peace Prize in 1970 and was depicted as "The Wizard" in the book by Charles C. Mann, *The Wizard and the Prophet*, which was published in 2018 [15]. Interestingly, Swaminathan did his PhD at Cambridge University, England, from 1950 to 1952 on "Species Differentiation, and the Nature of Polyploidy in certain species of the genus *Solanum*—section Tuberarium". Today most of India's potato production takes place as a winter crop in the sub-tropical plains, a success story resulting from a programme started in 1949 with the establishment of the Central Potato Research Institute (ICAR-CPRI), just 2 years after Indian independence [23].

1.5.4 European Union

The EU's Common Agricultural Policy (CAP) dates back to 1962 as "a partnership between agriculture and society, and between Europe and its farmers". Subsidies were paid to farmers to improve agricultural productivity, to guarantee producer prices, and also to ensure the stability both of markets for food production and of prices for consumers [24]. There were also tariffs on food imports. The consequences in the 1970s were "butter and beef mountains" and "wine and milk lakes" because the subsidies encouraged farmers to produce commodities with a guaranteed price. Surpluses were therefore "dumped" on world markets to the detriment of the livelihoods of farmers in poorer countries. In 1982, this prompted Richard Body, a British farmer and Member of Parliament for the former agricultural constituency of Holland with Boston, to write *Agriculture: The Triumph and The Shame*, in which he argued against subsidies and import tariffs and in favour of free trade [25]. Back in 1982, he was also able to express the view that globally, poverty was the cause of hunger, not scarcity of food. However, the response to the problem of surpluses was that from 1988 to 2008 farmers were paid to remove a percentage of their fields from production, so-called "set-aside", which proved a practical but unsatisfactory solution to stable and secure food production in Europe. The CAP was modified in 2005 with the Single Payment Scheme (SPS, or Single Farm Payment), in which farmers were paid a set amount per hectare of agricultural land maintained in a cultivatable condition, rather than by crop. Moving forward to the present day, CAP 2023–2027 is "designed to shape the transition to a sustainable, resilient and modern European agricultural sector. It will continue playing a key role in supporting Europe's agricultural sector as well as strengthening the efforts of European farmers to tackle climate change and protect the environment, in line with the Commission's Farm to Fork and Biodiversity strategies" [26]. However, at the beginning of 2024 farmers across Europe, including Belgium, France, Germany and the Netherlands, were protesting about the cost to them of the EU's environmental and net-zero policies. Furthermore, the June 2024 elections to the European Parliament saw the Greens lose seats, taking them down to 53 seats, whereas the centre-right European People's Party strengthened its position as the biggest group with 188 seats out of the 720. It remains to be seen how this will affect EU policies, particularly on agriculture.

1.5.5 United Kingdom

It is also interesting to look at the "New Approaches to Farm Funding in England" (they are slightly different in the other nations of the United Kingdom) after the United Kingdom left the European Union on 31 January 2020. They are complicated, so I am relying on an "Insight" published on 28 May 2024 by Sarah Coe, a researcher at the House of Commons Library specializing in agriculture [27]. The result of the Agriculture Act 2020 is that a new farm support policy is being phased in over the period 2021–2027 to replace previous support under the Common Agricultural Policy (CAP). Farmers will now be paid to provide "public goods" rather than to simply farm an area of land. The public goods come in the form of Environmental Land Management (ELM) schemes comprising: (1) the Sustainable Farming Incentive (SFI) for practices that protect and enhance the natural environment alongside food production (e.g., improving soil quality and planting hedgerows); (2) Countryside Stewardship targeted at specific locations, features and habitats and (3) Landscape Recovery for longer-term, larger-scale projects to enhance the natural environment. In financial terms, the direct basic payments to farmers in 2023 of £233.30 per hectare of land they farmed are being phased out by 2027. This is a big change, as under the CAP, direct payments made up more than 80% of government support to farmers. The change is promoting a lot of discussion in England about how much land should be taken out of production: for example, an essay in The Times by Emma Duncan on 1 June 2024 was headed "What do we want from our farms: food or flowers?" [28]. Furthermore, the front cover of *The Telegraph Magazine* for 17 August 2024 was a picture of a potato with the caption "Mashed, boiled, fried…finished? The race to save the British potato", in the context of a third poor harvest and hence falling production. Not a good advert for feeding the world! Inside, however, was an article by Richard Godwin explaining how the futuristic company B-hive is here to save us from a future without potatoes because they believe that without technological innovation, the British potato will die out [29]. We will meet B-hive again in the context of gene editing. Here, I want to make the point that whether or not farmers are paid subsidies and the nature of any subsidies paid have a big impact on food production, and the same is true for trade tariffs on food imports. Furthermore, the commitments being made by governments to net zero carbon emissions by 2050 will also impact food production.

1.6 United Nations "2030 Agenda for Sustainable Development"

Globally the way forward can be seen in the context of the United Nations "2030 Agenda for Sustainable Development", which was adopted by all 193 member states in 2015 and started on 1 January 2016 [30]. The agenda has 17 goals, the second of which is to end hunger, achieve food security and improved nutrition, and promote sustainable agriculture. The 2022 Sustainable Development Goals Report stated that in 2020 there were still between 720 and 811 million people worldwide suffering from hunger and a staggering 2.4 billion were moderately or severely food-insecure, lacking regular access to adequate food. The report also concluded that progress had stalled by 2022 as a result of a confluence of crises dominated by COVID-19, climate change and conflicts. The 2023 report gave figures for 2022 of 735 million facing chronic hunger and 2.4 billion as moderately or severely food insecure as a result of inadequate food [31]. Furthermore, the projection for 2030 was not zero hunger but 600 million people still facing hunger. The Secretary-General of the United Nations pleaded for a greater sense of urgency in 2022 and again in 2023. Ominously, the 2023 Special Edition Report had the subtitle "Towards a Rescue Plan for People and Planet", and the Secretary-General feared that unless we act now, "the 2030 Agenda could become an epitaph for a world that might have been" [31]. The figures for hunger and food insecurity were virtually unchanged in the 2024 report for 2023 [32]. The Global Hunger Index (GHI) for 2024 (data from 2019 to 2023) was equally pessimistic, stating that the GHI makes it clear that the world is far from meeting the goal of zero hunger by 2030 [33]. There had been little change in the index since 2016, with serious or even alarming hunger in 42 countries, particularly in Sub-Saharan Africa and South Asia. Furthermore, at the beginning of December 2024, the wars in Gaza and Sudan had led to exceptional food crises. The index for a country is based on the share of the population that are undernourished and stunted, wasting and mortality in children under 5. The 2024 report pointed out that women and girls are typically hardest hit by food insecurity and malnutrition and also suffer disproportionately from the effects of weather extremes and climate emergencies; in other words, gender inequality issues need to be addressed.

The world population is expected to reach 8.5 billion in 2030 and continue to increase to 9.7 billion in 2050 and possibly 10.4 billion in 2100 [34]. For major food staples, the need is to increase production and improve nutritional value, particularly to alleviate micronutrient deficiencies (hidden hunger), during a period of continuing human-made climate change. If the earth's average land and ocean surface temperature over the period 1880–1900 (13.70 °C) can be taken as a surrogate for pre-industrial levels, there has already been a rise of 1.09 °C to the average for 2014–2022 (14.79 °C), which were the hottest 9 years since 1880 [35]. The latest data have confirmed 2023 as even hotter at 15.09 °C, a rise of 1.39 °C [35]. Perhaps more encouragingly, 134 countries at COP28 in Dubai in December 2023 endorsed the landmark sustainable agriculture, resilient food systems and climate action declaration which put food systems transformation on the global climate agenda [36]. The 134 countries are home to over 5.7 billion people and almost 500 million farmers, produce 70% of the food we eat and are responsible for 76% of all greenhouse gas emissions from global food systems, or 25% of total emissions globally. The declaration recognized that "countries must put food systems and agriculture at the heart of their climate ambitions, addressing both global emissions and protecting the lives and livelihoods of farmers living on the front line of climate change". It went on to say that endorsement "will help in strengthening food systems, building resilience to climate change, reducing global emissions, and contributing to the global fight against hunger, aligned with the UN Sustainable Development Goals". Out of the 15 major potato producers listed in Table 1.1, just India and the Russian Federation have not yet signed the declaration.

The FAO observes the contribution potatoes can make to the 2030 Agenda, and beyond, on May 30 each year through an "International Day of Potato" (https://www.fao.org/international-potato-day/en). To quote from their website. "On 30 May, we spotlight the potato, *Solanum tuberosum (L.)* – a crop regularly consumed by billions of people. Potatoes are a key crop across diverse farming systems globally, ranging from smallholders producing diverse heirloom varieties by hand in the Andes to vast commercial, mechanized farms on different continents. Potatoes contribute to the food security and nutrition as well as the livelihoods and employment of people in rural and urban areas the world over. The Day is an opportunity to build on the International Year of the Potato, which was observed in 2008. The observance will also be used to underscore the importance of the crop to combating hunger and poverty and addressing environmental threats to agrifood systems. The roles of small-scale family farmers, a significant proportion of whom are women, in safeguarding the wide spectrum of the crop's diversity, will also be recognized, while the cultural and culinary dimensions of the crop's cultivation and consumption will be celebrated". An analysis of the current and future role of the potato as a food security crop can be found in a 2021 review by Devaux et al. on *The Potato of the Future: Opportunities and Challenges in Sustainable Agri-Food Systems* [37]. They argue that the contribution of the potato to global food security should be considered in the context of the continuing evolution of the agri-food systems from completely rural to more urbanized, industrial ones. In completely rural systems, landraces of potatoes are grown from farm-saved seed-tubers, with little use of machinery and agrochemicals, for fresh consumption as a staple crop which provides little income but has a low environmental impact. In contrast, in completely industrial systems, modern cultivars are grown from certified seed tubers, with extensive use of machinery and agrochemicals, for processing into products with added values but also higher environmental impacts. The authors also point out that there is scope to reduce food losses along potato value chains.

1.7 Consultative Group for International Agriculture Research

There is one other organization that needs to be introduced in the context of agriculture and a world without hunger. The Consultative Group for International Agriculture Research (CGIAR) was formed in 1971 as a global partnership of 15 International Research Centres, headquartered since 2011 at Agropolis International in Montpellier, France. It has a number of international sources of funding which are acknowledged on its website along with its 2030 Research and Innovation Strategy (https://www.cgiar.org). One of its main aims is the eradication of hunger and poverty and provision of food security through research. The research centres also make improved germplasm available to individual countries for release as cultivars or for further breeding through National Agricultural Research Systems (NARS). The International Potato Centre (CIP) was founded in Lima, Peru, in 1971. Peru was chosen because it is a centre of biodiversity for tuber-bearing *Solanum* species.

1.8 Junk Food Cycle

Given all of these concerns about hunger, it is ironic that the World Obesity Federation reported in 2023 that in 2020 there were 2.6 billion overweight (BMI ≥ 25 kg/m^2) people in the world, of which 988 million were obese (BMI ≥ 30 kg/m^2). Furthermore, the numbers were expected to increase to 3.51 and 1.56 billion, respectively, in 2030 [38]. The United Kingdom is an example of the problem and a warning to other countries. In 2017, we spent 16% of our income on food compared with 33% in 1957 but have created a "Junk Food Cycle", which needs to be addressed. Although 80% of our processed food is cheap and popular, it is unhealthy, being high in salt, refined carbohydrates, sugar and fats, and low in fibre [39]. Nevertheless, the emphasis in this book will be on concerns about hunger and malnutrition as the world's population increases and the ghost of Thomas Robert Malthus reappears.

1.9 Thomas Robert Malthus

In the first chapter of his 1803 enlarged edition of *An Essay on the Principle of Population*, Thomas Robert Malthus stated his subject to be the "ratios of the increase of population and food" [40]. He investigated one great cause "that had impeded the progress of mankind towards happiness", namely, "the constant tendency in all animated life to increase beyond the nourishment prepared for it". He pointed out that in a number of countries there were examples of human populations doubling in size in less than 25 years, at least for short periods when unchecked. In contrast, even if all of the fertile land for food production was occupied and improved, it could not feed a geometrically increasing population. He took England and Scotland as examples of countries where there was still a great deal of uncultivated land and where the science of agriculture was well established. He argued that if the average produce of the island (including Wales?) could be doubled in the first 25 years of effort, to achieve a further doubling in the next 25 years would be contrary to all available scientific knowledge of the land. The most optimistic outcome would be the same increase over the next 25 years. In practice the yearly additions would decrease. Hence, in the most favourable circumstances, he imagined the population of England and Scotland doubling from 11 to 22 million in 25 years and food production increasing to feed an additional 11 million people. Then in the next 25 years the population would double to 44 million, but the food production would only increase to feed an additional 11 million people; that is, 33 of the 44 million. After 100 years the population would be 176 million, but the means of subsistence would only be enough to support 55 million. The population of England and Scotland from the 1801 census was, in fact, 10 million and rose to 62 million in 2021, when we were not self-sufficient in food, relying on almost 50% from imports. Malthus extended his argument to the whole earth: the world population was 1 billion in 1803, so after 225 years (2028), the unchecked population would be 512 billion, but the means of subsistence would be enough to support only 10 billion people (a modern account of human demography can be found in the 2019 book *The Human Tide* by Paul Morland [41]). Although the human population has been checked and food production has been increasing, the subject of concern to Malthus is still with us because the land available for food production is finite, reaching capacity, and the human population is still increasing.

1.10 Land Use

It is useful to put the land cultivated for potatoes into a wider context. I have used "Land Use" from "Our World in Data" (https://ourworldindata.org/land-use), which was published in September 2019 [42].

The earth's surface is 71% ocean (361 million km^2) and 29% land (149 million km^2).

Land Surface (figures in brackets are percentage of total land area of 149 million km^2)

- 106 million km^2: Habitable

28 million km^2: Barren land (19%) (deserts, salt flats, exposed rocks, beaches and dunes)

15 million km^2: Glaciers (10%) (permanent snow and ice: 14 million km^2 is Antarctica)

Habitable Land

- 48 million km^2: Agriculture (1000 years ago, the area was only 4 million km^2)

40 million km^2: Forests (27%)

15 million km^2: Shrub (10%) (small and woody vegetation)

1½ million km^2: Freshwater (1%) (lakes and rivers)

1½ million km^2: Urban and built-up land (1%)

Agricultural Land

- 37 million km^2: Livestock (25%) (meat and dairy)

11 million km^2: Crops (7%) (excluding animal feed)

Although livestock account for 77% of agricultural land, they only produce 18% of the world's calories and 37% of the world's protein, the rest coming from plant-based food. Out of a crop plant area of 1.1 billion ha (11 million km^2), 588 million ha (55%) were used in 2022 for wheat (219.2), maize (203.5) and rice (165.0) combined, compared with just 17.79 million ha (1.6%) for potatoes (the figures for the period 2016–2022 do not vary much).

1.11 What This Means for Potatoes

The conclusion from 2008 was that "Let them eat potatoes" was a reasonable and serious proposition. It still is in the context of the United Nations "2030 Agenda for Sustainable Development". Throughout the world, the potato is seen as a crop that can contribute in varying degrees to food and nutritional security, despite the area cultivated being much less than that for grain crops. The extent to which this happens will depend on the decisions made by politicians, the economics of potato production compared with other crops and the consumer preferences of the societies in which their farmers operate. It will also depend on an understanding of the biology and history of the crop. Can more land be brought into cultivation for potatoes? Can potatoes replace existing crops on land already cultivated? Can potato yield per hectare be increased? Can the nutritional value of potatoes be improved? What are the risks of poor crops or even crop failure? In this book, I first consider the biology and history, together with the lessons learned, and then look to the future and the possibility of increasing potato production and nutritional value. There is also a more distant future.

1.12 Mars and Beyond

The Apollo 11 mission resulted in Neil Armstrong and Edwin Aldrin becoming the first humans to land on the moon on 20 July 1969. Three years later, in 1972, Apollo 17 became the last manned mission to the moon. Now, over 50 years later, there is renewed interest in human space travel, plans for a return to the moon, and a NASA (National Aeronautics and Space Administration) goal of sending people to Mars by 2030, with the possibility of establishing a long-term human presence on the Red Planet and a role for potatoes in a "bioregenerative food system" (BFS).

In 2009, NASA biologist Raymond Wheeler wrote a book chapter on potatoes for human life support in space [43]. He pointed out that as space missions travel greater distances and stay longer, methods for regenerating life support consumables will be required, namely, breathable air, clean water and food. Plant photosynthesis could remove carbon dioxide from the air and generate oxygen, plant transpiration could recycle wastewater to clean water, and a crop could provide food. Furthermore, microbial communities associated with the plant root system could break down organic compounds in the wastewater for plants to recycle as nutrients. The whole process would be driven by light from the sun or electric lamps. The book chapter reviewed the suitability of potatoes for this role, based on NASA-sponsored research from 1988 onwards into the growth of potatoes in controlled environment chambers. Some results were expected from previous physiological studies, whereas others were new and interesting, including the following: intolerance of some cultivars to continuous light, increased stomatal conductance at super-elevated carbon dioxide concentrations, ethylene production by whole canopies, successful growth in culture using a nutrient film technique, and yields of nearly 200 t/ha under high light intensities and carbon dioxide enrichment. Furthermore, small-scale space flight experiments revealed that tubers can form and sprout in weightlessness.

Then in 2016 there were headlines in the papers about NASA collaborating with CIP (International Potato Centre) scientists to test if potatoes will grow on the Red Planet, with parallels being drawn with the Matt Damon film *The Martian* [44]. The research involved growing potatoes in a greenhouse in Lima, Peru (12.08°S, 76.95°W, 244 masl) in "Mars-like" soil from the northern Atacama Desert that lacked basic nutrients, was devoid of organic matter, and was characterized by extreme salinity. The potatoes were fertilized and watered and grown for 134 days. The results were published in the *International Journal of Astrobiology* in 2017 [45]. Out of 65 genotypes tested, 40% survived and yielded from 0.3 to 5.2 g of tubers per plant. The best two clones were from CIP's breeding programme for adaptation to the subtropical lowlands and tolerance to abiotic stress. The fresh weight yields of their tubers were 9.3% and 8.8% of those achieved in control soil. The main conclusion was that appropriate genotypes and soil management are crucial to withstanding extreme salinity, which was also an important result for agriculture back here on Earth.

More recently, in 2021, a group of Chinese scientists proposed a "Whole-Body Edible and Elite Plant (WBEEP)" strategy, involving genetic transformation, to develop crops with more edible parts (i.e., a greater harvest index), richer nutrient content, higher yields and higher mineral nutrient use efficiencies for space farms [46]. They considered potatoes a top contender for space agriculture but identified three problems to overcome using WBEEP. First, elimination of toxic glycoalkaloids from the foliage so that it can be eaten, as they are not required for pest and pathogen resistance on space farms. Second, increasing yield through photosynthetic efficiency and nutrient accumulation by genetic biofortification to fully meet the nutritional requirements of humans. Third, reducing fertilizer consumption by increasing fertilizer use efficiency through improved nutrient utilization efficiency. The authors also thought that the reproductive biology of potatoes was an advantage. Asexual reproduction ensures the regeneration of food resources, whereas sexual reproduction results in more offspring (seeds) with lower storage and transportation costs.

Mars is not an attractive holiday destination. It is a cold, dry desert world with surface conditions unsuitable for life from Earth, including dust storms and high-energy radiation; gravity one-third of that on Earth; an atmosphere rich in carbon dioxide but lacking nitrogen and oxygen; and a solar intensity that is 43% of that on Earth. However, the choice of potatoes for research into space farms is providing results of relevance for life on Planet Earth. They confirm that potatoes are a high-yielding and nutritious crop with scope for improvement.

A crop that can help to feed the world in 2050 and beyond. A crop that can provide food security and help to eliminate hunger, malnutrition and poverty.

References

1. Rosenthal E (2008) Global food crisis lifts up the potato. International Herald Tribune, Monday, October 27, 2008, pp 1 and 7
2. McMahon P (2013) Feeding frenzy. Profile Books, London
3. Hall J (2008) The spectre of ration book Britain. The Sunday Telegraph, June 8 2008, Business B5
4. MacKenzie D (2008) Let them eat spuds. NewScientist 2 August 2008:30–33
5. International Year of the Potato. https://www.fao.org/agriculture/crops/thematic-sitemap/theme/hort-indust-crops/international-year-of-the-potato/en/
6. FAO statistics on 26 September 2024 (http://faostat.fao.org)
7. Welch C, Nwaubani AT, Larmer B, Zhang J (2023) Eight billion: a special report. National Geographic, April 2023:34–91
8. Worldometer. https://www.helgilibrary.com/indicators/potato-consumption-per-capita/ on 12 January 2023
9. Burgos G, Felde TZ, Andre C, Kubow S (2020) The potato and its contribution to the human diet and health. In: Campos H, Ortiz O (eds) The potato crop: its agricultural, nutritional and social contribution to humankind. Springer, ebook, pp 37–74
10. McIntosh TP (1927) The potato. Oliver and Boyd, Edinburgh
11. Bradshaw JE (ed) (2010) Root and tuber crops, Handbook of plant breeding, vol 7. Springer, New York
12. Muthoni J, Shimelis H (2023) An overview of potato production in Africa. In: Caliskan ME, Bakhsh A, Jabran K (eds) Potato production worldwide. Academic, London, pp 435–456
13. Hudu MI, Yusuf M, Saleh I (2018) Economics of potato production in Africa: a comparative analysis of Nigeria and Kenya for 2015. Niger J Agric Ext 19(3):21–29
14. Uche CO, Umar HS, Girei AA (2020) Assessing the profitability and constraints to Irish potato production in Plateau State, Nigeria. IJIRAS 7(5):13–20
15. Mann CC (2018) The wizard and the prophet. Picador, London
16. Applebaum A (2018) Red famine. Penguin, UK
17. Liefert WM, Liefert O (2022) Russia's development as a top player in world grain trade. In: Wegren SK, Nilssen F (eds) Russia's role in the contemporary international agri-food trade system. Palgrave Macmillan, ebook, pp 69–94
18. Nóia R, Ewert F, Webber H, Martre P, Hertel TW, Ittersum MK van, Asseng S (2022) Needed global wheat stock and crop management in response to the war in Ukraine. Glob Food Secur 35. doi: https://doi.org/10.1016/j.gfs.2022.100662
19. Dikötter F (2010) Mao's great famine. Bloomsbury, London

20. Dikötter F (2022) China after Mao. Bloomsbury, London
21. Earle R (2020) Feeding the people. Cambridge University Press, Cambridge
22. Jin L (2018) Potato production and breeding in China. In: Wang-Pruski G (ed) Achieving sustainable cultivation of potatoes volume 1: Breeding improved varieties. Burleigh Dodds, Cambridge, pp 259–270
23. Luthra SK, Gupta VK, Tiwari JK, Kumar V, Bhardwaj V, Sood S, Dalamu, Kaur RP, Kumar R, Vanishree G, Kumar D, Mhatre P, Chakrabarti SK (2020) Potato breeding in India. CPRI Technical Bulletin No 74 (revised). ICAR-Central Potato Research Institute, Shimla, Himachal Pradesh, India
24. Fiennes J (2022) Land healer. Witness Books, London
25. Body R (1982) Agriculture: the triumph and the shame. Maurice Temple Smith Limited, London
26. EU Common Agriculture Policy. https://agriculture.ec.europa.eu/common-agricultural-policy/cap-overview/cap-glance_en on 4 April 2023
27. Coe S (2024) Insight: new approaches to farm funding in England. House of Commons Library 28 May 2024 https://commonslibrary.parliament.uk/new-approaches-to-farm-funding-in-england/
28. Duncan E (2024) What do we want from our farms: food or flowers? An essay in The Times, 1 June 2024
29. Godwin R (2024) The fight to save Britain's favourite vegetable. The Telegraph Magazine 17 August 2024 pages 8–13
30. United Nations "2030 Agenda for Sustainable Development" (https://www.un.org/sustainabledevelopment/hunger/) on 14 December 2022
31. United Nations 2023 The sustainable development goals report special edition. https://sdgs.un.org/
32. United Nations (2024) The sustainable development goals report 2024. https://unstats.un.org/sdgs/report/2024/Goal-02/
33. Global Hunger Index October 2024, Synopsis. Retrieved on 2 December 2024 from https://www.globalhungerindex.org
34. United Nations Department of Economic and Social Affairs Population Division (2022) World population prospects 2022 summary of results. United Nations, New York
35. NOAA (National Oceanic and Atmospheric Administration) National Centers for Environmental Information, Monthly Global Climate Report for Annual 2023, published online January 2024, retrieved on May 28, 2024 from https://www.ncei.noaa.gov/access/monitoring/monthly-report/global/202313
36. COP28 UAE. https://www.cop28.com/en/news/2023/12/COP28-UAE-Presidency-puts-food-systems-transformation
37. Devaux A, Goffart JP, Kromann P, Andrade-Piedra J, Polar V, Hareau G (2021) The potato of the future: opportunities and challenges in sustainable agri-food systems. Potato Res 64:681–720
38. World Obesity Federation, World Obesity Atlas 2023. https://data.worldobesity.org/publications/?cat=19
39. England: The national food strategy: the plan-July 2021. nationalfoodstrategy.org
40. Malthus TR (1803) An essay on the principle of population. The 1803 Edition. Edited and with an Introduction by Stimson SC (2018). Yale University Press, New Haven
41. Morland P (2019) The human tide. John Murray, London
42. Ritchie H, Roser M (2019) Land use. Published online at OurWorldInData.org. Retrieved from https://ourworldindata.org/land-use. Consulted 7 November 2023
43. Wheeler RM (2009) Potatoes for human life support in space. In: Singh J, Kaur L (eds) Advances in potato chemistry and technology, 1st edn. Academic, Burlington, pp 465–495
44. Tegel S (2016) Life on Mars? Peruvian team tests if potatoes will grow on Red Planet. The Independent, Friday 22 January 2016, 29
45. Ramírez DA, Kreuze J, Amoros W, Valdivia-Silva JE, Ranck J, Garcia S, Salas E, Yactayo W (2017) Extreme salinity as a challenge to grow potatoes under Mars-like soil conditions: targeting promising genotypes. Int J Astrobiol 18 (1):18–24 (2019) doi:https://doi.org/10.1017/S1473550417000453
46. Liu Y, Xie G, Yang Q, Ren M (2021) Biotechnological development of plants for space agriculture. Nature Com 12:5998. https://doi.org/10.1038/s41467-021-26238-3

2 Wild Relatives

The wild tuber-bearing *Solanum* species never intended to feed us. They evolved swollen subterranean stems to survive from season to season as dormant storage organs in a wide range of habitats in the Americas, with centres of diversity in Mexico and Peru. Our knowledge of them has come from numerous collecting expeditions to North, Central and South America from the 1920s to the 1990s. They are a valuable but underutilized plant genetic resource for breeding new potato cultivars better adapted to providing global food security during a period of climate change. They need to be conserved, both in situ and ex situ in genebanks, studied, and used in breeding programs through knowledge of their taxonomy, reproductive biology and adaptive traits. The International Treaty on Plant Genetic Resources for Food and Agriculture (ITPGRFA) came into force on 29 June 2004, and by 16 July 2024, it had 152 contracting parties (out of 194 countries). The treaty recognized the contribution of farmers to the diversity of crops that feed the world, established a global system to provide farmers, plant breeders and scientists with access to plant genetic materials, and ensured that recipients shared the derived benefits with the countries of origin.

2.1 Wild Tuber-Bearing *Solanum* Species

Over the last 66 million years, the Earth's continents moved into their present positions, today's prominent mountain ranges formed, new ocean current systems became established, major changes occurred in the composition and distribution of the global flora, and from 50 million years ago the climate cooled [1]. The flowering plants (angiosperms) continued their expansion and evolution. The tuber-bearing *Solanum* species started to evolve in the Americas some 8 million years ago, the estimated date for the common ancestor of today's wild potatoes and tomatoes [2]. Recurrent glacial-interglacial cycles occurred over the past 2.5 million years. The last glacial period started 110,000 years ago, reached its maximum from 25,000 to 15,000 years ago, and ended 11,500 years ago [3]. Since then, there has been what we regard as a relatively benign interglacial period known as the Holocene Era. Some scientists, such as Paul J. Crutzen, argue that we have now entered the Anthropocene Era in which human dominance of Earth's biological, chemical and geological processes is an undeniable

J. E. Bradshaw, *Can Potatoes Feed the World?*, Sustainable Development Goals Series,
https://doi.org/10.1007/978-3-031-92890-1_2

reality [4]. We have already reached the point where 45% of habitable land is used for agriculture, and we need to feed over 8 billion people.

The wild tuber-bearing *Solanum* species never intended to feed us and, in fact, produce poisons to discourage their consumption by predators. They evolved swollen subterranean stems to survive from season to season as dormant storage organs in which energy is stored as starch. There is evidence that humans were eating wild potatoes some 13,000 years ago before starting to cultivate some of them high in the Andes [5–7]. Today, the wild relatives of modern cultivated potatoes are regarded as a valuable but underutilized plant genetic resource for breeding new cultivars better suited to providing food security during a period of climate change. It is therefore important to understand their distribution, biology and conservation, but first there is the question of recognizing and naming them.

2.2 The National Botanic Garden of Wales

Carl Linnaeus (1707–1778) proposed a naming system for plants in the eighteenth century, which was adopted by botanists [8, 9]. Each species has a Latin name consisting of two words, the genus and the species. These are the lowest ranks in what became a hierarchy of categories that result from a consideration of the similarities and differences among individuals and groups of plants. At each level, a number of smaller groups are brought together to make a larger one: species, genus (*Solanum*), subtribe, tribe, subfamily, family (Solanaceae), order (Solanales) and so on to the highest rank. However, after Charles Darwin (1809–1882) published *On the Origin of Species by Means of Natural Selection* in 1859 [10], systematists developed more natural classifications of species in terms of their evolutionary relationships and history (phylogeny). The outcomes are phylogenetic trees (chronological hypotheses) produced by cladistic analysis, in which a clade is a large group of organisms descended from a common ancestor and including the ancestor [11]. The overall result is the pedigree, or genealogical history, that we call the "Tree of Life".

The National Botanic Garden of Wales is a new garden which opened in May 2000 and has a double-walled garden of flowering plants (angiosperms) (Fig. 2.1). Its aim is to demonstrate the latest classification system being developed by the Angiosperm Phylogeny Group (APG) of botanists [12], using mostly molecular (DNA) data, as explained in an accompanying book with extensive colour photographs [13]. The classification reflects the phylogeny of flowering plants, based on the guiding principle of monophyly: "that taxonomic groups should hold all the descendants of a single ancestor, and only its descendants". The systematic beds are arranged to reflect the positions of the orders of flowering plants on the evolutionary tree with its three major branches (magnoliids, monocots and eudicots) radiating from the basal group of angiosperms. A fourth updated version of the system (APG IV) was published in 2016, 7 years after its predecessor (APG III) [12]. Within orders, families are listed alphabetically.

The directions for finding the wild potatoes as you progress up (forward in time) the evolutionary tree are as follows (Fig. 2.2). Take the Asterids branch (clade) from the Core Eudicots, then the Lamiids branch (Euasterids I clade) from the Asterids, and finally the Solanales branch (order) from the Lamiids. There you will find the family Solanaceae with an estimated 91 genera and 2450 species, including the genera *Atropa*, *Capsicum*, *Datura*, *Nicotiana*, *Petunia* and *Solanum*. A dated phylogenetic tree for 1075 species in the family Solanaceae was published in 2013, based on a cladistic analysis of molecular (DNA) rather than morphological data, and a published fossil record [2]. It provided minimum age estimates in million years ago (mya) for the major splits from a common ancestor, for example, at 8 mya, cultivated tomato (*S. lycopersicum*) and cultivated potato (*S. tuberosum*); at 14 mya, cultivated eggplant (*S. melongena*) and tomato/potato; and then at the genus level, 19 mya for *Solanum* and *Capsicum* (peppers) and 24 mya for *Solanum* and *Nicotiana* (tobacco). Taxonomy is always a work in progress.

Fig. 2.1 Systematics beds inside the walled garden at the National Botanic Garden of Wales, 25 June 2015

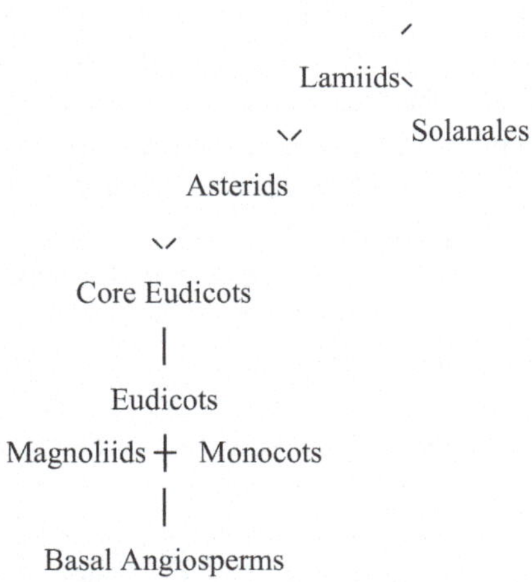

Fig. 2.2 Phylogenetic tree showing the order Solanales

2.3 Collecting Expeditions and Genebanks

Our knowledge of the potato's wild and cultivated relatives has come from numerous collecting expeditions to North, Central and South America from the 1920s to the 1990s [14–16]. As a result, wild species and cultivated landraces have been conserved and studied in situ, ex situ in genebanks and as herbarium specimens. Reference books have been written on the potatoes of Argentina, Brazil, Paraguay and Uruguay [17], Bolivia [18, 19], Peru [20] and North and Central America [21].

The first potato collecting expeditions were the Russian ones of Yuriy Voronov and Sergey Bukasov to Central and South America (Mexico, Guatemala, Cuba, Panama, Colombia, Venezuela, and parts of Chile and Peru) in 1925 and 1926 and that of Sergey Juzepchuk to Peru, Bolivia and Chile in 1928 [22]. Then Nikolay Vavilov (1887–1943) collected further samples of potatoes along with many other crops during his trips to the United States and Central and South America from 1930 to 1933 [22]. It was Vavilov who established our modern view of the origin, geography and evolution of cultivated plants from his research on the collected germplasm over the period 1920–1940 [23]. The extensive collections were housed and studied in "his institute", which was founded in Petrograd (Leningrad from 1924, St. Petersburg since 1991) in 1921 and bears his name, the Vavilov Research Institute of Plant Industry (VIR is the Russian abbreviation) [22]. The collections survived the 28-month-long siege of Leningrad (1941–1943) during WWII thanks to the sacrifices of a dedicated group of staff who died of starvation surrounded by collections of food. As for Vavilov, he also died of starvation, "a public enemy" of Stalin's Soviet Union, in Saratov city prison in January 1943. Stalin died in 1953, and from 1955, Vavilov's reputation was restored, and the value of scientific plant breeding in feeding the world's growing population was recognized [24]. Most Russian taxonomic descriptions of potato wild relatives were made from living plants of the collections grown in the experimental stations of VIR [25]. By 2010 the VIR potato collection had 2640 accessions of wild species, 3370 accessions of cultivated landraces, 2100 accessions of cultivars and 570 accessions of interspecific hybrids, making a total of 8680 accessions [25]. For comparison, it was estimated that in 2020 there were about 27,000 unique potato accessions conserved worldwide in genebanks [16], whereas simply adding the number of accessions held in genebanks gives a much larger number (over 80,000).

The pioneering Russian work encouraged collecting expeditions and the establishment of a number of potato germplasm collections (genebanks) by other European countries and the United States during the 1940s and 1950s (see websites for genebanks). The Commonwealth Potato Collection (CPC, Dundee, Scotland) began with the British Empire Potato Collecting Expedition of 1938–1939, which was led by a professional plant collector, E.K. Balls, and included J.G. Hawkes, who had just graduated in botany from Cambridge University. The Dutch-German Potato Collection (CGN, Wageningen, The Netherlands) dates back to a Dutch expedition in 1955 and a German one in 1959, and the German Groß Lusewitz Potato Collection (GLKS, IPK, Groß Lusewitz, Germany) dates back to 1949. Finally, the US Potato Genebank (NRSP-6, Sturgeon Bay, United States) was started by R.W. Hougas in 1950 and includes material collected by D. Correll in 1947, 1948, 1958 and 1960. Although these germplasm collections are made freely available to all bona fide research scientists, they were collected for the benefit of potato breeders and researchers in the countries where they became housed, not the countries where they were found.

This started to change in 1971 with the establishment of the International Potato Centre (CIP) in Lima, Peru. Since its foundation, CIP has been home to the world collection of potato germplasm and, through its breeding work, makes improved germplasm available to individual (developing) countries for release as cultivars or for further breeding through their own National Agricultural Research Systems (NARS). Today the world collection is maintained as a global public good under the International Treaty on Plant Genetic Resources for Food and Agriculture (ITPGRFA), and material is distributed with its Standard Material Transfer Agreement (SMTA). The treaty was adopted by the Thirty-First Session of the Conference of the Food and Agriculture Organization of the United Nations on 3 November 2001 and came into force on 29 June 2004. The treaty recognized "the enormous contribution of farmers to the diversity of crops that feed the world"; wanted to establish "a global system to provide farmers, plant breeders and scientists with access to plant genetic materials"; and wanted to ensure "that recipients share benefits they derive from the use of these genetic materials with the countries where they have been originated" [26]. On 16 July 2024, the treaty had 152 contracting parties (out of 194 countries), including one member organization (the European Union). Of the 15 major potato-producing countries listed in Chap. 1, China, Russian Federation, Ukraine and Belarus had not signed.

Information on the world collection can be found on the CIP genebank website [27]. Very briefly, CIP maintains a collection of 2595 accessions of wild species as botanical seed, backed up in the Svalbard Global Seed Vault, which opened on 26 February 2008 and is located for security deep in a remote mountainside close to the North Pole [28]. CIP's cultivated potato collection totals 4727 accessions, including 4354 traditional landrace cultivars, mainly but not exclusively from the Andean region, and improved cultivars. The entire clonal collection is conserved in vitro and distributed internationally as tissue-cultured materials. Furthermore, cryopreservation of the collection is now done routinely. Finally, CIP's herbarium collection has more than 90% of the wild and cultivated species represented from southern Colorado Canyon in the United States to southern Chile.

Although I have been to CIP, I must confess that I have not seen wild potato species and cultivated landraces growing in Peru, nor elsewhere in South America. My visual acquaintance came from plants growing in a glasshouse where I worked (Fig. 2.3). They formed part of the Commonwealth Potato Collection (CPC), which is currently held at what is now the James Hutton Institute in Dundee, Scotland [29, 30]. It resulted from expeditions to Mexico, the United States, Central and South America over the period 1938–1981, with a major involvement of Professor J.G. Hawkes of Birmingham University, United Kingdom [14]. He started his work on potatoes, with material from the British Empire Potato Collecting Expedition of 1938–1939, for his PhD thesis at Cambridge University under the supervision of Redcliffe N. Salaman (1874–1955). The CPC comprises 1500 accessions, two-thirds of which are wild species (80 of the ones recognized by Hawkes [14]) and one-third cultivated. It was fully converted to true seed populations in the 1960s. Each year 60 accessions are rejuvenated so that the entire collection is done every 25 years (Fig. 2.3). Sufficient seed for 20 seedlings of each accession is sown in spring in an insect-proof Quarantine Glasshouse, and the plants are tested for seed-transmitted quarantine organisms (viruses/viroids). The 20 plants are intercrossed at flowering by applying bulked pollen to all open flowers, repeating the process as

Fig. 2.3 Rejuvenation of part of the Commonwealth Potato Collection. (From Bradshaw JE, Bonierbale M (2010) Potatoes. Handbook of Plant Breeding 7. Springer, New York, pp. 1–52 with permission; 4 July 2008)

necessary. Seed is extracted from ripe berries, dried to 6% moisture, packeted, and kept in a dedicated store at low temperature (−20 °C) and humidity, conditions designed to retain seed viability for many years. A working collection store is maintained at 4 °C and 10% relative humidity. The collection is backed up in the Svalbard Global Seed Vault.

The germplasm collections have proved extremely useful to potato breeders as sources of raw materials for improving cultivated potatoes for traits such as disease and pest resistance, tolerance of abiotic stresses, and nutritional value [31]. However, to date, genes from relatively few accessions have made it into successful cultivars, but this may change to meet the goals of the "2030 Agenda for Sustainable Development" and to deal with global warming.

There is an important point to make from Fig. 2.3. The wild and cultivated species of the Andes flower in the long days of a Scottish summer, as do our cultivated potatoes. Hence, flowering time is not a barrier for hybridizations. However, unlike our cultivated potatoes, they do not produce tubers in long days, and this makes evaluation of tuber characteristics problematical. With shorter days, the reverse problem occurs. The wild and cultivated species of the Andes will both flower and produce tubers. In contrast, long-day-adapted cultivated potatoes do not flower but will produce tubers, although the plants may have an earlier maturity. In India, parents for hybridizations are planted outdoors in crossing blocks. In the cool hills around Shimla (about 30°N), potatoes are grown as a summer crop under natural day-length conditions. Flowering generally starts in the third week of June, with July the peak flowering time for doing the hybridizations. At Modipuram (also about 30°N) (Fig. 2.4), in the hot sub-tropical plains, potatoes are grown as a winter crop in short days. The crossing block is planted in the first week of October, and artificial lights are used to produce long days. Flowering starts in the last week of November with the peak in December [32].

Fig. 2.4 Crossing block at Modipuram centre of the ICAR-Central Potato Research Institute, India. (From Bradshaw JE (2021) Potato Breeding: Theory and Practice. Springer, Cham with permission; 13 December 2008)

2.4 Distribution of Wild Tuber-Bearing *Solanum* Species

In 1990, Professor Hawkes published an authoritative and comprehensive book on the potato, covering its evolution, biodiversity and genetic resources [14]. It has proved valuable to me and many others interested in potatoes for a variety of reasons. We learn that wild tuber-bearing *Solanum* species grow in a wide range of habitats and altitudes in 16 countries from the southwest of the United States (38°N), through Central America and the Andes, to Argentina and adjacent Chile (41°S). There are two clear centres of diversity: one in central Mexico and the other in Peru, with species occurring throughout the high Andes. Some species are widespread, whereas others are very restricted in their distribution, with no single species occurring throughout the whole geographical area. Although wild species are morphologically diverse, many resemble a smaller and more delicate version of the cultivated potato. One final and important point: as well as evolving to adapt to different habitats, wild species co-evolved with insect pollinators and also with pathogens and hence have genes for disease and pest resistance. Six species widely used by potato breeders are shown in Fig. 2.5.

2.5 Ploidy Levels in Wild Tuber-Bearing *Solanum* Species

The species shown in Fig. 2.5 provide examples of different ploidy levels. Potatoes, like all plants, have a cellular organization which today can be seen in great detail using powerful optical and electron microscopes and appropriate staining techniques [33]. Nevertheless, by the end of the nineteenth century, fibre-like threads called chromosomes could be seen in the cell nucleus and observed during the nuclear and

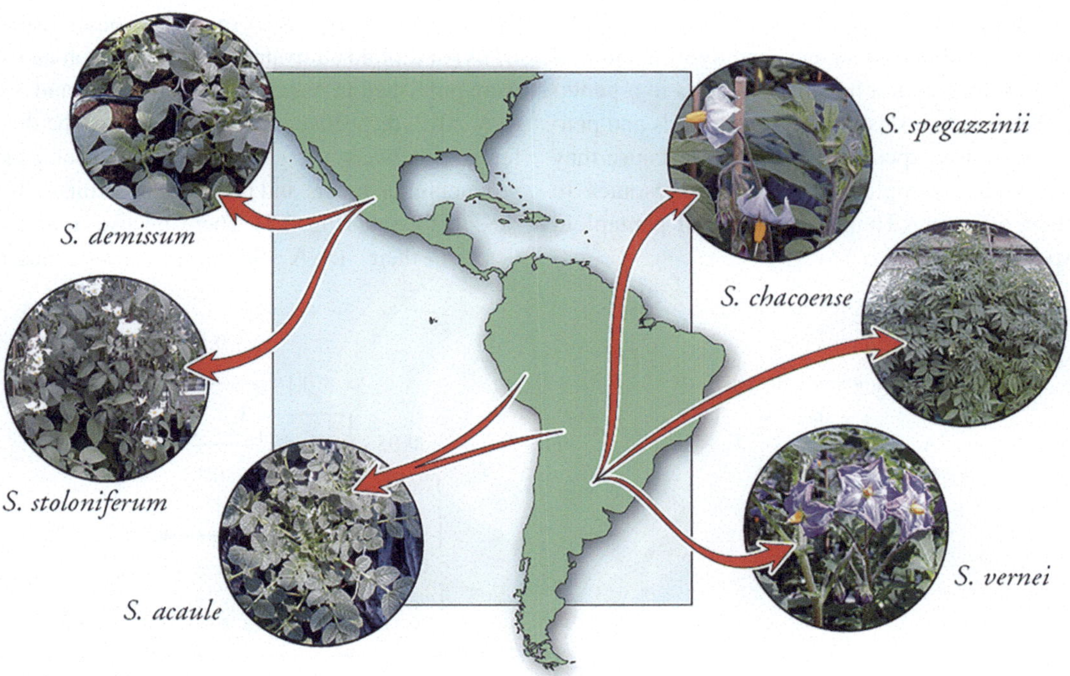

Fig. 2.5 Six species widely used by potato breeders (*S. spegazzinii* is now included in *S. brevicaule*). (From Bradshaw JE, Ramsay G (2009) Potato Origin and Production. In: Singh J, Kaur L (eds) Advances in Potato Chemistry and Technology. Elsevier, Amsterdam, pp. 1–26 with permission)

cell divisions that gave rise to new daughter cells. In 1903, Walter Sutton [34] correctly explained Mendel's laws of inheritance in terms of chromosome behaviour (pairing and separation) during two critical nuclear and cell divisions that occur during the formation of the male (sperm) and female (egg) reproductive cells (gametes) in the anthers and ovules, respectively. In other words, the chromosomes are the physical carriers of hereditary information from one generation to the next. Wild potato species have a basic set of 12 chromosomes ($x = 12$). About 80% have two sets of the 12 chromosomes, one inherited from the male parent and the other from the female parent, and are therefore diploid ($2n = 2x = 24$). Some species are tetraploid with four sets ($2n = 4x = 48$), two from each parent, and some are hexaploid with six sets ($2n = 6x = 72$), three from each parent [35, 36]. Out of the six species shown in Fig. 2.5, one is hexaploid (*S. demissum*), two are tetraploid (*S. stoloniferum* and *S. acaule*), and three are diploid (*S. spegazzinii* = *S. brevicaule*, *S. chacoense* and *S. vernei*). Hybridization of a diploid with a tetraploid results in a triploid with three sets of chromosomes, and hybridization of a tetraploid with a hexaploid results in a pentaploid with five sets. However, triploids and pentaploids are expected to be sterile because they can't pass complete sets of chromosomes to their offspring. Hence, triploid and pentaploid species are rare.

2.6 Molecular Phylogenetics and the Concept of Species

The latest taxonomy of wild tuber-bearing *Solanum* species requires an understanding of molecular phylogenetics and the concept of species. The places to find out about the construction of phylogenetic trees are books on molecular evolution and mathematical models in biology [37–39], preferably including the word introduction in their titles. They are likely to include the following kind of example of data and phylogenetic tree (Fig. 2.6). The tree attempts to show the evolutionary history of four extant species (A–D) from a common ancestor (G) at the root of the tree, via other hypothetical ancestors (E and F).

It is important to understand the phylogenetic distances, which can be explained with an example. Suppose that A and B are potatoes (*S. tuberosum*) and tomatoes (*S. lycopersicum*), two species in the genus *Solanum*. DNA is the chemical basis of heredity, with the genetic information coded in the sequence of four bases (abbreviated A, T, G, C) in each of its complementary strands. The first draft sequences of the genomes (total DNA) of diploid cultivated potatoes and tomatoes were published in *Nature* on 14 July 2011 and 30 May 2012, respectively [40, 41]. Some of the differences between the genomes result in potatoes producing the tubers and tomatoes the fruits (berries) that we eat (Fig. 2.7). There are also similarities in their DNA that result in enough

Phylogenetic distances

	A	B	C	D
A	-	-	-	-
B	0.2	-	-	-
C	0.4	0.4	-	-
D	0.5	0.5	0.5	-

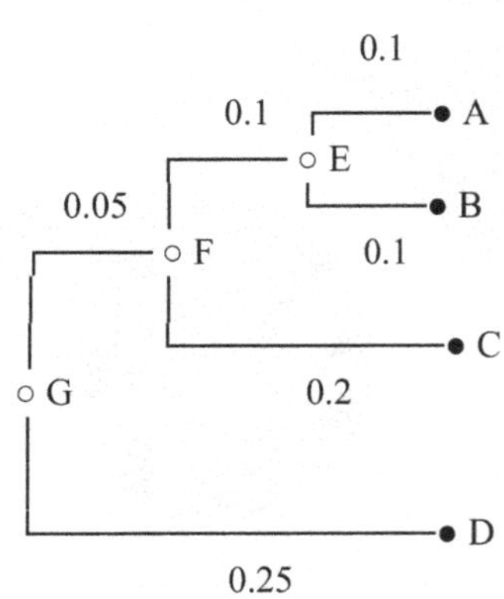

Fig. 2.6 Phylogenetic distances (Jukes-Cantor) between extant species (taxon) A to D and the derived phylogenetic tree: ● terminal node (leaf) for species A to D, ○ hypothetical ancestor for root (G) and two internal nodes (E and F), and branch (edge) lengths (numbers on horizontal lines) between nodes (vertical lines are just for clarity), which add to give the phylogenetic distances

morphological similarity for us to place both species in the genus *Solanum*. Mutations resulting in base changes in other sections of their DNA have been found but appear to be of no consequence and have accumulated over time. Hence, the number of such base differences between the extant species tells us how far they have diverged (phylogenetic distances) from a common ancestor. Furthermore, if the mutation rate is constant for the lineages, the accumulation of base changes can be interpreted as a molecular clock. If we can calibrate this clock through a fossil record, say of seeds, we can estimate in years the time from the start of divergence to the present day. We can also compare species in different genera, families and orders and so on through the higher ranks of classification. Where whole genome sequences are not available for the species of interest, the sequences of a number of genes common to all the species can be used (molecular data), as happened in constructing modern phylogenetic trees.

The tree in Fig. 2.6 is constructed from DNA sequence data where each species has provided a DNA sequence from a gene (or section of DNA) common to all the species. The sequences need to be aligned, which is not a trivial task. The lengths of the branches represent the amount of mutation (base changes in the DNA) that occurred between bifurcations of the lineage. They are not simply the number of differences in the base sequences, as back mutations to the original base can occur, which might deceive us into thinking no mutations have occurred. The amount of mutation along any branch is proportional to the elapsed time of evolution, provided the assumption about the molecular clock holds. The tree is built from the phylogenetic distances derived from the sequence information (base substitutions) using what is known as the Jukes-Cantor model [37–39]. The method used is known as the unweighted pair-group method with arithmetic means (UPGMA). As A and B are the closest species at 0.2 apart, we start with equal branch lengths from E of 0.1. We next combine A and B into a group and work out the average distance of the group to species C (0.4) and to species D

Fig. 2.7 Potatoes and tomatoes in a market in Sanlucar de Barrameda, Spain, 8 February 2023

(0.5). As C is the shortest distance, we assign equal branch lengths of 0.2 from F to A, B and C, which means that F and E are 0.1 apart. Finally, we combine A, B and C into a group and work out the average distance of the group to species D (0.5). Hence, D is 0.25 from the root of the tree, and G and F are 0.05 apart.

If all of the members of species A look similar and clearly different from B, C and D, and the same applies to these species, we have a taxonomic classification and taxonomic species. If all of the members of species A are interfertile but will not hybridize with members of B, C and D, and the same applies to these species, we have biological species. If the four species have a common ancestor G, then we have an evolutionary history and evolutionary species. If there are no inconsistencies in these three concepts of species, we have an intellectually satisfying outcome. If not, we have problems to resolve, as happens with potatoes and their wild relatives. We will shortly discover that there are 107 wild tuber-bearing taxonomic species in the genus *Solanum* which fall into three crossability groups and three or four phylogenetic groups (clades).

As the number of species (or other groups) increases, the number of trees that can relate the terminal nodes in the search for the "true" one becomes astronomical (over 34 million rooted trees for 10 species and a lot more with 107 species!). Furthermore, in practice, a tree will not fit the data exactly. So, constructing a phylogenetic tree is not a trivial task and requires an understanding of the following topics: probability and conditional probability, matrix models of base substitution (e.g., Jukes-Cantor model and Kimura models), methods for determining phylogenetic distances (e.g., Jukes-Cantor distance and Kimura distances), methods for tree construction (e.g., the Neighbour-Joining algorithm or Maximum Parsimony or Maximum Likelihood), the inclusion of a more distantly related taxon (outgroup) to ensure a rooted tree (e.g. species D in Fig. 2.6), assessing the reliability of a tree (e.g., bootstrapping) and lastly the algorithms used to produce the available computer software.

2.7 Taxonomy of Wild Tuber-Bearing *Solanum* Species

Professor Hawkes provided a detailed taxonomic description of the 219 wild tuber-bearing species which he recognized, all of which he placed in the *Solanum* section *Petota*, along with nine species that do not produce tubers [14]. Furthermore, he grouped the 219 species into two superseries, *Stellata* (star-shaped flowers) and *Rotata* (round-shaped), containing nine and ten series, respectively, which he considered a meaningful evolutionary sequence, with *Stellata* the more primitive and Mexico the possible centre of origin of tuber-bearing species. Twenty-four years later, in 2014, we find that a new generation of taxonomists has reduced the number to 107 species [42]. How could this be? Professor Hawkes adopted the morphological species concept, albeit in modified form. Although differences can be seen between members of a species, they are small compared with those seen between individuals from different species. Professor Hawkes wanted to distinguish one species from another by associating breaks or gaps in the total pattern of morphological diversity with ways of maintaining the integrity of species, such as eco-geographical separation and differences in flowering time. Put simply, plants in different geographical areas or habitats, or with different flowering times, can't hybridize. Since 1990, however, morphological studies have been complemented by ones at the molecular level, using material from extensive new collecting expeditions. The results provided the taxonomic rationale for the drastic reduction in the number of recognizable species to 107, all in section *Petota*, but excluding the non-tuber-bearing ones. This reduction can be confusing (*S. spegazzinii* is now included in *S. brevicaule*) but is considered a more meaningful taxonomy than the series of Hawkes. Both taxonomies remain in use!

The 107 species are not biological ones. Their crossability has been determined by artificial pollinations done over many years, with the result that three main groups can be identified. Hybridizations between species within the groups are usually successful, whereas those between species in different groups are usually unsuccessful. This is because species within groups have evolved in different locations but have not developed mechanisms to prevent their hybridization, whereas genetic barriers to hybridization have evolved between species in different groups. Therefore, hybridizations between species in different groups are not successful, even if the species occur in the same location, although there are exceptions. What is the main barrier to hybridization?

Viable seeds that will germinate and grow into a healthy plant require a normal endosperm for nourishing the embryo. In some crosses between wild species the endosperm degenerates and thus provides a barrier to success. Whether or not this happens depends on the genetic composition of the endosperm, which develops from the fusion of a sperm nucleus from the male parent with two (polar) nuclei from the female parent to give a triple fusion nucleus. Hybridizations are successful if the species have the same Endosperm Balance Number (EBN). This means that the two polar nuclei and the one sperm nucleus will have an EBN ratio of 2 to 1. EBN is determined experimentally relative to species assigned an arbitrary value of 1 [43, 44]. The three main crossability groups are therefore EBN = 1, EBN = 2 and EBN = 4.

Examples of successful and unsuccessful hybridizations are given in Table 2.1. It can be seen that an unreduced sperm nucleus (2EBN pollen) from a 2*x* 2EBN male parent can successfully fertilize a 4*x* 4EBN female parent with two 2EBN polar nuclei, and likewise for an unreduced egg (2EBN). This is important in breeding work because most wild species are 2*x* 2EBN (Table 2.2), and the principal cultivated potato is 4*x* 4EBN. Potato breeders can, in fact, use various methods of ploidy manipulation to produce wild species with the same EBN as cultivated species and hence transfer desirable genes from wild to cultivated species [45].

Finally, the new molecular data were used for cladistic analyses of the 107 species. When based on plastid (chloroplast) genes, the result was four clades (1–4), whereas nuclear genes gave three clades (1 + 2, 3 and 4 or more simply 1, 3 and 4, as shown in Table 2.2).

The taxonomy of the wild species published in 2014 [42] includes a list of the 107 species in alphabetical order, along with their countries of occurrence, ploidy level, endosperm balance number and finally their clade (1, 3 or 4). A summary is shown in Table 2.2 of the 74 species for which all of this information was available.

Although wild species are found in all 16 countries listed, there are two clear centres of diversity: one in central Mexico and the other in Peru, with species occurring throughout the high Andes of Colombia, Ecuador, Peru, Bolivia and Argentina. Sixty of the 74 species are diploid and fall into four groups based on EBN number and clade. The 1EBN species are both geographically (Mexico and Peru) and genetically separated (Clade 1 and Clade 3 with one species in Clade 4). The 2EBN species occur from Colombia to Argentina with one exception *(S. verrucosum* occurs in Mexico) with a genetic separation

Table 2.1 Endosperm balance in triploid fusion nucleus of endosperm

Female × Male	Two polar nuclei	One sperm nucleus	Female: Male EBN
2*x* 1EBN × 2*x* 1EBN	2 × ½EBN	½EBN	2:1 yes
2*x* 2EBN × 2*x* 2EBN	2 × 1EBN	1EBN	2:1 yes
2*x* 1EBN × 2*x* 2EBN	2 × ½EBN	1EBN	1:1 no
2*x* 2EBN × 2*x* 1EBN	2 × 1EBN	½EBN	2:½ no
4*x* 4EBN × 4*x* 2EBN	2 × 2EBN	1EBN	4:1 no
4*x* 4EBN × 2*x* 2EBN	2 × 2EBN	1EBN	4:1 no
4*x* 4EBN × 2*x* 2EBN pollen	2 × 2EBN	2EBN	4:2 yes
2*x* 2EBN egg × 4*x* 4EBN	2 × 2EBN	2EBN	4:2 yes

Table 2.2 A classification for 74 of the 107 wild tuber-bearing *Solanum* species for which information was available on country of occurrence, ploidy level (*x*), endosperm balance number (EBN) and, for diploid species, their clade (1, 3 or 4) in brackets after the number of species (<1 means species accounted for in another country)

	2*x* 1EBN	2*x* 2EBN	4*x* 2EBN	6*x* 4EBN	All
United States	<1		<1		
Mexico	7(1)	1(4)	4	5	17
Guatemala	<1		<1	<1	
Honduras	<1		<1		
Costa Rica					
Panama					
Venezuela			<1		
Brazil		<1			
Paraguay		<1			
Uruguay		<1			
Colombia		1(3)	1		2
Ecuador	1(3)	3(3)	<1	<1	4
Peru	10(3), 1(4)	11(3), 14(4)	3	1	40
Bolivia		6(4)	<1		6
Argentina		5(4)	<1		5
Chile		<1			
All	19	41	8	6	74

Information from [42]

(Clade 3 and Clade 4) that divides them into a northern and southern group. The tetraploid species are divided between Mexico and Peru, whereas the hexaploid species occur mainly in Mexico. Clades are not given for the tetraploid and hexaploid species because they are genetically more complicated, in effect containing genes from more than one clade. A more recent (2018) and comprehensive taxonomy, based on a phylogenetic analysis of 201 accessions of *Solanum* section *Petota* species, has confirmed the main features of Table 2.2 whilst revealing a northern (Peru) and southern (Bolivia southwards) geographic division of Clade 4 [46]. As always, taxonomy is a work in progress.

2.8 Self-Incompatibility (SI)

There is one other important aspect of reproductive biology that I need to mention. Self-pollination does not normally occur in plants of the diploid species, with the notable exception of *S. verrucosum*, which is self-compatible (SC). They possess what we now regard as a classical gametophytic self-incompatibility (GSI) system. The genetics of the system and much of its molecular basis are understood, but I am not going into the details [47–52]. Here I just want to mention that pollen grains from the same plant (self-pollen) and from a different plant (cross-pollen) of the same species germinate on the stigma, and pollen tubes start to grow down the style, but only those of cross-pollen reach and fertilize the eggs. Reactions occur in the first third of the style that allow it to recognize the tubes of self-pollen and stop them from growing. Can this happen in crosses between different species? One example involves *S. verrucosum.* If an SI diploid 1EBN species is the female parent in a cross with SC *S. verrucosum* (2EBN), the GSI system operates and pollen tube growth is stopped in the style. However, if *S. verrucosum* is the female parent, pollen tubes reach and fertilize the eggs, but seed set is poor because endosperm collapse occurs, with a few exceptions, because of the difference in EBN [53, 54]. The relationship between GSI and interspecies incompatibility is still not fully understood [55] but may be resolved as more genomes (total DNA) of wild species are sequenced. I assume that the pollen of more distantly related non-tuber-bearing *Solanum* species either does not reach the stigmas of tuber-bearing species, or fails to germinate, or the pollen tubes fail to penetrate the stigma.

2.9 Genomes of Wild Tuber-Bearing *Solanum* Species

Over the period 2015–2022, the genomes of three wild tuber-bearing *Solanum* species were sequenced: *S. chacoense*, *S. commersonii* and *S. verrucosum* [56–58]. *S. verrucosum* and *S. chacoense* are both diploid 2EBN species in Clade 4 but geographically well-separated, with the former in Mexico and the latter widely distributed in Peru, Bolivia, Argentina, Brazil, Paraguay and Uruguay. In contrast, *S. commersonii* and *S. chacoense* are both diploid species which occur in the same geographical area, Argentina, Brazil and Uruguay, but the former is 1EBN and the latter is 2EBN. Hence, they do not hybridize. In 2023, the number of sequenced genomes dramatically increased to 23 as a result of a large research project which produced a whole-genome phylogeny of 92 *Solanaceae* species [59]. The analysis estimated that the common ancestor of all *Solanum* species occurred 25 million years ago (mya) and of all *Solanaceae* species 80 mya. As even more genomes are sequenced, there will be further insights into the evolution and adaptations of the wild tuber-bearing *Solanum* species.

2.10 The Need for Conservation

The wild tuber-bearing *Solanum* species are a valuable but underutilized plant genetic resource for breeding new potato cultivars better adapted to providing global food security during a period of climate change. They need to be conserved, both in situ and ex situ in genebanks, studied, and used in breeding programs through knowledge of their adaptive traits, taxonomy and reproductive biology. More details can be found in the 2018 book chapter by Bamberg et al. on *Ensuring the genetic diversity of potatoes* [60], the 2020 one by Ellis et al. on Ex situ *conservation of potato [Solanum Section Petota (Solanaceae)] genetic resources in genebanks* [16], and the 2021 book by Bradshaw on *Potato Breeding: Theory and Practice* [61].

References

1. Willis KJ, McElwain JC (2014) The evolution of plants, 2nd edn. Oxford University Press, Oxford
2. Särkinen T, Bohs L, Olmstead RG, Knapp S (2013) A phylogenetic framework for evolutionary study of the nightshades (Solanaceae): a dated 1000-tip tree. BMC Evol Biol 13:214. https://doi.org/10.1186/1471-2148-13-214. PMID: 24283922; PMCID: PMC3850475
3. Murphy DJ (2007) People, plants & genes: the story of crops and humanity. Oxford University Press, Oxford
4. Schwägerl C (2014) The anthropocene: the human era and how it shapes our planet. Synergetic Press, Sante Fe
5. Pyne L (2020) In search of prehistoric potatoes. Archaeology, March/April 2020, Archaeological Institute of America
6. Louderback LA, Pavlik BM (2017) Starch granule evidence for the earliest potato use in North America. Proc Natl Acad Sci USA 114:7606–7610. https://doi.org/10.1073/pnas.1705540114
7. Ugent D, Dillehay T, Ramirez C (1987) Potato remains from a late Pleistocene settlement in south central Chile. Econ Bot 4:17–27
8. Linnaeus C (1735) Systema Naturae (in Latin)
9. Linnaeus C (1753) Species Plantarum (in Latin)
10. Darwin C (1859) On the origin of species by means of natural selection. The 150th Anniversary Landmark Edition: Bynum W (ed.) Penguin Group, London
11. Niklas KJ (2016) Plant evolution. The University of Chicago Press, Chicago/London
12. Angiosperm Phylogeny Group IV (2016). http://www.mobot.org/MOBOT/research/APweb/
13. Edwards D, Spears P, Channing A (2011) Flowering plant families at the national botanic garden of Wales. FIRST, London
14. Hawkes JG (1990) The potato: evolution, biodiversity and genetic resources. Belhaven Press, London
15. Spooner DM, Hijmans RJ (2001) Potato systematics and germplasm collecting, 1989–2000. Am J Potato Res 78:237–268
16. Ellis D, Salas A, Chavez O, Gomez R, Anglin N (2020) Ex situ conservation of potato [*Solanum* Section *Petota* (Solanaceae)] genetic resources in genebanks. In: Campos H, Ortiz O (eds) The potato crop: its agricultural, nutritional and social contribution to humankind. Springer, ebook, pp 109–138
17. Hawkes JG, Hjerting JP (1969) The potatoes of Argentina, Brazil, Paraguay and Uruguay. a biosystematic study. Oxford University Press, Oxford

18. Hawkes JG, Hjerting JP (1989) The potatoes of Bolivia. Their breeding value and evolutionary relationships. Oxford University Press, Oxford
19. Ochoa CM (1990) The potatoes of South America: Bolivia. Cambridge University Press, Cambridge
20. Ochoa CM (2004) The potatoes of South America: Peru. 1: the wild species. International Potato Center, Lima
21. Spooner DM, van den Berg RG, Rodriguez A, Lara-Cabrera SI (2004) Wild potatoes (Solanum section Petota; Solanaceae) of North and Central America. Systematic Botany Monographs Volume 68
22. Loskutov IG (1999) Vavilov and his institute. A history of the world collection of plant genetic resources in Russia. International Plant Genetic Resources Institute, Rome. ISBN 92-9043-412-0
23. Vavilov NI (1920–1940) English translation of papers on origin and geography of cultivated plants. Cambridge University Press, Cambridge, Digitally Printed Version 2009
24. Love D (1992) Translator's foreword of Origin and Geography of Cultivated plants by N.I. Vavilov. Cambridge University Press, Cambridge, Digitally Printed Version 2009
25. Gavrilenko T, Antonova O, Ovchinnikova A, Novikova L, Krilova E, Mironenko N, Pendinen G, Islamshina A, Shvachko N, Kiru S, Kostina L, Afanasenko O, Spooner DM (2010) A microsatellite and morphological assessment of the Russian National cultivated potato collection. Genet Resour Crop Evol 57:1151–1164
26. International Treaty on Plant Genetic Resources for Food and Agriculture (ITPGRFA). www.fao.org/plant-treaty/overview/en/
27. CIP Genebank. https://cipotato.org/genebankcip/
28. Hawtin G, Fowler L (2011) The Global Crop Diversity Trust. In: Frison C, López F, Esquinas-Alcázar JT (eds) Plant genetic resources and food security. FAO, Bioversity International and Earthscan, Abingdon, pp 209–221
29. Commonwealth Potato Collection. https://ics.hutton.ac.uk/germinate-cpc/#home
30. Bradshaw JE, Ramsay G (2005) Utilisation of the Commonwealth Potato Collection in potato breeding. Euphytica 146:9–19
31. Bradshaw JE, Bryan GJ, Ramsay G (2006) Genetic resources (including wild and cultivated *Solanum* species) and progress in their utilisation in potato breeding. Potato Res 49:49–65
32. Luthra SK, Gupta VK, Tiwari JK, Kumar V, Bhardwaj V, Sood S, Dalamu, Kaur RP, Kumar R, Vanishree G, Kumar D, Mhatre P, Chakrabarti SK (2020) Potato breeding in India. CPRI Technical Bulletin No 74 (revised), ICAR-Central Potato Research Institute, Shimla, Himachal Pradesh, India
33. Blackmore S (2012) Green universe. Royal Botanic Garden Edinburgh and Papadakis Publisher, Winterbourne
34. Sutton WS (1903) The chromosomes in heredity. Biol Bull Mar Biol Lab, Woods Hole 4:231–248
35. Fukuda Y (1927) Cytological studies on the development of pollen grains in different races of *S. tuberosum* with special reference to sterility. Botanical Magazine (Tokyo) 61:459–476
36. Smith HB (1927) Chromosome counts in the varieties of *S. tuberosum* and allied wild species. Genetics 12:84–92
37. Graur D, Li WH (2000) Fundamentals of molecular evolution, 2nd edn. Sinauer Associates, Sunderland
38. Allman ES, Rhodes JA (2004) Mathematical models in biology. Cambridge University Press, Cambridge
39. Yang Z (2014) Molecular evolution: a statistical approach. Oxford University Press, Oxford
40. The Potato Genome Sequencing Consortium (2011) Genome sequence and analysis of the tuber crop potato. Nature 475:189–195
41. Tomato Genome Consortium (2012) The tomato genome sequence provides insights into fleshy fruit evolution. Nature 485:635–641
42. Spooner DM, Ghislain M, Simon R, Jansky SH, Gavrilenko T (2014) Systematics, diversity, genetics, and evolution of wild and cultivated potatoes. Bot Rev 80:283–383
43. Johnston SA, der Nijs TPM, Peloquin SJ, Hanneman RE Jr (1980) The significance of genic balance to endosperm development in interspecific crosses. Theor Appl Genet 57:5–9
44. Hawkes JG, Jackson MT (1992) Taxonomic and evolutionary implications of the endosperm balance number hypothesis in potatoes. Theor Appl Genet 84:180–185
45. Ortiz R (1998) Potato breeding via ploidy manipulation. Plant Breed Rev 16:15–86
46. Li Y, Colleoni C, Zhang J, Liang Q, Hu Y, Ruess H, Simon R, Liu Y, Liu H, Yu G, Schmitt E, Ponitzki C, Liu G, Huang H, Zhan F, Chen L, Spooner D, Huang B (2018) Genomic analyses yield markers for identifying agronomically important genes in potato. Mol Plant 11:473–484
47. Cipar MS, Peloquin SJ, Hougas RW (1964) Inheritance of incompatibility in hybrids between *Solanum tuberosum* haploids and diploid species. Euphytica 13:163–172
48. Dodds KS (1965) The history and relationships of cultivated potatoes. In: Hutchinson JB (ed) Essays in crop plant evolution. Cambridge University Press, Cambridge, pp 123–141
49. Dzidzienyo DK, Bryan GJ, Wilde G, Robbins TP (2016) Allelic diversity of S-RNase alleles in diploid potato species. Theor Appl Genet 129:1985–2001
50. Gebhardt C, Ritter E, Barone A, Debener T, Walkemeier B, Schachtschabel U, Kaufmann H, Thompson RD, Bonierbale MW, Ganal MW, Tanksley SD, Salamini F (1991) RFLP maps of potato and their alignment with the homoeologous tomato genome. Theor Appl Genet 83(1):49–57
51. Kaufmann H, Salamini F, Thompson RD (1991) Sequence variability and gene structure at the

self-incompatibility locus of *Solanum tuberosum*. Mol Gen Genet 226:457–466
52. McClure B, Cruz-García F, Romero C (2011) Compatibility and incompatibility in S-RNase-based systems. Ann Bot (Lond) 108:647–658. https://doi.org/10.1093/aob/mcr179
53. Hermsen JGTh (1977) General considerations on interspecific hybridization. In: Sanchez-Monge E, Garcia-Olmedo F (eds) Proceedings 8th Congress of Eucarpia, Madrid, Spain, pp 299–304
54. Jansky S, Hamernik A (2009) The introgression of 2×1EBN *Solanum* species into the cultivated potato using *Solanum verrucosum* as a bridge. Genet Resour Crop Evol 56:1107. https://doi.org/10.1007/s10722-009-9433-3
55. Behling WL, Douches DS (2023) The effect of self-compatibility factors on interspecific compatibility in *Solanum* Section *Petota*. Plants 12(8):1709. https://doi.org/10.3390/plants12081709
56. Leisner CP, Hamilton JP, Crisovan E, Manrique-Carpintero NC, Marand AP, Newton L, Pham GM, Jiang J, Douches DS, Jansky SH, Buell CB (2018) Genome sequence of M6, a diploid inbred clone of the high-glycoalkaloid-producing tuber-bearing potato species *Solanum chacoense*, reveals residual heterozygosity. Plant J 94(3):562–570
57. Aversano R, Contaldi F, Ercolano MR, Grosso V, Iorizzo M, Tatino F, Xumerle L, Molin AD, Avanzato C, Ferrarini A, Delledonne M, Sanseverino W, Cigliano RA, Capella-Gutierrez S, Gabaldón T, Frusciante L, Bradeen JM, Carputo D (2015) The *Solanum commersonii* genome sequence provides insights into adaptation to stress conditions and genome evolution of wild potato relatives. Plant Cell 27(4):954–968
58. Hosaka AJ, Sanetomo R, Hosaka K (2022) A de novo genome assembly of *Solanum verrucosum* Schlechtendal, a Mexican diploid species geographically isolated from other diploid A-genome species of potato relatives. G3, 12(8):jkac166. https://doi.org/10.1093/g3journal/jkac166. PMID: 35775942; PMCID: PMC9339273
59. Wu Y, Li D, Hu Y, Li H, Ramstein GP, Zhou S, Zhang X, Bao Z, Zhang Y, Song B, Zhou Y, Zhou Y, Gagnon E, Sarkinen T, Knapp S, Zhang C, Stadler T, Buckler ES, Huang S (2023) Phylogenomic discovery of deleterious mutations facilitates hybrid potato breeding. Cell 186:2313–2328. https://doi.org/10.1016/j.cell.2023.04.008
60. Bamberg J, Jansky S, del Rio A, Ellis D (2018) Ensuring the genetic diversity of potatoes. In: Wang-Pruski G (ed) Achieving sustainable cultivation of potatoes. Burleigh Dodds, Cambridge, pp 57–79
61. Bradshaw JE (2021) Potato breeding: theory and practice. Springer, Cham

3 Domestication and Cultivation in South America

Andean farmers, without our modern understanding of botany and genetics, utilized naturally occurring variation from mutation and sexual reproduction to establish landraces which they propagated from tubers. The origins and spread of potato cultivation through the Andes and into Southern Chile are relevant to the genetic base of modern cultivars and the need for genes from extant landraces to provide increased food security in many countries. As archaeological evidence is sparse and historical records are limited, we need to consider the current distribution of cultivated potatoes and their wild relatives and also analyse their DNA. Three key conclusions emerge. (1) Potatoes were first cultivated and domesticated from a few wild species in the high Andes of southern Peru, where the summers are cool and wet and daylengths are 12–13 h. (2) Although the first cultigen was a diploid potato, Group Stenotomum, it was the cultivation of tetraploid Group Andigena that spread throughout the Andes, and the genetically distinct tetraploid Group Tuberosum that became established in the longer days of Southern Chile. (3) There were natural hybridizations between cultivated potatoes and their wild relatives which could not subsequently occur elsewhere in the world in the absence of wild relatives.

3.1 Wild Relatives as Food

Potatoes have soft tissues that disintegrate over time so that they are difficult to find in the archaeological record. Nevertheless, our knowledge of the prehistory of wild tuber-bearing *Solanum* species (potatoes) being used as food comes from archaeology. A good introduction is a letter by Lydia Pyne in the March/April 2020 magazine *Archaeology*, a publication of the Archaeological Institute of America [1]. The title is "In Search of Prehistoric Potatoes: Native Peoples of the American Southwest Dined on a Little-Known Spud at least 10,000 Years Ago". There are two surprises in the title: the earlier-than-expected date and the place, which is not in one of the centres of diversity of the wild species (Mexico and Peru). The initial archaeological dig took place between 2004 and 2008 at North Creek Shelter (38°N), located in southern Utah at the base of a sandstone cliff at an elevation of 1900 masl. Subsequently, in 2017, archaeologist Lisbeth Louderback (University of Utah) and botanist Bruce Pavlik (Natural History Museum of Utah) published their findings in the Proceedings of the National Academy of Sciences of the United States of America (PNAS) [2]. They had been able to extract well-preserved starch granules (40–50 μm

J. E. Bradshaw, *Can Potatoes Feed the World?*, Sustainable Development Goals Series,
https://doi.org/10.1007/978-3-031-92890-1_3

Table 3.1 Today's monthly weather averages for Lake Titicaca, Peru, from weather-atlas.com

Month	Low temperature (°C)	High temperature (°C)	Rainfall (mm)	Daylength (hours mins)
January	3.6	16.7	133.3	12 57
February	3.5	16.7	108.7	12 37
March	3.2	15.5	98.5	12 11
April	0.6	16.8	43.3	11 44
May	−3.8	16.6	9.9	11 22
June	−7.0	16.0	3.1	11 12
July	−7.5	16.0	2.4	11 17
August	−5.4	17.0	5.8	11 36
September	−1.4	17.6	22.1	12 01
October	0.3	18.6	41.1	12 28
November	1.5	18.8	55.3	12 52
December	3.0	17.7	85.9	13 04

long) from stone tools used to grind raw foods and to use five strictly defined diagnostic characteristics to identify them as the species *S. jamesii* (2*x*, 1EBN, Clade 1). The excavated stones came from a substratum of the site dated to between 10,100 and 10,900 years ago. *S. jamesii* is known as the "Four Corners potato" because it is found where Utah, Colorado, Arizona and New Mexico meet. Dormant tubers, which can persist in the ground for at least eight years, produce shoots after the onset of summer monsoon rain in July. They grow into small plants with white (yellow tinge) flowers and produce small tubers, about the size of walnuts, in the autumn. The tubers are highly nutritious, having twice the protein, zinc and manganese content of cultivated *S. tuberosum* and three times the calcium and iron content. They are, however, naturally high in glycoalkaloids, which are extremely bitter and potentially toxic compounds. Native American communities still eat these potatoes to this day, having boiled them in white clay to reduce their bitterness. Given their long prehistory and history, Louderback and Pavlik [2] speculate over whether or not some populations of *S. jamesii* were actually cultivated and their produce transported to other locations. Currently, we do not know.

There is also some evidence for a different wild potato being used as food in the southern extreme of the distribution of wild tuber-bearing *Solanum* species. Starch grains were found attached to preserved tuber skins in buried cultural remains at a human settlement in Monte Verde (41°S), near the city of Puerto Montt in south-central Chile, dated to 13,000 years ago [3]. The grains were identified as *S. maglia*, a white-flowered plant with small tubers that occurs in coastal forests and is the only wild species of this southern temperate region. Today local inhabitants are said to eat these wild, edible potatoes. Again, the authors of the scientific paper speculate about whether or not they have been cultivated during the last 13,000 years. However, in his 1990 book, Professor Hawkes [4] points out that *S. maglia* does not occur anywhere near the site of the human settlement. He did, however, accept that some sort of wild potato was being eaten at the time, and that may be as much as we can safely conclude.

3.2 Domestication: Which Species, Where and When?

Domestication is more than just the cultivation of wild species. It is selecting them for better adaptation to cultivation and other human needs. Hence, we need to ask which species were domesticated, where and when? But there is no written history to answer these questions. We do know where the wild species occur today. We also know that today some 3000 different types of potato (landraces) are grown by indigenous (Indian) farmers in South America, throughout the Andes (3000–3500 masl) from western Venezuela to northern Argentina, and also in lowland south-central Chile [5]. The area of greatest diversity of these cultivated potatoes is in the central Andes of Peru and Bolivia and hence, according to the ideas of N. I. Vavilov, the most likely centre of origin of the cultivated crop, assuming diversity increases over time with crop evolution [6, 7]. Today, however, we can also look for answers in the similarities and differences between the wild and cultivated potatoes at the DNA level.

More precise answers to our first two questions came from phylogenetic analyses of DNA markers (AFLPs) from 362 individual wild (261) and landrace (98) potatoes, with three outgroup non-tuber-bearing *Solanums* [8]. The analyses revealed a cladistic split of members of a complex set of species (clade 4) into a northern (Peru) and southern (Bolivia and Argentina) component. The data supported a monophyletic origin (descendants of a single ancestor) of the landraces from the northern component in Peru. Put another way, the authors were saying that there was a single domestication in the highlands of southern Peru, from the northern group of members of the *S. brevicaule* complex of morphologically similar diploid species typified by *S. bukasovii* (now included in *S. candolleanum* [5] along with 26 other species in the scheme of Hawkes [4]). The above-ground characteristics of the *S. candolleanum* complex and the cultivated potatoes now growing in the same area are similar, although the latter are taller and more vigorous.

It has proved difficult to date the earliest cultivation of potatoes and hence answer the when question. Hawkes [4] concluded that the potato is an ancient domesticate which has been cultivated for at least 7000 years, based on preserved food plant remains found at various excavated sites on the arid coast of Peru and one site in the high Chilca canyon, south of Lima.

We now need to turn our attention to the area around Lake Titicaca in Peru at latitude 15°S and a height of just over 3800 masl. It is instructive to start by looking at the current monthly weather averages (Table 3.1). There is little variation in daylength and the highest average temperature, whereas there is a large variation in the rainfall and the lowest average temperature: summer (October to March) is cool and wet, but winter (April to September) is cold and dry. Hence, summer is the growing season for potatoes. However, between 15,000 and 5000 years ago, conditions were very different, as the high mountains of the Andes (Cordillera) slowly emerged from the last ice age [9].

Rumold and Aldenderfer, in PNAS in 2016 [10], provided evidence for the early use of potatoes at Jiskairumoko, an early village site in the western Titicaca Basin of the high south-central Andes (3890 masl), dating to 5420–3620 years before present. Out of 141 starch, micro remains recovered from 14 ground stone tools, 50 were identified as consistent with cultivated or domesticated potatoes, based on reference to published materials and a study of wild and cultivated potato starch morphology. However, the researchers couldn't say with certainty that the 50 micro remains were from cultivated/domesticated potatoes and not from wild potato species such as *S. candolleanum*, which is native to the area. They were, however, able to cite references to evidence for small-scale farming being in place by 3620 years before present. We can therefore reasonably conclude that domestication was complete by this date. In his book *The Incas and their ancestors*, M. E. Moseley [9] argued that the herding of animals and the domestication of llamas and alpacas probably preceded the settled cultivation of crops by early Andean farmers. The herding would have involved moving up and down the mountains with the changing seasons. Under these circumstances seasonal visits to favoured sites for food plants would have meant that domestication proceeded gradually, with favoured types being re-planted to ensure harvests at subsequent re-visits. The llamas would be used to transport produce as well as providing fertilizer in the form of dung! In other words, cultivation became a necessity in the harsh conditions of the mountains of the Andes (Cordillera), but more extensive farming did not start to take hold until the arrival of more favourable conditions about 3620 years ago.

Ceramics depicting potatoes, recovered from burial sites, indicate that they were widely known to Peruvian people from the beginning of the common era (CE, now preferred to AD). Much later, during the fifteenth century, the Incas developed sophisticated agriculture in the high-altitude, cold and dry climate around their Peruvian capital of Cuzco. They grew crops of maize, potatoes and other food plants during summer on terraces built of dressed stone which trapped the heat of the sun and aided irrigation, then stored the produce for use during winter [9]. Later, historical accounts from the time of the Spanish conquest of the Americas (1537 CE onwards) reveal that the potato was widely grown throughout the Andes, from Colombia to Peru, and also in southern Chile [4]. Did the cultivation of potatoes spread during the fifteenth century as the Incas expanded and governed their empire with a network of roads covering northern Chile, upland Argentina, Bolivia, Peru, Ecuador and the south of Colombia, or did the Incas conquer indigenous people beyond Peru and Bolivia who were already growing potatoes? We shall probably never know.

After the International Potato Centre (CIP) was founded in Lima, Peru, in 1971, their scientists collected more than 17,000 potato cultivars (landraces) native mainly to South America. Subsequently, duplicate accessions were identified and eliminated, as well as mixtures, atypical plants, and those with virus symptoms, leaving the current collection of over 4000 traditional landrace cultivars as a valuable genetic resource for future breeding [11]. They comprise four species, of which one is divided into two genetically distinct groups (Gp or Group) (Table 3.2), according to the now widely accepted classification of cultivated potatoes made by Spooner and colleagues in 2007 [12]. This was based on a phylogenetic tree constructed from molecular marker data from 742 cultivated landraces and eight closely related wild species. Also shown in Table 3.2 are the groups originally proposed by Dodds back in 1962 [13], along with the ploidy levels of the species and groups. Personally, I still find the groups of Dodds useful for some purposes and will be referring to them. Clearly ploidy changes occurred during the spread of potato cultivation by the indigenous farmers of South America, but so did other genetic changes.

3.3 Reproductive Biology of Potatoes

The indigenous farmers of the Andes successfully exploited the naturally occurring variants that resulted from the reproductive biology of potatoes without understanding any modern botany or genetics. The result of the domestication of the wild diploid species *S. candolleanum* was a diploid cultigen *S. tuberosum* Stenotomum

Table 3.2 Cultivated landraces of South American potatoes in the CIP collection [11], classified according to Spooner and colleagues [12] and Dodds [13]

Species	Group-Spooner	Group-Dodds	Ploidy	Number of landraces	Where grown
S. ajanhuiri			$2x$	14	Peru, Bolivia
S. juzepczukii			$3x$	36	Peru, Bolivia
S. curtilobum			$5x$	6	Peru, Bolivia
S. tuberosum	Andigenum				
		Stenotomum	$2x$		Peru, Bolivia
		stenotomum		289	
		goniocalyx		102	
		Phureja	Mainly $2x$	206	Peru, Bolivia, Ecuador, Colombia, Venezuela
		Chaucha	$3x$	121	Peru, Bolivia
		Andigena	$4x$	3233	Andes from Venezuela to Argentina
	Chilotanum	Chilean Tuberosum	$4x$	179	Coastal lowlands of southern Chile

Group ($2n = 2x = 24$), which is now grown from central Peru to central Bolivia [4]. Goniocalyx (diploid) has been recognized as a northern subgroup of Stenotomum, noted for tubers with a bright yellow flesh. The wild species and its subsequent cultigen both reproduce sexually through botanical seed and asexually through tubers. On the arrival of suitable growing conditions, the tubers in the ground sprout and grow into potato plants which flower and then produce more underground tubers before the plant senesces and dies. The tubers are the swollen ends of underground stems which can be quite long. This results in the new plants being separated from one another. Cultivation of potatoes requires dividing the harvested tubers into those that are going to be eaten and those that will be planted into prepared ground in time for the next growing season and hence the next crop. Both tubers for eating and planting will need to be stored during the cold winter months.

Flowering is also important (Fig. 3.1). Potato plants are hermaphrodites, so their flowers have both male and female sex organs, and self-pollination within a flower is a possibility. The flowers in Fig. 3.1 have five purple petals (collectively, the corolla). The male sex organs are the stamens, five in number, each of which is a yellow pollen-producing anther on a stalk called a filament. They surround the female pistil, which comprises a pollen-receptive area, the stigma, on the end of a stalk called the style, which extends from the ovary where the ovules are found. Inside each ovule is an egg-containing embryo sac.

In the potatoes we are considering, there is a mechanism which allows cross-pollination between plants but prevents self-pollination. The genetics and biochemistry of the process are now largely understood, and we refer to it as a gametophytic self-incompatibility system, but the details need not concern us [14]. What is important is that the mechanism of Mendelian inheritance, coupled with cross-pollination, generates a lot of genetic variation by producing new combinations of the genetic variants, which arise by natural mutation. The cross-pollination is done by insects that are capable of releasing pollen from potato anthers by vibration, such as some bee species [15]. After successful pollinations, the ovules develop into dormant botanical seeds, and the ovary becomes the berry (fruit), usually with over 100 seeds, that falls to the ground when ripe [16]. Next season the seeds can germinate and grow into potato plants among those that arise from tubers.

Thus, in the wild, a population of potato plants will be a complex mixture of ones from tubers from the same plant (genetically identical), ones from a genetically different plant, and new genetically unique ones from botanical seed. Under cultivation, the tubers from one plant can be planted close together and kept separate from those of another plant, thus establishing recognizable clones; but landraces may comprise a group of similar-looking clones. Under these circumstances, it should be possible to identify any new types of plants arising from botanical seed. Farmers could spare these volunteer plants when weeding and examine those that survive to harvest. They could then incorporate some of them into their stock of tubers, either assigning them to the landraces they most resemble or keeping them separate. If the farmer is selective in which harvested plants provide the next crop and encourages new cultivators in neighbouring areas, cultivation spreads and crop evolution takes place. We do not know how often Andean farmers incorporated the offspring of sexual reproduction into their stocks of tubers, nor can we actually rule out the deliberate harvesting of berries and sowing of true botanical seeds. We can, however, look at what they eventually achieved.

3.4 Domestication and Glycoalkaloids

Potato tubers contain small quantities of steroidal glycoalkaloids, whereas other plant parts have higher concentrations, probably to provide protection against insects and pathogens. Tuber concentrations above 15 mg 100 g^{-1} FW (fresh weight) give them a bitter taste, and above 20 mg are regarded as unsafe for human consumption, resulting in symptoms typically associated with

Fig. 3.1 Cultivar Mayan Gold, a long-day-adapted Group Phureja potato (top), flowering and being pollinated by a bumble bee in July and (bottom) berries and harvested tubers in September, from planting in mid-April in Edinburgh, Scotland

food poisoning. Today potato cultivars are tested to ensure that they are very unlikely to exceed the safe limit, despite levels being affected by other factors such as climate, storage environment, maturity, damage, temperature and exposure to light (a good reason not to store potatoes in the light) [17]. In contrast, wild species have higher contents, and it is usually assumed that domestication involved selection for less bitter-tasting tubers and hence lower levels of glycoalkaloids. But did the first cultivators deliberately choose species that were low in glycoalkaloids, or were they just lucky? The results of research published in 1990 provide some clues [18]. Genebank accessions of *S. bukasovii* (= *S. candolleanum*), the most likely progenitor of cultivated potatoes, had tuber glycoalkaloid levels (9–29 mg 100 g^{-1} FW), which were close to those found in many clones of *S. tuberosum* Stenotomum Group, the first cultigen. Hence, domestication would have involved little selection for lower levels of glycoalkaloids. In contrast, other former candidates for domestication, namely *S. canasense*, *S. leptophyes* and *S. sparsipilum*, had levels of 42–164 mg 100 g^{-1} FW. Only the first of these species is now included in *S. candolleanum*, the other two being in *S. brevicaule* (the former southern group of the *S. brevicaule* complex). High levels were also found in *S. jamesii* (115–128 mg 100 g^{-1} FW), so perhaps grinding of tubers at North Creek Shelter did lower the levels and result in some detoxification [10]. The only examined accession of *S. maglia* also had moderate levels (56 mg 100 g^{-1} FW), but was from Argentina, so we do not know its likely levels in Monte Verde, Chile (see Wild Relatives as Food). More recently, a comparison of genes in wild species, landraces and modern cultivars (diversity panel) provided evidence of farmer selection for lower levels of steroidal glycoalkaloids on two key genetic loci required for their synthesis: one in the chemical pathway that converts acetyl-CoA to cholesterol and the other in the pathway that converts cholesterol to steroidal glycoalkaloids [19]. In conclusion, I think that it is fair to credit the early Andean farmers with making the potato an edible crop.

The CIP collection of landraces also reveals other decisions made by Andean farmers. They selected plants with shorter stolons so that all of the tubers were close together, thus making their harvest easier. So too did more easily detachable tubers. They also retained a much wider variety of tuber shapes and skin and flesh colours than seen in wild species and also greater than in modern cultivars [20–22]. The wild species have small tubers, round or oval in shape, with white flesh and skins that are white or have a faint bluish-purple flush. Presumably many small tubers rather than a few large ones are the best survival strategy. In contrast, landraces have larger tubers, and the shapes are either regular, from compressed through elliptic to long, or unusual, such as coiled and concertina-shaped. Skin colour is pink, red, blue or purple, due to anthocyanins, or white. The distribution of pigments can result in pigmented eyes or eyebrows, and splashed, scattered, spectacled or stippled tubers. Flesh colour is yellow or orange, due to carotenoids, or white, and anthocyanin pigmentation can also occur. Interestingly, potato tuber diversity was one of the examples given by Charles Darwin of the principle that the valuable and selected parts of all cultivated plants show the greatest amount of modification [23]. I think that it is reasonable to assume that Andean farmers also selected for higher yields and harvest index (weight of tubers as proportion of weight of whole plant), as well as for resistance to pests and diseases and tolerance of abiotic stresses.

3.5 A Detour to Higher Altitudes

Three frost-tolerant species are cultivated in the Andes: *S. ajanhuiri*, *S. juzepczukii* and *S. curtilobum* (Table 3.2). These "bitter potatoes" are grown at high altitudes (up to 4500 m for *S. juzepczukii*) in the central Andes of Peru and Bolivia [4]. It is not known when they arose. *S. ajanhuiri* is the diploid hybrid of the wild frost-resistant diploid species *S. megistacrolobum* (now *S. boliviense*) (2*x*, 2EBN) and *S. tuberosum* Stenotomum Group (2*x*, 2EBN). *S. juzepczukii* is the triploid hybrid of the wild frost-resistant tetraploid species *S. acaule* (4*x*, 2EBN) and *S. tuberosum* Stenotomum Group (2*x*, 2EBN). *S. curtilobum* is a pentaploid hybrid, most likely between *S. tuberosum* Andigena Group (4*x*, 4EBN) and *S. juzepczukii* (3*x*, 2EBN). You might expect triploid *S. juzepczukii* to be sterile, but it can produce some gametes with three sets of chromosomes which combine with gametes with the normal two sets of chromosomes from Group Andigena. Of particular interest with the three species is the way in which their tubers were handled to produce a dehydrated product (chuño). This could be stored for several years and used when fresh potatoes were scarce. In the high, cold mountains, the potatoes were in effect freeze-dried using the very low night temperatures and intense sunlight of the day [4]. The frozen potatoes were trampled by foot to eliminate any remaining water and then washed and dried. As a bonus, the washing removed the toxic glycoalkaloids. Chuño production has continued to the present day, yielding a product highly prized by Andean people.

3.6 Spread of Potato Cultivation Northwards

The other diploid cultigen in Table 3.2 is Group Phureja. Hawkes, in his 1990 book [4], explains that Group Phureja evolved from Group Stenotomum when cultivation moved into the lower, eastern frost-free Andean valleys. Selection by the indigenous cultivators resulted in faster maturity and lack of tuber dormancy so that two or even three crops could be grown each year. Another consequence of this was that Group Phureja could spread northwards through the lowland gap in northern Peru and southern Ecuador and then on into northern Ecuador, Colombia and Venezuela. Today, it is the second most widely cultivated type of potato in South America. However, an even more widely cultivated type of potato arose from Group Stenotomum in the central Andes of Peru and Bolivia, namely, tetraploid Group Andigena (4*x*, 4EBN), whose cultivation spread both northwards and southwards throughout the Andes. Group Andigena also hybridized with Group Stenotomum to give Group Chaucha triploid potatoes, which are still grown in the same areas as Group Stenotomum.

3.7 Group Andigena Potatoes

There has been a lot of argument and discussion over the origin of Group Andigena. I think that the most likely explanation is the one put forward in 2006 by two Japanese researchers, Sukhotu and Hosaka [24]. Group Andigena arose many times at many places in the fields of cultivated potatoes from the fusion of diploid pollen and diploid eggs of Group Stenotomum plants. Such pollen and eggs are referred to as unreduced gametes, and the process is referred to as sexual tetraploidization because a tetraploid plant is produced from two diploid ones. Gametes with two sets of chromosomes instead of one arise through errors, caused by gene mutations, in two critical cell divisions leading to pollen and eggs. The evidence came from comparing modern Andigena landraces with closely related cultivated and wild species using chloroplast and nuclear DNA markers. The results were consistent with Group Stenotomum being the sole ancestor of Group Andigena (i.e., a monophyletic origin). As a result of having four sets of chromosomes, the inheritance of traits is complicated in Group Andigena, which is regarded as an autotetraploid which displays tetrasomic inheritance. Furthermore, self-pollination is possible in Group Andigena plants because the gametophytic self-incompatibility system breaks down in tetraploids [25, 26]. Nevertheless, 40% (range 21–74%) natural cross-pollination was estimated to occur in the Andes [27] and 20% (range 14–30%) in an artificially constructed Andigena population [28]. This level of cross-pollination is sufficient to maintain genetic variation in a population, with the offspring from cross-pollination expected to be more vigorous and fertile than those from self-pollination (hybrid vigour versus inbreeding depression).

The cultivation of Andigena potatoes spread both northwards and southwards and eventually could be found throughout the Andes from Venezuela to Argentina. As mentioned earlier, we do not know if this occurred before or during the Inca period (1200–1532 CE). One reason why Andean farmers might have retained the tetraploid potatoes in preference to the diploid ones can be found in modern-day comparisons. Tetraploid potatoes are often higher yielding because their tubers are larger and heavier [29].

At this point, it is worth mentioning two modern-day studies. The first was a field study by Professor Hawkes and two colleagues from CIP that was published in 1980 [30]. The site was the village of Cuyo-Cuyo in a 500 m deep valley in southern Peru (15°S), at 3300 masl. Above was an upland plateau, the altiplano at 4000 masl; below was the Sandia Gorge, which quickly dropped to the humid eastern slopes of the Andes. The sides of the valley were covered by an extensive system of (Inca) terraces where traditional agriculture was practised. Two fields were examined, one with 846 plants in 76 rows (1–2.5 m in length), of which 10 were sampled, and the other with 326 plants in 20 rows, all of which were examined and harvested. Plants were grown along the rows in pairs in which the two plants could be different varieties. Furthermore, the distribution of varieties in each field was complicated (i.e., not following one after another). Although 25 varieties were identified by one of the Quechua Indian farmers (Fig. 3.2) involved in the study, four accounted for 61% of the plants: Yana wiriquilla, Pucca palta, Yana papa and Patoruntu. There were also differences between the two fields; for example, all but one plant of

Fig. 3.2 Cusco Potato Park, Peru, where Quechua people grow landraces of potato (kindly supplied by John Marshall for Bradshaw JE (2021) Potato Breeding: Theory and Practice. Springer, Cham with permission)

Pucca palta was in field 1, and there were other fields of only Yana wiriquilla. Over 95% of the plants were tetraploids, the remainder being diploid and triploid, based on chromosome counts made from root-tip preparations.

The farmers involved in the study assessed Yana wiriquilla as having good quality based on flavour and a high dry matter content, giving a floury potato. The best quality potatoes were eaten boiled or mashed, whereas the poorer ones were made into soup. Finally, the study found that potato plants could be seen in the oca crop that followed potatoes in the rotation, but it was not known if they were from true seed or tubers. Today the indigenous Quechua people work with CIP (The International Potato Centre) on the in-situ conservation of valuable germplasm by growing 1367 landraces of potato in the Cusco Potato Park (20 miles from Pisac), a 10,000-ha reserve high in the Andes of southern Peru (Fig. 3.2) [31, 32]. The conservation work is supported by the Asociación Andes, a Non-Governmental Organization (NGO).

The second study was an experiment done in 2002 at the Instituto Nacional de Investigación Agraria (INIA) Experimental Station Salcedo, outside the Peruvian city of Puno, beside Lake Titicaca at 16°S and 3800 masl [15]. The field-plot experiment demonstrated that three closely related wild species (2EBN) that grow around farmers' fields can hybridize at low frequency (about 4%) with Group Andigena (4EBN) and that some hybrid progeny would be selected by present-day Andean farmers. However, this one experiment does not tell us what actually happened in the Andes. It is DNA from extant landraces that provides us with the evidence that Andean farmers did select some naturally occurring hybrids. In other words, the crop evolved as cultivation spread through the Andes, and it is the southward journey that is of particular interest. So, let us return to the comparison of genes in the diversity panel of 20 diploid wild species, 20 South American landraces (Groups Andigena, Phureja, Stenotomum and Chilotanum), 23 post-1850 North American cultivars and four out-

Table 3.3 Today's monthly weather averages for Castro, Chiloé Archipelago, Chile, from weather-atlas.com

Month	Low temperature (°C)	High temperature (°C)	Rainfall (mm)	Daylength (hours mins)
January	10.9	17.6	37	14 52
February	10.9	17.6	45	13 44
March	10.1	16.0	64	12 20
April	9.2	14.0	88	10 55
May	8.2	11.8	126	9 43
June	6.7	9.8	124	9 7
July	5.6	9.3	115	9 25
August	5.8	9.6	120	10 28
September	5.9	10.5	65	11 50
October	6.6	12.0	79	13 16
November	8.0	13.5	74	14 34
December	9.7	15.6	68	15 14

groups, mentioned in an earlier section [19]. A phylogenetic analysis of DNA sites in conserved genes subdivided the panel into three primary populations: wild species, Andean landraces (Groups Phureja, Stenotomum and Andigena) and Group Chilotanum landraces plus the cultivars. For simplicity, I am going to report quantitative results using descriptive words rather than numbers and proportions. All of the cultivated groups contained a significant contribution of genetic variants from their diploid progenitor, *S. candolleanum*. However, out of the Andean landraces, it was tetraploid Andigena that had the highest number of genetic variants from the diploid wild species. Whereas the diploid landraces had genetic variants from diploid species native to Peru, tetraploid Andigena also had genetic variants from species native to Bolivia and Argentina (e.g., *S. berthaultii*). It has therefore been argued that these additional variants helped in the adaptation of Group Andigena to the environments it encountered as cultivation spread to the southern end of the Andes [19].

3.8 Landraces of the Coastal Lowlands of Southern Chile

Now we need to cross to the coastal lowlands of southern Chile and consider the Chilotanum (Chilean Tuberosum) landraces (Table 3.2). The analysis of the diversity panel revealed the following interesting information [19]. Not only did Chilean Tuberosum landraces have all of the genetic variants present in the Andean landraces, but also unique ones from diploid species found only in the Andes of southern Bolivia and northwest Argentina: *S. gourlayi* (now *S. brevicaule*), *S. kurtzianum*, *S. microdontum* and *S. spegazzinii* (now *S. brevicaule*); all 2*x*, 2EBN. Other researchers have provided good evidence for the contribution of *S. berthaultii* (= *S. tarijense*; 2*x*, 2EBN), based on the presence or absence of a 241-bp deletion in plastid (chloroplast) DNA [33, 34]. They suggest that Andigena populations in southern Bolivia or northern Argentina pollinated populations of *S. berthaultii* with the plastid deletion. The result was landraces of Andigena with the deletion, which have been found in northern Argentina. It is assumed that tubers of these landraces were transported from the Andes to coastal Chile, a distance of over 500 km. Furthermore, it is assumed that some of the genetic variants which had accumulated were important in producing a crop that would produce tubers in the long summer days of coastal Peru. There, pollinations by the sole wild species of the area, *S. maglia* (deletion absent), made further genetic contributions to Chilean Tuberosum (deletion usually present), including tuber production in long days. Phylogenetic analyses using DNA markers (microsatellites) showed that Chilean Tuberosum and *S. maglia* are closely related [33]. Today, the deletion is present in around 88% of

Chilean Tuberosum landraces but only in around 1% of Andigena landraces, mainly from southern Bolivia and Argentina [5].

Table 3.3 shows the present-day climate for Castro (43°S, 130 masl), the capital of the Chiloé Archipelago, which is home to a wide variety of landraces of Chilean Tuberosum. It can be seen that the summer months (October to March) are marked by daylengths of 12–15 h, cool average low and high temperatures, and modest rainfall. In contrast, the winter months are colder and wetter with daylengths of 9–12 h.

Of particular interest in the diversity panel were variants of a gene known to affect tuberization in different daylengths (*StCDF1*) [35]. Almost all (23/25) of the long-day clones in the diversity panel had variants such as *StCDF1.2*, which result in truncation of a key protein involved in the control of tuberization, early vine maturity and the ability to form tubers in long days. Strictly speaking, such potato plants are facultative short-day plants because tuber formation is accelerated following transfer of the plants to shorter daylengths. The variants in most of the North American cultivars were most closely related to those in *S. microdontum*, but this is not proof of source without further research being done, given the contribution of *S. maglia* to Chilean Tuberosum. Finally, it is important to note that Chilean Tuberosum and Andigena landraces are morphologically distinct in their vine and leaf structures, as well as being genetically distinct [36].

3.9 The First Europeans to See Potatoes

The first Europeans to see potatoes being cultivated in South America were the Spanish conquerors of the Inca Empire, who were helped in this endeavour by the spread of smallpox and other diseases among the indigenous people rather than by military might [9].

In 1532, Francisco Pizarro and a small group of mercenaries disembarked on the desert coast of Peru and ascended the Andes to the town of Cajamarca. There the Inca emperor was kidnapped, ransomed for gold and silver, and then killed before the soldiers marched on Cuzco [9]. They must have seen potatoes, but there is no record until that of Vicente de Valverde in 1539, after he had been made bishop of Cuzco by Pope Paul III in 1537 [37]. He likened them to Spanish truffles. The first recorded sighting was in 1537 in the Valle de la Grita, Colombia, during the expedition of Jiménez de Quesada, from Santa Marta on the Caribbean coast, southwards along the valley of the Rio Magdalena, and then up into the valley of the Opón river [4].

The Spanish had come to South America looking for gold and silver, but they also found the sophisticated terraced agriculture of the Incas, including potatoes as an important food crop. The ships that returned to Spain via the Canary Islands not only carried gold and silver in their stores but also potatoes, living tubers that survived the journey and could be planted and grown in new lands.

References

1. Pyne L (2020) In search of prehistoric potatoes. Archaeology, March/April 2020, Archaeological Institute of America
2. Louderback LA, Pavlik BM (2017) Starch granule evidence for the earliest potato use in North America. PNAS 114:7606–7610. https://doi.org/10.1073/pnas.1705540114
3. Ugent D, Dillehay T, Ramirez C (1987) Potato remains from a late Pleistocene settlement in south central Chile. Econ Bot 4:17–27
4. Hawkes JG (1990) The potato: evolution, biodiversity and genetic resources. Belhaven Press, London
5. Spooner DM, Ghislain M, Simon R, Jansky SH, Gavrilenko T (2014) Systematics, diversity, genetics, and evolution of wild and cultivated potatoes. Bot Rev 80:283–383
6. Vavilov NI (1940) The theory of the origin of cultivated plants after Darwin. Nauka [Science] 2:55–75
7. Vavilov NI (1920–1940) English translation of papers on origin and geography of cultivated plants. Cambridge University Press, Cambridge, Digitally Printed Version 2009
8. Spooner DM, McLean K, Ramsay G, Waugh R, Bryan GJ (2005) A single domestication for potato based on multilocus amplified fragment length polymorphism genotyping. PNAS 102:14694–14699
9. Moseley ME (2001) The Incas and their ancestors: the archaeology of Peru, 2nd edn. Thames and Hudson, London

10. Rumold CU, Aldenderfer MS (2016) Late Archaic-Early formative period microbotanical evidence for potato at Jiskairumoko Basin of southern Peru. PNAS 113(48):13672–13677. https://doi.org/10.1073/pnas.1604265113
11. CIP Genebank. https://cipotato.org/genebankcip/process/potato/potato-cultivated/
12. Spooner DM, Nunez J, Trujillo G, Herrera MR, Guzman F, Ghislain M (2007) Extensive simple sequence repeat genotyping of potato landraces supports a major re-evaluation of their gene pool structure and classification. PNAS 104:19398–19404
13. Dodds KS (1962) Classification of cultivated potatoes. In: Correll DS (ed) The potato and its wild relatives. Texas Research Foundation, Renner, Texas, pp 517–539
14. Dodds KS (1965) The history and relationships of cultivated potatoes. In: Hutchinson JB (ed) Essays in crop plant evolution. Cambridge University Press, Cambridge, pp 123–141
15. Scurrah M, Celis-Gamboa C, Chumbiauca S, Salas A, Visser RGF (2008) Hybridization between wild and cultivated potato species in the Peruvian Andes and biosafety implications for deployment of GM potatoes. Euphytica 164:881–892
16. Gopal J, Ortiz R (2006) True potato seed. In: Gopal J, Khurana SMP (eds) Handbook of Potato Production, Improvement and Postharvest Management. The Haworth Press, Binghamton, pp 557–585
17. Friedman M, Levin CE (2016) Glycoalkaloids and calystegine alkaloids in potatoes. In: Singh J, Kaur L (eds) Advances in potato chemistry and technology, 2nd edn. Elsevier, Amsterdam, pp 167–194
18. Johns T, Alonso JG (1990) Glycoalkaloid change during the domestication of the potato, *Solanum* Section *Petota*. Euphytica 50:203–210
19. Hardigan MA, Laimbeer FPE, Newton L, Crisovan E, Hamilton JP, Vaillancourt B, Wiegert-Rininger K, Wood JC, Douches DS, Farré EM, Veilleux RE, Buell CR (2017) Genome diversity of tuber-bearing *Solanum* uncovers complex evolutionary history and targets of domestication in the cultivated potato. PNAS 114(46):E9999–E10008. https://doi.org/10.1073/pnas.1714380114
20. Glendinning DR (1983) Potato introductions and breeding up to the early 20th century. New Phytol 94:479–505
21. Simmonds NW (1995) Potatoes. In: Smartt J, Simmonds NW (eds) Evolution of crop plants, 2nd edn. Longman Scientific & Technical, Harlow, pp 466–471
22. Ortiz R, Huaman Z (1994) Inheritance of morphological and tuber characteristics. In: Bradshaw JE, Mackay GR (eds) Potato genetics. CAB International, Wallingford, pp 263–283
23. Darwin C (1868) The variation of animals and plants under domestication, volume 1. Digitally printed 2010. Cambridge University Press, Cambridge
24. Sukhotu T, Hosaka K (2006) Origin and evolution of Andigena potatoes revealed by chloroplast and nuclear DNA markers. Genome 49:636–647
25. Lewis D (1943) The physiology of incompatibility in plants. III. Autopolyploids. J Genet 44:171–185
26. Lewis D (1954) Comparative incompatibility in angiosperms and fungi. Adv Genet 6:235–285
27. Brown CR (1993) Outcrossing rate in cultivated autotetraploid potato. Am Potato J 70:725–734
28. Glendinning DR (1976) Neo-Tuberosum: new potato breeding material. 4. The breeding system of Neo-Tuberosum, and the structure and composition of the Neo-Tuberosum gene pool. Potato Res 19:27–36
29. Stockem J, De Vries M, Van Nieuwenhuizen E, Lindhout P, Struik P (2020) Contribution and stability of yield components of diploid hybrid potato. Potato Res 63(3):345–366
30. Jackson MT, Hawkes JG, Rowe PR (1980) An ethnobotanical field study of primitive potato varieties in Peru. Euphytica 29:107–113
31. Ellis D, Salas A, Chavez O, Gomez R, Anglin N (2020) Ex situ conservation of potato [*Solanum* Section *Petota* (Solanaceae)] genetic resources in genebanks. In: Campos H, Ortiz O (eds) The potato crop: its agricultural, nutritional and social contribution to humankind. Springer, ebook, pp 109–138
32. Collyns D (2019) Spuds we like. How Peru's potato scientists could avert food crisis. The Guardian 30 November 2019, p 41
33. Spooner D, Jansky S, Clausen A, del Rosario HM, Ghislain M (2012) The enigma of *Solanum maglia* in the origin of the Chilean cultivated potato, *Solanum tuberosum* Chilotanum Group. Econ Bot 66:12–21
34. Hosaka K (2004) Evolutionary pathway of T-type chloroplast DNA in potato. Am J Potato Res 81:153–158
35. Kloosterman B, Abelenda JA, Carretero Gomez MDM, Oortwijn M, de Boer JM, Kowitwanich K, Horvath BM, van Eck HJ, Smaczniak C, Prat S, Visser RGF, Bachem CWB (2013) Naturally occurring allele diversity allows potato cultivation in northern latitudes. Nature 495(7440):246–250
36. Grun P (1990) The evolution of cultivated potatoes. Econ Bot 44(3 Suppl):39–55
37. Earle R (2020) Feeding the people. Cambridge University Press, Cambridge

4 South America to the World

Potatoes were introduced from the Andes to the Canary Islands around 1562 and from there to Europe, where they were first cultivated in Spain, then in Portugal and Italy. They were short-day-adapted Group Andigena potatoes, which were able to form tubers in the short days of autumn. The first illustrations and descriptions of these potatoes appeared over the period 1597–1601. From 1600, it proved possible to find regions in most countries of the world with seasons and climates to which potatoes could be adapted for cultivation, particularly between 44°N and 58°N as a lowland temperate summer crop and between 23°N and 34°N as a lowland subtropical winter crop. Potatoes were initially grown on small plots of land for home consumption (e.g., cottage gardens) but increasingly from 1750 in larger fields for a range of markets and end uses, as well as for a staple food. DNA extracted from herbarium specimens revealed that since about 1800, new potato introductions came from coastal Chile, and today their offspring predominate in potato cultivation. The period from 1800 onwards also saw the birth of modern potato breeding with implications today for the role of potatoes as a food security crop.

4.1 Canary Islands and Spain

Seville is a Spanish city on the Guadalquivir River, which flows into the Atlantic Ocean 119 km away at Sanlúcar de Barrameda. This was the starting and finishing point of the first voyage around the world from 1519 to 1522 (Fig. 4.1). Magellan set sail from Sanlúcar with five ships on 20 September 1519. Elcano returned with just one ship on 6 September 1522. The voyage can be regarded as the birth of our modern, interconnected world. An account of the expedition can be found in a book by Antonio Pigafetta [1]. The ships reached Tenerife in the Canary Islands on 26 September 1519. They crossed the Atlantic, reaching the most easterly part of Brazil, north of Recife, on 29 November and Rio de Janeiro on 13 December 1519. In Brazil, they ate potatoes, but these were sweet potatoes. They were also too far south on the Pacific Coast of Chile to see the potatoes of the Chiloé Archipelago. Eventually, the one remaining ship, Victoria, sighted Tenerife on 28 July 1522, arriving back in Sanlúcar de Barrameda on 6 September 1522, having never encountered a potato!

J. E. Bradshaw, *Can Potatoes Feed the World?*, Sustainable Development Goals Series,
https://doi.org/10.1007/978-3-031-92890-1_4

Fig. 4.1 Sanlúcar de Barrameda: starting and finishing point of the first voyage around the world from 1519 to 1522, 8 February 2023

Table 4.1 Today's monthly weather averages for Seville, Spain, from weather-atlas.com

Month	Low temperature (°C)	High temperature (°C)	Rainfall (mm)	Daylength (hours mins)
January	5.7	16.0	66	9 56
February	7.0	18.1	50	10 50
March	9.2	21.9	36	11 59
April	11.1	23.4	54	13 11
May	14.2	27.7	31	14 11
June	18.0	32.2	10	14 42
July	20.3	36.0	2	14 27
August	20.4	35.5	5	13 34
September	18.2	31.7	27	12 25
October	14.4	26.0	68	11 14
November	10.0	20.2	91	10 11
December	7.3	16.6	99	9 38

There is no known historical account of potatoes being brought to Europe. Furthermore, it appears that potatoes (patata in Spanish), *Solanum tuberosum* (Fig. 4.2), were sometimes confused with sweet potatoes (batata in Spanish), *Ipomoea batatas*.

However, careful examination of historical records by J.G. Hawkes and J. Francisco-Ortega revealed that barrels of potatoes were exported from Gran Canaria to Antwerp (Belgium) in November 1567 and from Tenerife via Gran Canaria to Rouen (France) in April 1574 [2]. They therefore concluded that the potato was being grown as an export crop in Gran Canaria and Tenerife in 1567 and 1574, respectively, and hence could have been introduced to the Canary

Fig. 4.2 Potatoes (patatas) on sale in a market in Sanlúcar de Barrameda, Spain, 8 February 2023

Islands around 1562 and from there to mainland Europe. They also concluded that the November shipment came from an autumn crop, whereas the April one came from a spring crop.

Earlier, Hawkes and Francisco-Ortega had visited Seville to study the account books of the Hospital de la Sangre for the period 1546–1601 [3]. They found that potatoes were bought regularly in the Seville market from 1580 onwards, with the first record appearing in 1573. Most purchases were in December and January, with none from April to October. They concluded that potatoes had become established as a crop in the area by about 1570 and were being harvested mainly in December and January for local markets. Potatoes are just too bulky for long-distance travel as a food crop. The initial potato introductions probably came from the Canary Islands rather than directly from South America, but we do not know for certain. Hawkes and Francisco-Ortega also concluded that they were Andigena potatoes, which were able to form tubers in the short days at the end of the year.

Table 4.1 shows the present-day Mediterranean climate for Seville (37°N, 9 masl). The summer months (June, July and August) are too hot and dry for potato cultivation. Furthermore, the day-length is over 14 h in May, June and July. Today in the area around Seville, the potato can be grown as a spring crop (Fig. 4.3), planted in January and harvested in April, or as an autumn crop, planted in September and harvested in December.

According to Hawkes, in his 1990 book [4], the potato was also known in Italy before 1587. This conclusion comes from a book by Father Magazzini de Vallombrosa published posthumously in 1623. The Father says that potatoes were grown at Vallombrosa, near Florence, having been brought there from Spain and Portugal by Camelite Friars. Hawkes concludes that potatoes were first cultivated in Spain, then in Portugal and Italy.

4.2 Botanists and Herbalists

The evidence for the potato becoming well-known to European botanists and herbalists at the end of the sixteenth century is also summarized by Hawkes in his 1990 book [4]. Three famous names are the Flemish botanist Carolus Clusius (Charles d'Ecluse), the Swiss botanist Gaspard Bauhin and the English herbalist John Gerard.

Clusius provides information on potatoes, including an illustration, in his *Rariorum Plantarum Historia*, which was published in 1601, when he was a professor at the University of Leiden in the Netherlands and the first Prefect of their new botanical garden, the Hortus Academicus. He recounts that when he was in Vienna in 1588, he was sent two tubers and a fruit from Philippe de Sivry, Prefect of Mons in Belgium, who in turn had received them from a friend of the Papal Legate in Belgium in 1587, thus indicating an Italian origin. Then in 1589, Philippe de Sivry also sent him a watercolour painting of a potato, dated 1588. This potato was late maturing with long stolons and had deep-

Fig. 4.3 Spring potato crop (left) and reservoir for irrigation, near Seville, Spain, 7 February 2023

red-skinned tubers of irregular shape with deep eyes. More details can be found in the summary produced by R. N. Salaman in his 1926 book *Potato Varieties* [5].

Gaspard Bauhin provided the first botanical description of the potato in 1596 in his *Phytopinax seu enumeration plantarum*. He named the potato *Solanum tuberosum* but subsequently changed it to *Solanum tuberosum esculentum.* Much later, Linnaeus used *Solanum tuberosum.* Gaspard Bauhin is also credited with the earliest known herbarium specimen of the potato, made around 1620. Hawkes [4] says that it is an obvious specimen of an Andigena potato.

John Gerard provided the first printed illustration of a potato in his *Herball* of 1597, which was followed by Bauhin in 1598 and Clusius in 1601. Gerard's potato, which he grew in his garden, had long stolons but was not as late maturing as that of Clusius and had white-skinned tubers of irregular shape with deep eyes. Again, more details can be found in Salaman's book. How this potato came into Gerard's possession has been the subject of much speculation (e.g., links with Sir Walter Raleigh and Sir Francis Drake) but remains unknown [4].

In summary, the first European potatoes were short-day-adapted, Andigena types from the Andes, presumably sent to Europe either from the port of Cartagena in Colombia (most likely) or from Peru (possible). They were being cultivated in Spain, Portugal and Italy during the latter part of the sixteenth century. By 1600, the potato was known to botanists and herbalists in Austria, Belgium, England, the Netherlands and Switzerland, and presumably in France and Germany. However, as we will shortly see, potatoes were not widely cultivated on a field scale until the 1750s onwards. What happened between 1600 and 1750? Did potatoes remain a botanical curiosity confined to physic gardens, or were they grown on the estates of the wealthy or in the cottage gardens of the poor? How, when, and by whom were they selected to tuber in the longer days of northern Europe? When did Russia, Poland and Germany become the big producers in Europe? Were there more introductions from South America? Many questions, but are there any answers?

4.3 The Period 1600–1750 in Europe

In her book *Feeding the People*, published in 2020, Rebecca Earle [6] argues that during the period from 1600 to around 1750, potato cultivation was driven by villagers, peasants and labourers, albeit on a small scale in their cottage gardens. As a result, it spread in Austria, Belgium, Britain, France, Germany, Ireland, the Netherlands and Switzerland, as well as in Spain, Portugal and Italy. Had it been spread by wealthy landlords or state policy, we would have detailed historical records, which are lacking. What we do have are recorded disputes over ecclesiastical taxes where the cottagers argued that they should pay a reduced tax because the potatoes were intended for their own use, rather than for sale, and that this had occurred for many years: i.e., before the recorded disputes arose from the late 1600s onwards. We also have agricultural manuals from the early 1600s explaining how to grow potatoes. Earle mentions the book written by Frederick Morton Eden in 1797, *The State of the Poor Or, An History of the Labouring Classes in England, from the Conquest to the Present Period* [7]. In Volume I, Book II, Chapter II, there is a section on potatoes where Eden says that in England, potatoes were first grown in fields in small patches in Lancashire in 1684 and in open fields in Scotland on the Kilsyth Estate (Lanarkshire) in 1728, although a Parish History by the Reverend Peter Anton (1893) gives the date as 1739. Before that, they had been grown in cottage gardens in England and Scotland. We must assume, without written evidence, that the potato plants in cottage gardens throughout Europe were flowering and producing berries and that some of the resulting seedlings were being selected in the following seasons, as happened in the Andes (I mentioned in Chap. 2 that potatoes from the Andes will flower in long days). In fact, without written evidence, we cannot rule out cottagers actually harvesting the berries and sowing the true botanical seeds or possibly exchanging seeds with their neighbours and more distant cottagers. As Salaman says in his 1926 book *Potato Varieties*, practically every variety, which has been raised since the introduction of the potato to Europe around 1570 has come from true seed [5]. It is just that we don't have a historical record of what actually happened. The result, nevertheless, was potatoes being selected for their ability to produce tubers in the longer, cooler summer days of Europe above latitude 45°N.

Interestingly, experiments done in Great Britain (GB) and the United States from 1959 onwards showed that adaptation to long days can occur quickly [8–10]. The starting material for the experiment was Andigena landraces, held in the Commonwealth Potato Collection in GB, comprising approximately 45% Bolivian, 35% south Peruvian, 10% north Peruvian and 10% Colombian. Five cycles of selection for tuber production in the long days of GB increased yield per plant from 0.33 to 1.15 kg compared with 0.76 kg for a Tuberosum control population, albeit from a late harvest [9]. Likewise in the United States (Cornell University, Ithaca, New York), with some of the same starting material, six cycles of selection increased yield per plant from 0.32 to 1.37 kg [10]. However, in GB, the long-day-adapted Andigena population (often referred to as Neotuberosum) still resembled the unadapted one in having plants that were twice as large (haulm size) as the Tuberosum control population and with twice as many plants producing flowers. Furthermore, leaf shape was unchanged. Three further generations of selection took place in the Cornell population for Tuberosum haulm (not excessive) and leaf shape (large broad leaflets, few dissections, and a wide angle of leaf insertion), collectively referred to as vine type. The outcome was plants that looked like Tuberosum [11]. One further characteristic requires mention. When individuals from the unadapted Andigena population were used to pollinate Tuberosum, most of the offspring (>90%) were male sterile; but this did not occur if Andigena plants were the female parent. This was explained as an unfavourable interaction between the cytoplasm (mitochondria or chloroplasts) of Tuberosum and the nuclear genes of Andigena. Unexpectedly, during the selection of long-day-adapted Andigena in the Cornell population, there was a reduction in this unfavourable interaction [12]. In other words, pollination of Tuberosum with adapted Andigena (Neotuberosum) produced more plants than expected that were male fertile and sometimes numbers approaching that seen in Tuberosum-by-Tuberosum crosses (about 70%). The researchers tried to explain this in terms of selection for genes that restored male fertility, but not convincingly. Furthermore, there was a final twist to this story. In 2009, researchers [13] were studying Cornell's Neotuberosum and derived material with molecular markers for their nuclear DNA (simple sequence repeats, SSRs), as well as the presence or absence of a 241-bp deletion in the plastid (chloroplast) DNA. They found that the deletion, which is characteristic of Chilean Tuberosum, was present in one of six Neotuberosum clones, seven out of 12 clones with Neotuberosum as the direct female parent, and almost half of the derived material. Furthermore, the nuclear markers grouped the six Neotuberosum clones and all Neotuberosum-derived material with the Chilean and not the Andigena landraces included in the study. They concluded that Neotuberosum is not derived solely from Andigena landraces, possibly because of misclassification of landraces included in the original population and possibly because of selection of plants from the tubers of Tuberosum cultivars left in the Neotuberosum selection field from previous years (groundkeeper plants). The presence of such groundkeepers was acknowledged in one of the papers on the GB Neotuberosum population [14]. Nevertheless, Shelagh, the one cultivar bred from this population, has Andigena cytoplasm, as we shall see later in this chapter.

My own conclusion from all of this work is that adaptation to long days can occur quickly in Andigena potatoes, but they remain distinct from Chilean Tuberosum potatoes unless the latter are deliberately or unintentionally incorporated into the selection program. Rapid adaptation to long days has also been achieved in the progenitor of Andigena potatoes, namely diploid Phureja/Stenotomum [15, 16].

I mentioned earlier that the 1588/1601 potato of Clusius had a deep red skin, whereas the 1597 potato of Gerard had a white skin. Over 100 years later, in 1730, in his book *Considerations on Agriculture*, George Rye in Ireland mentions an outstanding potato with black skin and white flesh, as well as four others, of which one had white skin and another yellow flesh [17, 18]. Salaman, in his 1926 book, explains that black-skinned potatoes can arise when ones with deep red (beetroot red) skins are mated to other types, including white ones [5]. However, there is no written record of where, when and how frequently different types of potatoes arose and were maintained.

4.4 The Period 1750–1800 in Europe

The potato's potential as a food crop, grown on a field scale, was first seen in Ireland at the end of the seventeenth century and throughout the eighteenth century. We will explore the reasons and consequences in the next chapter. Here I want to focus on the rest of Europe during the second half of the eighteenth century. A number of authors have argued that the stimulus to potato cultivation throughout this period in Europe was a combination of food shortages and the need to feed the armies and cheap labour required for military and economic strength, particularly during the Industrial Revolution, which began in the second half of the eighteenth century and gained momentum in the first half of the nineteenth century. We can see this through some well-known historical figures who have been portrayed as key players or influencers.

Frederick II (the Great) of Prussia (1712–1786) has been called "The Potato King" and credited with being a key player in encouraging potato cultivation in Prussia and beyond. Furthermore, the War of Bavarian Succession (1778–1779) has been called "The Potato War", as the opposing armies fed themselves on the region's potato crop. According to John Reader in his 2008 book *Propitious Esculent* [18], Frederick the Great had learnt the value of keeping peasants and soldiers alive with potatoes during the early battles of the War of the Austrian Succession (1740–1748). In 1744, he had ordered the distribution of free seed potatoes throughout Prussia (now northern Germany and Poland) and instructions on how to grow them. Then at the beginning of the Seven Years' War (1756–1763), he issued a "Potato Order" advising the planting of potatoes as a very nutritious food wherever space permitted. Potatoes came to be seen by poor families as a way of staying alive when wheat was seized by invading armies. Another consequence of the war was that the rulers of the invading countries, France, Austria and Russia, saw the value of potatoes and encouraged potato growing by their farmers.

Catherine II (the Great) of Russia (1729–1796) is credited with the spread of potato cultivation from around St Petersburg in 1764 to many parts of the country following a royal decree in 1765 [19]. Before that, potatoes had been used as a delicacy at royal banquets, following their introduction from the Netherlands at the end of the seventeenth century by Peter the Great, at least according to legend. However, cultivation in peasants' gardens in western Russia had probably spread from what is now Poland at about the same time.

Antoine-Augustin Parmentier (1737–1813), a French pharmacist, is considered by a number of authors to be one of the great potato influencers of the eighteenth century [6, 17, 19]. Whilst in the French army, he was taken prisoner by the Prussians during the Seven Years' War (1756–1763). He survived 3 years of captivity on a diet consisting mainly of potatoes. On his release, he returned to Paris in remarkably good health, convinced of the nutritional value of the potato. He researched the chemical composition of potatoes and cooking and storage methods, as well as promoting potato cultivation in France and beyond, particularly after food shortages following the failure of the wheat harvest in 1770. In 1773, he won a competition for the best essay on alternatives to cereal grains for use in times of scarcity in which he proposed the potato! In 1785, he succeeded in having the potato served at the royal court. He also received royal approval from Louis XVI to grow 16 hectares of potatoes on the western edge of Paris. He convinced sceptical locals that they were valuable and worth eating by having soldiers guard them during the day but not at night so that they could be stolen. He survived the French Revolution (1789–1799) and was made a member of the Légion d'Honneur by Napoleon. Examples of his work are as follows: he made dried potato biscuits from boiled potatoes for use by sailors, he developed methods of drying cooked potatoes in forms that could be reconstituted by adding boiling water and he made potato breads.

Benjamin Thompson, Count Rumford (1753–1814), a British physicist and inventor, has been credited with providing the first real soup kitchens for feeding the poor when he was advisor to Karl (Charles) Theodor, the elector of Bavaria, from 1785 to 1796 [6]. Potatoes constituted the core of his soup recipes for Munich's poorhouse, and subsequently his soup kitchens spread to many European cities (more information on soup recipes can be found in Earle's book [6]). In 1799, he and Joseph Banks established the Royal Institution of Great Britain and appointed Humphry Davy its first lecturer. His assistant was Michael Faraday. Today the Royal Institution is famous for its Christmas Lectures for Children, which reach a large audience through their TV coverage.

Adam Smith (1723–1790), a key figure of the Scottish Enlightenment, published *An Inquiry Into the Nature and Causes of the Wealth of Nations* in 1776 [20]. He briefly mentioned potatoes in Chapter XI, Of the Rent of Land, Part I: Of the Produce of Land Which Always Affords Rent. The context is that of landlord (landowner), farmer, labourer and consumer, in which the farmer pays rent to the landlord and also hires labourers to do the work, which results in food (produce) to sell at market to consumers for profit (Fig. 4.4).

Consideration is given to progress in improvements in farming and their effects on the rent and profit of what is improved; for example, grass for cattle and their produce (meat) and wheat as the common food (bread) of the people. Vineyards (in wine countries) and well-watered kitchen gardens are considered part of the farm. Smith's yield figures, when the correct allowance is made for water content, show that potatoes have a 41% higher yield than wheat, compared with the 44% for the period 2016–2020, which I quoted in Chap. 1. He also considered potatoes to have a nourishing quality as food. However, he concluded, "It is difficult to preserve potatoes through the year and impossible to store them like wheat for 2–3 years together. The fear of not being able to sell them before they rot discourages their cultivation and is perhaps the chief obstacle to their ever becoming great in any country, like bread, the principal vegetable food of all the different ranks of the people." Interestingly, Eden, in his 1797 book, quotes Smith's reservations but is more optimistic about the future of the potato, albeit as a food for the poor, which can be stored underground through winter and spring [7]. What is clear is that Smith appreciated that economics determined whether wheat was favoured over potatoes or vice versa, and the same was true of grass for animal products. The landlords may grow potatoes in their veg-

Landowner		**Labourer**
Vegetable garden		Garden for own food
Rent for land	← **Farmer** →	Hires labour
	↓	
	Grows potatoes in field	
	Sells to **Consumer** at market	

Fig. 4.4 Adam Smith (1723–1790): statue in the High Street, Royal Mile, Edinburgh, 8 December 2022

etable gardens to feed kings and queens; the labourers may grow potatoes in their cottage gardens to feed their families, but the farmers grow potatoes to sell them at market: potatoes, like other foods, had become a commodity to trade. Today, in the modern world, we also have to consider the economics and logistics of potato supply chains and, for large farming enterprises, the risks attached to growing potatoes rather than other crops.

By way of contrast, in Part II of Chapter XI of The Wealth of Nations, Smith considers the gold and silver mines of Peru, the reason that the Spanish went to South America.

Despite the reservations of Smith, by 1800 the potato was accepted as an important food throughout Europe, not just in Ireland, with a further European expansion of cultivation during the nineteenth century. But what about the rest of the world?

4.5 Europe to the Rest of the World

During the seventeenth (1600s) and eighteenth (1700s) centuries, a number of European countries developed widespread political and commercial interests in the rest of the world, and British, Dutch, French, Portuguese and Spanish colonists and missionaries took with them their common crops, including potatoes [21]. Accounts tend to disagree on exact dates, but a very brief outline of the spread of potato cultivation worldwide is as follows. Potatoes were taken from England to Bermuda in 1613 and from there to Virginia on the east coast of North America in 1621 [4]. They were being grown in China, India and Japan during the 1600s [4], but probably later, in the 1700s (and as late as the 1800s), in various parts of Africa. Finally, the potato reached New Zealand. The Maori have traditionally credited Captain James Cook with introducing the potato to New Zealand, possibly in 1769, but more likely in 1773 [4, 18]. Hence, by 1800, many parts of the world were known to have seasons suitable for growing potatoes, and as in Europe, selection must have taken place for tuber formation in the daylengths of the new locations. We do not know the number of introductions into these new locations nor the details of their incorporations into the local farming practices.

Two hundred years later, at the beginning of the twenty-first century, the potato was a truly global crop, as shown by the global geo-referenced database of the distribution of the potato published by R.J. Hijmans in 2001 [22]. Potatoes were being grown in 149 of the 227 countries listed in the FAO database, from latitudes 65°N to 47°S, and at altitudes from sea level to 4000 m, with 25% located above 1000 m; a remarkable achievement for a crop that was unknown outside of South America just 500 years ago. There were two main peaks in distribution by latitude. The major peak (52% of total potato area) was between 44°N and 58°N and represented the temperate climates where the potato was a summer crop. The other peak (19%) was between 23°N and 34°N and mainly represented the subtropical lowlands where the potato was a winter crop. The reduction in the temperate peak and increase in the subtropical one since 2001 was presented in Chap. 1 and discussed through the 15 countries that accounted for nearly 75% of world potato production (364 million tonnes fresh weight of tubers) for the period 2016–2020.

Today, each potato-growing country will have its own history of potato introduction(s) and cultivation. For us, however, it is more informative to focus on the environments and climates where potatoes can be successfully cultivated. As potatoes are bulky, those for eating need to be grown close to where they are going to be consumed. This means that if potatoes are going to "feed the world", they need to be grown globally in a wide range of environments.

Potatoes don't like it too hot (average daily temperatures above 21 °C) or too cold (below 5 °C) [23], and their maximum light use efficiency occurs at 18 °C as a result of the effect of temperature on the balance between photosynthesis and respiration [24]. As a result, they are grown as follows.

1. As a summer crop in the tropical highlands of Bolivia, Peru and Mexico, and in the lowland temperate regions of the world (North America, Western and Eastern Europe, Northern China and Australia and New Zealand) (Fig. 4.5). In the high Andes, the growing season can be as long as 180 days, whereas in the lowland temperate regions, crop duration is typically 120–150 days and yields are potentially high as a result of long daylengths, moderate temperatures, and a relatively long growing season.
2. All year round in parts of China and Brazil, in the equatorial highlands of South America (e.g., Ecuador and Colombia) and East Africa (e.g., Kenya and Uganda), and in South Africa.
3. As spring and autumn crops in the Mediterranean (e.g., North Africa and southern Spain).
4. As a winter crop in the lowland subtropics (e.g., northern India, Bangladesh and southern China) (Fig. 4.6). Here the growing season can be as short as 75 days, with 90–120 days the norm, which is too short for really high yields.

In a country like Japan that goes from 24°N to 46°N, the potato is grown as a summer crop in the north and a winter crop in the south, with double cropping in spring and autumn in some regions.

As potatoes cannot be grown all of the year round in most parts of the world, it is normal to store both seed tubers for planting the next crop and ware tubers for consumption. Hence, post-harvest infrastructure in terms of road transport and cold storage facilities is an important aspect of successful potato production, and hence for feeding the world.

Potatoes also require adequate water, either from rain or irrigation. The latter is essential for summer crops in the North-West of the United States, autumn crops in the Mediterranean, and winter crops where rice is grown in the rainy summer season. In other areas, crops may still benefit from some irrigation [24]. Hence, two major factors in determining where potatoes can be grown and their yields are annual monthly minimum and maximum air temperature and mean precipitation, and these are changing. High-resolution Köppen-Geiger climate classification maps are being used to analyse historical climate conditions from 1901 and to make predictions about the future to 2099, based on the most plausible CO_2-induced warming rates, with new maps published in 2023 [25]. Under the "middle-of-the-road" socio-economic scenario SSP2-4.5, where CO_2 emissions start to fall from mid-century but do not reach net-zero by 2100,

Fig. 4.5 Cultivar Yukon Gold growing in mid-summer near Skagway (59°N), Alaska (left, 29 June 2011) and at Dumfries House (55°N), Ayrshire, Scotland (right, 22 June 2018)

Fig. 4.6 Potato trial (left, 8 April 2008) at the Sichuan Academy of Agricultural Sciences, Chengdu, Southwest China (31°N), planted early December for late April harvest, and potato trial (right, 13 December 2008) at the Modipuram Campus of the ICAR (Indian Council of Agricultural Research)-Central Potato Research Institute, Uttar Pradesh, India (29°N), planted 20 September and harvested 12 December (83 days)

the authors project that from 1991–2020 to 2071–2099, 8% of the land surface will transition to a different major Köppen-Geiger class. Furthermore, they estimate that 2.6 million km^2 of the global land surface will transition from polar to cold, 2.4 million km^2 from cold to temperate, 1.1 million km^2 from cold to arid and 2.8 million km^2 from temperate to tropical. We will return to the consequences of climate change for potatoes in Chap. 7.

We have now seen how potato cultivation spread from Europe to the rest of the world but have avoided the question of further introductions from South America. It is time to look at recently acquired information from the DNA of herbarium specimens.

4.6 DNA from Herbarium Specimens

DNA is a stable macromolecule. The genetic information (four bases) is held on the inside of its double helix structure, with the two complementary strands held together by hydrogen bonding. Nevertheless, it is remarkable that in favourable circumstances DNA can be extracted from herbarium specimens and sequenced. The first potato paper to be published was in 2008 by M. Ames and D. M. Spooner, with the focus on the presence (Chilean T-type cytoplasm) or absence (Andean cytoplasm) of a 241-bp deletion in the plastid (chloroplast) DNA [26]. They successfully amplified and analysed the DNA from 49 out of 64 of their herbarium leaf samples. These came from 13 European countries [Austria (1), Belgium (2), England (3), France (10), Germany (1), Hungary (1), Italy (2), the Netherlands (6), Poland (1), Portugal (1), Spain (5), Sweden (6), Switzerland (2) and Unknown (8)]. The results were clear: 20 samples from 1720 (one given as around 1700) to 1892 had Andean cytoplasm, and 29 samples from 1811 to 1910 had the Chilean cytoplasm. Both types of cytoplasm were found in the samples from England, France, the Netherlands, Sweden and Switzerland. The five Spanish samples (1827–1910) all had Chilean cytoplasm. The authors were happy about the ability of the plastid deletion marker to discriminate the likely geographical sources of the potato introductions being studied through the herbarium samples (>0.81 probability for Chile and >0.94 probability for the Andes). The authors concluded that the original introductions of potatoes came from the Andes, but introductions from Chile appeared from 1811 and came to predominate, presumably because they were better adapted to the long summer days of Europe. However, one cannot rule out the possibility that the Chilean potatoes were also pollinated by the Andigena potatoes and the resulting hybrids with Chilean cytoplasm were selected, despite usually being male sterile.

The second paper to be published was in 2019 by a group of eight researchers [27]. They sequenced whole chloroplast genomes and parts of the nuclear genomes of 88 potato plants. These included 29 historical herbarium specimens spanning the years 1660–1896, obtained from various European institutions. Three were Chilean, including two collected by Charles Darwin in 1834 during the voyage of the Beagle; the third was collected later in 1862. One was from Canada (1896). Twenty-five were from Europe: France (1, 1680) and Spain (4, 1781–1843); Ireland (2, 1846, 1847) and the United Kingdom (7, 1660, 1845–1875, another before 1900); the Netherlands (1, 1738); Austria (1, 1879), Germany (4, 1863–1877) and Poland (3, 1868–1891); and Unknown Europe (2, both 1706). The other potatoes were 43 South American landraces (post-1970) and 16 modern (post-1962) European cultivars. As whole chloroplast genomes had been sequenced, six types of cytoplasm (chlorotypes) could be recognized. The results were clear-cut. The Andean landraces of *S. tuberosum* from Colombia, Ecuador, Peru, Bolivia and Argentina had Andean cytoplasm. The three herbarium specimens from Chile had Chilean cytoplasm. The six herbarium specimens from Europe between 1660 and 1781 had Andean cytoplasm. Of the 19 herbarium specimens from Europe between 1810 and 1891, six had Andean cytoplasm (1847–1879) and 13 had Chilean cytoplasm (1810–1891). The analysis of the nuclear genomes also showed that European potatoes became progressively more similar to the historical Chilean potatoes. Finally, the researchers looked at the *StCDF1* gene that affects tuberization in different daylengths. The form (allele) *StCDF1.1*, which occurs in the Andean landraces, encodes a normal-length protein, which results in tuberization in daylengths of around 12 h. It was found to be present in the six herbarium specimens from Europe between 1660 and 1781. Three variants (alleles), *StCDF1.2*, *StCDF1.3* and *StCDF1.4*, were found in addition to *StCDF1.1* in 15 of the 19 herbarium specimens from Europe between 1810 and 1891, with *StCDF1.4* the most frequent. These alleles have insertions in their DNA (7 bp in alleles 2 and 4 and a putative transposon in allele 3), which result in the coded protein being truncated and the ability to form tubers in longer days. They are what we call dominant alleles; only one of the four alleles at a genetic locus in a tetraploid plant needs to be the variant. The three herbarium specimens from Chile all had the *StCDF1.1*, but three is too small a number to conclude that the other alleles were absent from Chilean landraces and arose by mutation in Europe. But this does mean that there are other genes, which influence the ability to form tubers in long days in Chile.

The conclusion from these two studies with herbarium specimens is that during the expansion of potato cultivation in Europe from 1800 to 1900, potato introductions from both the Andes and coastal Chile were present, with the latter becoming more important. Furthermore, genetic variants were present, which aided the ability to form tubers in long days. However, the number of new introductions from South America is unknown.

4.7 Extant Cultivars

Extant cultivars also provide some interesting information, and four examples are worth consideration.

Let's start in the Canary Islands, where potatoes were first grown outside of South America from about 1562 [2]. The first documented introduction was from Peru to Tenerife in 1622 [28]. By 1776, potatoes were the most important crop in the Canary Islands after wine grapes. In 1868, descriptions were available of 20 landraces, with most names still in use. However, from 1800 seed potatoes had also been imported from continental Europe. Today (2023) between 600 and 800 ha of 50 ancient varieties are grown in the Islands, with about 95% on Tenerife, where they are planted at 300–1500 masl in terraces in areas of dense cloud cover. Much of the cultivation involves hand labour by women farmers, with some use of animals. Animal manure is used as fertilizer. The main planting period is January for harvest in June and July (spring/summer potatoes) (Figs. 4.7 and 4.8), but some varieties are planted from June to November for harvest from December to April (autumn/winter potatoes). Yields are highest from the January planting.

Currently, the major threat to the crop is the Guatemalan potato moth (*Tecia solanivora* Povolny), which was inadvertently introduced from South America in 1999 and can destroy more than 50% of the production [28]. Late blight, viruses and nematodes can also be problems, as they are in other countries.

A phylogenetic analysis of 19 Canary Islands landraces and 25 South American landraces from the Andes and lowland south-central Chile, using genetic markers (SSRs) at 24 nuclear loci, separated them into a Chilean and an Andean cluster [28]. The separation corresponded to the presence and absence of the now-familiar 241-bp deletion in the chloroplast DNA, with three exceptions. The exceptions occurred among the 19 Canary Islands landraces. These landraces comprised 11 Group Andigena (tetraploids without deletion), 2 Group Chaucha (triploids that clustered with the Andean landraces but surprisingly had the deletion) and 6 Group Chilotanum (tetraploids with deletion, one of which clustered with the Andean landraces). The two Group Chaucha landraces could have arisen on the Canary Islands from unknown hybridizations, as all 150 Chaucha landraces in the collection held at CIP lacked the deletion. In summary, the analysis showed that Andean- and Chilean-derived landraces occur on the Canary Islands but from unknown sources. Therefore, it is not possible from this work to infer the origin(s) of the first European potatoes; hence the importance of the work on herbarium specimens dealt with in the previous section.

The situation in Japan is of particular interest, as potatoes were being grown there in the early seventeenth century, before Japan cut itself off from the outside world until the middle of the nineteenth century, a period of around 200 years. Then North American and European cultivars were successfully introduced, and publicly funded breeding work began in 1902. An analysis was published in 1994 of the genetic relationships (based on molecular markers called RAPDs) of Japanese and North American cultivars, including four old purple-skinned Japanese cultivars (Murasaki-imo, Nemuro-murasaki, Rankoku 3 and Rankoku 5) and two Andigena

Fig. 4.7 Cloud cover around Teide (28°N, 3718 m) on Tenerife, taken from Gomera, and terraces in steep valleys on Gomera, which are also typical of Tenerife, 27 February 2023

Fig. 4.8 Papas arrugadas (Canary potatoes with skin and salty crust) served with red mojo, 1 March 2023

landraces from Bolivia and Peru [29]. The old cultivars and Andigena landraces were closely related and distinct from the other material. Furthermore, they had the A-type chloroplast DNA typical of Andigena landraces. The authors concluded that the old cultivars were descended from one or a few introductions from Europe in the early seventeenth century before the introduction of foreign cultivars in the mid-nineteenth century.

In contrast, results from similar genetic analyses, with nuclear DNA markers (SSRs, simple sequence repeats) and the 241-bp chloroplast deletion, on 32 Indian cultivars, 12 Andean landraces and 5 Chilean landraces failed to provide evidence of remnant populations of Andean landraces among the Indian cultivars [30]. All Indian cultivars clustered with the Chilean landraces, although five lacked the 241-bp deletion and hence could be hybrids of Andean landraces and Chilean clones or more advanced cultivars.

Finally, in 1999, a group of scientists from the Scottish Crop Research Institute [31] examined 178 cultivars, which had been grown in the United Kingdom during the nineteenth (mainly late) and twentieth centuries, including cultivars from North America and elsewhere in Europe. They looked at variation in both chloroplast and nuclear DNA using SSR markers. Using seven chloroplast SSRs, they were able to identify 26 categories (haplotypes) of chloroplast DNA. Interestingly, 151 out of the 178 cultivars had the haplotype, which was diagnostic of the Chilean T-type cytoplasm mentioned earlier (241-bp deletion). In contrast, 25 haplotypes were found among the other 27 cultivars. These included Lumpers (introduced 1806), Myatt's Ashleaf (1847) and Pink Fir Apple (1850), which

T × T	T × A	A × T	A × A
↓	↓	↓	↓
T cytoplasm	T cytoplasm	A cytoplasm	A cytoplasm
Moderate tuber yields	Very high tuber yields	High tuber yields	Low tuber yields
Moderate fruit set	Low fruit set	High fruit set	High fruit set
Moderate male sterility	High male sterility	Low male sterility	Low male sterility

Fig. 4.9 Offspring of hybridizations involving Group Tuberosum (T) and long-day-adapted Group Andigena (A) potatoes where the first parent is the female and fruit set is natural in the field

can all be considered derived from the early Andean introductions to Europe. Others, such as Maris Piper (1966) and Shelagh (1986), are the products of modern breeding programs where the female lineage traces back to an Andean landrace. The researchers also used eight nuclear SSRs to determine the amount of genetic variation within the two groups of cultivars with and without T-type cytoplasm. There was little difference between the two groups in contrast to their chloroplast DNA, where the variation was within the group without the T-type cytoplasm. The researchers concluded that there were few maternal lineages in the breeding of modern cultivars. They attributed this to the selection of potatoes with T-type cytoplasm because they were higher yielding [32] but which had also been pollinated by Tuberosum rather than Andigena pollen because such plants had a higher fruit set [32, 33] (Fig. 4.9). This was also possibly due to the interaction between Tuberosum cytoplasmic (organellar) genes and dominant Andigena nuclear genes that result in male sterility [33].

These four results are consistent with the view that the first introductions from South America were Andigena potatoes, but during the expansion of potato cultivation from 1800 to 1900, introductions from both the Andes and coastal Chile were present, with the latter becoming more important. We do not know their actual distributions nor how often Andean and Chilean potatoes were grown in the same field. What we do know is that the potato cultivars being grown today are dominated by derivatives of Chilean Tuberosum and are the products of breeding (modern) programs involving artificial hybridizations, which trace back into the nineteenth century. Such cultivars are also being grown in the South Agricultural Council (SAC) countries of South America (Brazil, Paraguay, Argentina, Uruguay and Chile), possibly from late nineteenth and early twentieth century introductions by European immigrants, whereas Andean landraces are confined to the Andean countries (Venezuela, Colombia, Ecuador, Peru and Bolivia) [34].

The prerequisites for this modern breeding were an understanding of sexual reproduction in flowering plants, which came from the work of botanists and microscopists over the period 1682–1847 [35] and extensive experiments in plant hybridization from about 1750–1900, first by naturalists and later by breeders. The results of the breeding work are cultivars adapted to a wide range of environments around the world, thus allowing potatoes to contribute to food security in many countries.

References

1. Pigafetta A (1525) The first voyage around the world 1519–1522. Cachey Jr TJ (ed) (2007), University of Toronto Press, Toronto
2. Hawkes JG, Francisco-Ortega J (1993) The early history of the potato in Europe. Euphytica 70:1–7
3. Hawkes JG, Francisco-Ortega J (1992) The potato in Spain during the late 16th century. Econ Bot 46:86–97
4. Hawkes JG (1990) The potato: evolution, biodiversity & genetic resources. Belhaven Press, London
5. Salaman RN (1926) Potato varieties. Cambridge University Press, Cambridge

6. Earle R (2020) Feeding the people. Cambridge University Press, Cambridge
7. Eden FM (1797) The state of the poor or, an history of the labouring classes in England, from the conquest to the present day. Volume 1. Digitally printed 2011, Cambridge University Press, Cambridge
8. Simmonds NW (1969) Prospects of potato improvement. Scottish plant breeding station forty-eighth annual report 1968–69, pp 18–38
9. Glendinning DR (1975) Neo-Tuberosum: new potato breeding material. 2. A comparison of Neo-Tuberosum with unselected Andigena and with Tuberosum. Potato Res 18:343–350
10. Munoz FJ, Plaisted RL (1981) Yield and combining abilities in Andigena potatoes after six cycles of recurrent phenotypic selection for adaptation to long day conditions. Am Potato J 58:469–479
11. Huarte MA, Plaisted RL (1984) Selection for tuberosum likeness in the vines and in the tubers in a population of neotuberosum. Amer Potato J 61:461–473
12. Vilaró FL, Plaisted RL, Hoopes RW (1989) Comparison of cytoplasmic male sterilities in progenies of Tuberosum × Andigena and Tuberosum × Neo-Tuberosum crosses. Amer Potato J 66:13–24
13. Ghislain M, Núñez J, del Rosario HM, Spooner DM (2009) The single Andigenum origin of Neo-Tuberosum potato materials is not supported by microsatellite and plastid marker analyses. Theor Appl Genet 118:963–969. https://doi.org/10.1007/s00122-008-0953-6
14. Glendinning DR (1976) Neo-Tuberosum: new potato breeding material. 4. The breeding system of Neo-Tuberosum, and the structure and composition of the Neo-Tuberosum gene pool. Potato Res 19:27–36
15. Carroll CP (1982) A mass-selection method for the acclimatization and improvement of edible diploid potatoes in the United Kingdom. J Agric Sci Camb 99:631–640
16. Haynes KG, Lu W (2005) Improvement at the diploid species level. In: Razdan MK, Mattoo AK (eds) Genetic improvement of solanaceous crops, volume I: potato. Science Publishers Inc, Enfield, pp 101–114
17. Wilson A (1993) The story of the potato through illustrated varieties. Balding and Mansell, England
18. Reader J (2008) Propitious esculent. William Heinemann, London
19. Hawkes JG (1992) History of the potato. In: Harris PM (ed) The potato crop, 2nd edn. Chapman & Hall, London, pp 1–12
20. Smith A (1776) An inquiry into the nature and causes of the wealth of nations. Amazon, Great Britain (with minor editorial revisions to the original text)
21. Burton WG (1989) The potato, 3rd edn. Longman Scientific & Technical, Harlow
22. Hijmans RJ (2001) Global distribution of the potato crop. Am J Potato Res 78:403–412
23. Govindakrishan PM, Haverkort AJ (2006) Ecophysiology and agronomic management. In: Gopal J, Khurana SMP (eds) Handbook of potato production, improvement, and postharvest management. Food Products Press, New York, pp 179–229
24. Haverkort AJ, Struik PC (2015) Yield levels of potato crops: recent achievements and future prospects. Field Crop Res 182:76–85
25. Beck HE, McVicar TR, Vergopolan N, Berg A, Lutsko NJ, Dufour A, Zeng Z, Jiang X, van Dijk AIJM, Miralles DG (2023) High-resolution (1 km) Köppen-Geiger maps for 1901–2099 based on constrained CMIP6 projections. Sci Data 10:724. https://doi.org/10.1038/s41597-023-02549-6
26. Ames M, Spooner DA (2008) DNA from herbarium specimens settles a controversy about origins of the European potato. Am J Bot 95:252–257
27. Gutaker RM, Weiß CL, Ellis D, Anglin NL, Knapp S, Fernández-Alonso JL, Prat S, Burbano HA (2019) The origins and adaptation of European potatoes reconstructed from historical genomes. Nat Ecol Evol 3:1093–1101
28. Ríos D, Devaux A, de Galarreta JIR (2023) Ancient potato varieties of the Canary Islands: their history, diversity and origin of the potato in Europe. Potato Res. https://doi.org/10.1007/s11540-023-09672-2
29. Hosaka K, Mori M, Ogawa K (1994) Genetic relationships of Japanese potato cultivars assessed by RAPD analysis. Am Potato J 71:535–546
30. Spooner DM, Nunez J, Rodriguez F, Naik PS, Ghislain M (2005) Nuclear and chloroplast DNA reassessment of the origin of Indian potato varieties and its implications for the origin of the early European potato. Theor Appl Genet 110:1020–1026
31. Provan J, Powell W, Dewar H, Bryan G, Machray GC, Waugh R (1999) An extreme cytoplasmic bottleneck in the modern European cultivated potato (*Solanum tuberosum*) is not reflected in decreased levels of nuclear diversity. Proc R Soc Lond B Biol Sci 266:633–639
32. Maris B (1989) Analysis of an incomplete diallel cross among three ssp. *tuberosum* varieties and seven long-day adapted ssp. *andigena* clones of the potato (*Solanum tuberosum* L.). Euphytica 41:163–182
33. Grun P (1990) The evolution of cultivated potatoes. Econ Bot 44(3 Suppl):39–55
34. Caldiz DO (2023) Potato production in South America. In: Caliskan ME, Bakhsh A, Jabran K (eds) Potato production worldwide. Academic, London, pp 409–433
35. Bradshaw JE (2016) Plant breeding: past, present and future. Springer, Cham

Late Blight, Crop Failure and Famine

5

In 2008, there was a caveat to "How the humble potato could feed the world". Reliance on a single crop for food security is dangerous if widespread crop failures can occur and result in famine, as happened in Ireland with the late blight epidemics of 1845, 1846, 1848 and 1849. Grain and potatoes need to complement one another for food security, and we need to prevent crop failures while also recognizing political issues. *Phytophthora infestans* (Mont.) de Bary, the cause of late blight, is an oomycete with two mating types and the potential for exponential production of pathogenic spores. Effective control requires an understanding of its biology, epidemiology and ability to evolve. Weather conditions suitable for blight need to be forecast. Fungicides with different chemistries need to be used in spraying regimes to prevent the spread of blight strains with insensitivity to particular modes of action. Cultivars with different broad-spectrum resistance genes are required to delay the evolution of virulent blight strains. Blight outbreaks need to be analysed for knowledge of the prevailing pathogen populations. Good farm hygiene is important because diseased tubers left lying around and volunteer plants from diseased tubers left in the ground are sources of inoculum.

5.1 Let Them Eat Potatoes

Let them eat potatoes hasn't always worked. The potential of the potato as a major food crop in Europe was first seen in Ireland at the end of the seventeenth century and throughout the eighteenth century. In his 2008 book [1], John Reader argued that the golden age of the potato in Ireland was the second half of the eighteenth century. Ireland had a suitable climate and adequate soils for growing potatoes, but there were also societal and economic factors that increased their importance as a staple food of the "labouring poor", as witnessed by the English farmer Arthur Young on his tour in Ireland over the period 1776–1779 [2]. One hectare of land could produce all the potatoes needed by a poor family of peasants and labourers for a monotonous but nutritious diet, particularly if complemented with a cup of milk at each meal. This conclusion can be derived from simple but realistic assumptions and sums made by Reader [1]: for example, if a hectare of land produces a modest 11,500 kg of potatoes and an adult needs 4.5 kg a day, a family of seven could be fed for a year. Two parents and five children each generation would mean an expanding human population went hand in hand with an

J. E. Bradshaw, *Can Potatoes Feed the World?*, Sustainable Development Goals Series,
https://doi.org/10.1007/978-3-031-92890-1_5

increasing area of potato cultivation. As a consequence, the population of Ireland increased from 2 million in 1700 to 8.5 million in 1845, when one million hectares of potatoes had been planted (i.e., a hectare needed to feed 8.5 people) [1]. Then disaster struck with the potato murrain, as it was called. The late blight epidemics of 1845, 1846, 1848 and 1849, of which 1846 was the worst, destroyed the potato plants and rotted their tubers, resulting in crop failures and famine in Ireland with profound societal consequences, both at the time and to the present day [3, 4]. The devastation was the result of the arrival from eastern North America via Belgium of a new, virulent (ability to produce disease) and aggressive pathogen, which encountered a widely grown susceptible cultivar (Irish Lumper, grown for its yield) and favourable weather conditions in the days before fungicides. One million people died, and two million emigrated to the Americas and further afield [4], and the population of Ireland has never fully recovered, nor has the area of potatoes (8794 hectares producing 344,000 tonnes per year in the Republic of Ireland from 2016 to 2020, which is enough for 188 g per person per day [5]). Today the population stands at 6.8 million for the island of Ireland, with 5 million in the Republic of Ireland and 1.8 million in Northern Ireland [6]. Whether famine could and should have been prevented has been debated ever since, remembering that the Union of Great Britain and Ireland took place on the 1st of January 1801 and that 50 years later, in 1851, the Great Exhibition of London was a demonstration of Britain's wealth. Late blight had also caused devastation of potato crops in eastern North America after its arrival from Mexico in 1841–1842 and in Northwest Europe after its arrival in 1842, but famine is associated with Ireland and the policies of the British government. David Nally argues in his 2011 book that "the government showed no inclination to initiate anything resembling the traditional anti-scarcity measures used to mitigate the worst horrors of famine around the world" [4].

If, like me in 2010, you visit Grosse Île, the Canadian Quarantine Station for Immigrants from 1832 to 1937, you will discover that in 1847 there were 100,000 immigrants, mainly Irish but some Scottish Crofters, and that 5000 of them died of typhus and other diseases. You will also see the names on the glass memorial, as well as the Irish Memorial overlooking the St. Lawrence River (Fig. 5.1). In Scotland, the West Highlands and Islands and Caithness in the northeast escaped late blight in 1845 but not in 1846, so there was famine in early 1847, which was also the worst year of famine in Ireland. James Hunter tells the story in his book *Insurrection: Scotland's Famine Winter* [7]. Since 1800, crofters (tenants) in the West Highlands and Islands had adopted potatoes as their staple food, in preference to oats and barley, which were difficult to grow in the harsh environments. The Whig government thought that the landlords should help their crofters, but instead, the landowners of Lewis, Barra, South Uist and Benbecula paid to have a total of 3700 crofters shipped to Canada. George Douglas, the medic in charge of Grosse Île, said of them, "I never, during my long experience at the station, saw a body of emigrants so destitute of clothing and bedding. Many children of nine and ten years old had not a rag to cover them". The Quebec authorities never received any money from the landowners towards the cost of caring for their former tenants. In Caithness, although the potatoes succumbed to late blight in 1846, there was a good cereal harvest. However, early in 1847, the landowners tried to ship their grain to the Scottish Lowlands and England for more money than the local population could afford, so there were riots in Wick.

Exports and the cost of food were issues during the famine in Ireland. With potatoes in the late eighteenth century adequately feeding an expanding but poor population of Irish agricultural workers, the latter were expected to spend more time working on the production of commodities for their landlords to export, such as grain and linen from flax. When late blight destroyed their potatoes, the poor could not afford to buy other food. Books such as the ones by John Reader [1] and David Nally [4] deal with the politics, economics and societal atti-

Fig. 5.1 Grosse Île, the Canadian Quarantine Station from 1832 to 1937, in the St. Lawrence River. (From Bradshaw JE (2016) Plant Breeding: Past, Present and Future. Springer, Cham with permission; 8 October 2010)

tudes that contributed to the human misery that was the famine (e.g., the Irish Poor Law of 1838, employment from the government's public works programmes rather than food aid, and workhouses). Here, I just want to draw attention to a point that is still relevant today. In 1845, the Corn Laws of 1815 were still operating. They imposed a customs duty on imported grain (wheat and other cereals such as maize) so that it could never be cheaper than the home-grown product. Hence, the cost of food was kept high for the benefit of landowners. Sir Robert Peel, the British Prime Minister in 1845 and 1846, thought that part of the solution to alleviate the growing famine was the repeal (abolition) of the Corn Laws. He succeeded in May 1846 despite two-thirds of his own Conservative MPs voting against the repeal because they were representing the interests of the landowners. In contrast, the new industrialists wanted the repeal, as it would lead to free trade and the industrial economy they sought for their benefit. Subsequently, the government fell, and the new Whig Prime Minister, Lord John Russell, assumed office on 30 June 1846. He, and the controversial assistant secretary to the treasury (i.e., a career civil servant), Charles Trevelyan, inherited and failed to deal with the problem of the famine, which resulted in a million deaths [4]. Today, as I continue to write this book in April 2023, as a result of the Russian invasion of Ukraine, there is a one-year suspension (June 2022 to June 2023) of tariffs on grain imports from Ukraine into the European Union, which is disliked by Polish and Hungarian farmers. Food security, tariffs and free trade remain political issues.

The Irish potato famine serves as a warning not to rely on potatoes as the main crop to feed a population. Grain and potatoes need to complement one another, and we need to take steps to prevent crop failures. So, what progress has been made with controlling late blight?

5.2 The Destroyer of Plants: *Phytophthora infestans* (Mont.) de Bary

The first question that needed to be answered was simply, what is the cause of late blight? The first volume of the *Journal of the Horticultural Society of London* was published in 1846. The second paper was by the Rev. Miles J. Berkeley on "Observations, Botanical and Physiological, on the Potato Murrain (plague)" [8]. He reminded readers that late blight first appeared in Belgium in 1842 and spread from there to the rest of Europe, with 1845 being the first year of devastation. He also mentioned that late blight was prevalent in Canada and the United States in 1844. Having weighed up all of the ideas and evidence, he concluded that late blight was caused by a fungus, as suggested by Dr Charles Morren and others in Belgium, and not independent of parasitic fungi as others thought. He reported the botanical description of the fungus made by the French cryptogamist Dr Jean F. C. Montagne, who thought that it was a Botrytis mould. It was therefore named *Botrytis infestans*, Mont. However, it was not until 1861 that the German mycologist Heinrich Anton de Bary provided proof of the causal agent by infecting a healthy plant with spores of the fungus and watching the disease develop while an uninfected plant remained healthy [1]. His seminal book of 1861 on late blight was in German [9]; but later, in 1876, he described the asexual life history (cycle) of the fungus in English [10]. By then, the fungus was known as *Phytophthora infestans* (Mont.) de Bary: the destroyer of plants (Fig. 5.2).

Fig. 5.2 Late blight on susceptible cultivar Vales Sovereign in a garden in Currie, Edinburgh, Scotland, on 27 August 2023 (July and August had been warm, wet months)

Since then, a vast literature comprising thousands of publications has accumulated on the pathogen and its disease. Two introductions are still relevant. In 1991, *Advances in Plant Pathology Volume 7* [11] was devoted to *Phytophthora infestans*, and a much shorter but still detailed account was published by William Fry in 2008 [12]. These are scientific reviews, which means that they contain many technical terms; for example, they will tell you the following (my comments in brackets). "*Phytophthora infestans* is a heterothallic (sexual reproduction requires two mating types, A1 and A2) oomycete (algal-like rather than fungal-like), and it is a near-obligate hemi-biotrophic pathogen (usually needs to obtain its nutrients from living tissues of its host) under natural and agricultural conditions". Here, I will try to minimize such terms in giving an explanation of how blight epidemics occur and the attempts to control them.

5.3 Blight Epidemics

In the absence of both mating types, the situation in Europe and North America in 1845, a blight epidemic starts from infected tubers. In the early 1840s, it is assumed that infected tubers of the A1 mating type were unintentionally brought by ship from Mexico, a centre of diversity of wild tuber-bearing *Solanum* species and also of *P. infestans*, where both mating types occur.

Crops are infected from diseased tubers, which had been discarded (not destroyed) and left lying around, either at lifting the previous season (cull piles) or after being stored (indoors or in clamps) but not used. When these sprout, *P. infestans* can grow in the shoots and sporulate to give spores that are dispersed to nearby newly planted crops. Likewise, if infected tubers are planted, they can give rise to infected shoots which can produce spores. Either way, under moist conditions, *P. infestans* produces what are called sporangiophores, which in turn produce and release sporangia (spores) containing zoospores. When dispersed, aerially or in water splashes, the sporangia can initiate new infections on the leaves and stems of the growing potato plants, which in due course produce more sporangiophores in cycles that lead to the observed epidemic. The spores can also be washed down into the soil and infect the developing tubers. Furthermore, tubers can be infected at harvest from infected foliage bearing sporangia. Any tubers that are not harvested can survive in the ground and produce infected volunteer plants the next year. Hence the need for good farm hygiene.

A zoospore is motile for up to about an hour after release. Then after encystment and germination on host tissue, its germ tube penetrates the leaf or stem tissue. A necrotic lesion becomes visible about 2 days later. As the lesion grows, there is an outer zone of mycelium working its way into healthy tissue. Behind it, after another 2 days or so, if the humidity is high enough, is a zone in which sporangiophores are produced with up to 300,000 sporangia per lesion. In turn, behind this sporulating zone is a sterile, inner zone. Thus, there is a latent period before newly infected tissue becomes infectious. After the infectious period, the tissue is removed from the epidemic because it neither produces spores nor is it susceptible to re-infection. Hence, a short latent period and a long infectious period result in the rapid development of an epidemic over a number of cycles of infection and sporulation.

Sexual reproduction is more complicated, requiring the presence of both the A1 and A2 mating types and the formation of female (oogonia) and male (antheridia) sex organs. It was therefore confined to Mexico until the late 1970s and the global spread of new populations of *P. infestans*. The outcome is the formation of oospores, which can persist for several years in soil and start epidemics earlier in the potato-growing season than those from asexual zoospores. Furthermore, as a result of sexual reproduction, the oospores are genetically variable.

5.4 Mathematical Models

As a result of the spread of the coronavirus SARS-CoV-2 from China at the end of 2019, we have all experienced the COVID-19 pandemic. One consequence is that we are now all familiar with the mathematical modelling of epidemics. A simple model has proved extremely useful in studying late blight epidemics. The model, with many examples, can be found in J. E. Van der Plank's 1968 book *Disease Resistance in Plants* [13], which influenced the thinking of my generation of plant breeders. B. E. J. Wheeler presents shorter accounts in two books published in 1975 and 1976 [14, 15].

We are going to consider a group of potato plants, which could be a small plot in a breeder's nursery or a large field of potatoes. We are going to relate the infection rate of the epidemic to the proportion of infected rather than infectious tissue as the simpler but adequate option. We will assume that an epidemic is started by a low concentration of spores so that they act independently and the number of lesions produced is proportional to the number of spores. The infection rate (r) of the epidemic is defined by the following differential equation:

$$dy / dt = ry(1 - y)$$

where y is the proportion (0–1) of infected tissue (disease) at time t (days) and (1-y) is the proportion of tissue still healthy and available for infection. In other words, we have an epidemic in which y increases with t, and the rate of increase depends on both the amount of diseased tissue present and the amount of healthy tissue remaining to be infected. We solve the differential equation as follows.

$$\int dy / [y(1 - y)] = r \int dt$$

$$\int dy / y + \int dy / (1 - y) = r \int dt$$

Therefore, $\ln(y) - \ln(1\text{-}y) = rt + k$ where k is a constant

$$\ln\left(y/(1-y)\right) = rt + k$$
$$y/(1-y) = e^{rt+k}$$
$$y = e^{rt+k}/\left(1+e^{rt+k}\right)$$

Hence, the graph of ln(*y*/(1-*y*)) against *t* is a straight line of slope *r* and intercept *k*, whereas the graph of *y* against *t* is a curve, the disease progress curve (e ≈ 2.718). Early in an epidemic, *y* will be small and (1-*y*) will be very close to 1, so that $y \approx e^{rt+k}$.

At the very beginning of the epidemic, $y = y_0$ and $t = 0$ so that $y_0 = e^0e^k = e^k$ (and $k = \ln(y_0)$) and $y \approx y_0e^{rt}$. In other words, initially we have exponential growth (compound interest) in which infections are independent of one another. Furthermore, *k* is the natural logarithm of the initial proportion of infected tissue, which depends on the number of lesions, which in turn depends on the number of spores initiating the epidemic (initial inoculum).

I am going to give a theoretical example based on my own experimental results [16]. Each year we had a disease nursery in Ayrshire on the western seaboard of Scotland where the moist weather was conducive to the development of late blight (Fig. 5.3). The year 1998 was fairly typical. We planted the nursery on 13 May and 23 July pots of glasshouse-grown plants of the susceptible cultivar King Edward, which had been inoculated 3 days earlier with a zoospore suspension of *P. infestans*, were placed in the nursery at intervals along spreader drills of King Edward. The amount of blight present on the experimental material in the nursery was scored on a 1–9 scale of increasing resistance [17] on 30 July and on 3, 7, 11 and 14 August (Fig. 5.3). The 1–9 scale corresponded to the proportion of necrotic tissue (1, 0.9, 0.8, 0.7, 0.6, 0.4, 0.25, 0.1 and 0). The mean

Fig. 5.3 Experimental material in disease nursery, Ayrshire, Scotland, August 1998: foliage on spreader drills of susceptible cultivar King Edward is dead. (From Bradshaw JE (2016) Plant Breeding: Past, Present and Future. Springer, Cham with permission; source: SCRI)

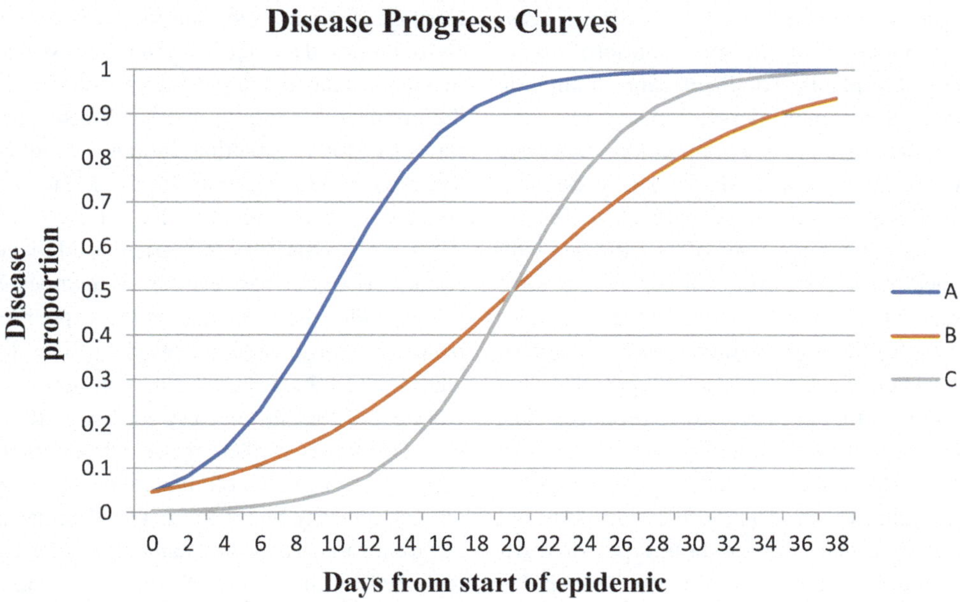

Fig. 5.4 Theoretical Disease Progress Curves differing in r and k (A: $r = 0.3$ and $k = -3$; B: $r = 0.15$ and $k = -3$; C: $r = 0.3$ and $k = -6$)

scores of susceptible controls (y = 0.24, 0.36, 0.66, 0.88 and 0.94) allowed me to plot $\ln(y/(1\text{-}y))$ against t in days from 23 July. The result was a good fit to a straight line with slope (r) 0.2725 and intercept (k) −3.299. It took about 3 weeks to completely destroy the susceptible plants. In contrast, r was 0.1603 for the resistance cultivar Stirling (y = 0.03, 0.10, 0.10, 0.175 and 0.35).

I have therefore decided to use values of $r = 0.3$ and $k = -3$ for my theoretical disease progress curve A shown in Fig. 5.4. Descriptive keys are available for working out the proportion of diseased (necrotic) tissue in a field of potatoes, rather than a disease nursery, and can be found in the books by Wheeler [14, 15].

In curve B the rate r is halved so that y (the proportion of disease) reaches a value of 0.5 on day 20 of the epidemic, compared with day 10 for curve A. In contrast, in curve C the rate remains the same as in curve A, but k is doubled from −3 to −6. The curve is pushed to the right by 10 days so that y is the same in curves A and C on days 2 and 12, 4 and 14, 6 and 16, etc. This can be confirmed as follows:

$$\ln\left(y_A/(1-y_A)\right) = rt_A + k_A \text{ and}$$
$$\ln\left(y_C/(1-y_C)\right) = rt_C + k_C$$
$$\text{Hence when } y_A = y_C, rt_A + k_A = rt_C + k_C$$
$$\text{and } (k_A - k_C)/r = t_C - t_A$$
$$\text{In our example, } (k_A - k_C)/r = (-3+6)/0.3 = 10 \text{ days.}$$

Furthermore, at the start of the epidemic, we saw that $k_A = \ln(y_{0A})$ and $k_C = \ln(y_{0C})$ so that

$$k_A - k_C = \ln(y_{0A}) - \ln(y_{0C}) = \ln(y_{0A}/y_{0C})$$
$$\text{and } y_{0A}/y_{0C} = e^{(kA-kC)}.$$

Hence, in our example, we have $k_A - k_C = 3$ and $y_{0A}/y_{0C} \approx 20$. In other words, there are 20 times more spores of A than C at the start of the epidemic.

The value of these disease progress curves is that they help us to think about ways of controlling an epidemic or, even better, preventing it in the first place. They also help us think about types of disease resistance in potatoes. Curve A in Fig. 5.4 is typical of a susceptible cultivar. In contrast, curve C is typical of a cultivar with a resistance gene(s) that can stop the infection process for most but not all of the spores initially attacking the potato plants. The *P. infestans* population comprises a common avirulent race and a rare virulent one that arose by mutation. Systemic fungicides can have a similar effect by stopping the infection process at a very early stage. In contrast, curve B is typical of a cultivar with resistance that reduces the number of spores (of both races) that can establish an infection, slows the spread of the lesions, increases the length of time to when the lesions start to sporulate, and reduces the number of new spores produced. The outcome is a reduced infection rate. Protectant fungicides that inhibit the germination of spores have a similar effect. Finally, I should point out that a cultivar can combine both types of resistance and also comment on the susceptible "spreader" cultivar King Edward (strictly King Edward VII). The parentage of King Edward is unknown [18]. It was raised by a gardener in Northumberland, England, under the name Fellside Hero and later renamed King Edward and put on the market in 1902. It was still being grown for supermarkets in 2024, 122 years later, having been one of the most popular cultivars in Britain for most of that period. Another successful rebranding of a susceptible cultivar was International Kidney (unknown parentage) to Jersey Royal (1879) after it was introduced to the Island of Jersey for export to Britain [18]. It is still a popular addition to supermarket shelves in Britain in May and June, or earlier if planted in glasshouses or polythene tunnels in December. Clearly, durable commercial success and durable resistance to late blight are rather different! But what actually happened after the late blight epidemics of the 1840s?

5.5 Bordeaux Mixture

The Gardeners' Chronicle was first published in London as a weekly newspaper in 1841 with John Lindley as its first editor. By 1851, it was more popular than the Observer and The Economist. It continued to be published until 1986. It can be found online in the Biodiversity Heritage Library. On Saturday 26 September 1846, it printed a letter from Matthew Moggridge of Swansea dated 4 September 1846 [19]. During the eighteenth century, a new "Welsh process" for efficiently smelting copper resulted in Swansea producing one third of the world's smelted copper over the period between the 1770s and 1840s [20]. Moggridge said that on 31 August 1846 he had examined potatoes being grown under the influence of the copper smoke from a nearby smelting works. "The leaves, haulm, and tubers improve as you approach the works, and that the nearest gardens, little more than 200 yards ($\approx$ 200 metres) from them, are entirely free from the blight, and the crop good in quality, quantity and flavour. The Potatoes are of different sorts. These last-named gardens, as I am informed by the proprietor, entirely escaped the disease in 1845, and have borne Potatoes for 40 years". Reader also reports this letter in his 2008 book [1] and explains that it was another 40 years before anyone would investigate the use of copper to control late blight. In contrast, copper was already used to protect wheat seed from the smut (bunt) fungus *Tilletia tritici*, named after the French botanist Mathieu Tillet, who investigated it in 1755. Experiments by the Swiss naturalist Isaac-Bénédict Prévost, published in his memoir of 1807, had shown that wheat seeds could be protected by soaking them in a solution of copper sulphate. However, nobody investigated its effect on the foliage of plants. It was a visit to a vineyard in 1882 by the French professor of botany at Bordeaux University, Pierre Marie Alexis Millardet, that led to the breakthrough. He noticed that the vine leaves had not been destroyed by the downy mildew fungus *Plasmopara viticola*, as had happened elsewhere. He discovered that it was the custom in Médoc to cover the leaves with copper sulphate mixed with lime when the grapes were ripening because the copperish spots had proved a deterrent to theft of the fruit. In May 1885, he published a recipe for his light-blue Bordeaux Mixture of copper sulphate and lime, which produces copper hydroxide [$Cu(OH)_2$] as the main active ingredient. By 1888, an effective recipe and spraying schedule had been developed in France for potatoes. It was soon in common use in Europe and North America and is still used today, particularly by organic growers. Crops must be sprayed before the arrival of *P. infestans*, as it is the mixture's copper ions that prevent the spores from germinating. However, Bordeaux Mixture has been banned in some countries because copper ions can build up in the soil and cause pollution and, depending on their concentration, can be toxic to a wide range of soil- and aquatic-borne organisms.

5.6 Fungicides

Fungicides have proved to be one of the most effective ways of controlling late blight. Useful historical accounts can be found by Schwinn and Margot in *Advances in Plant Pathology Volume 7*, published in 1991 [21], and *A Short History of Fungicides*, written in 2008 for the American Phytopathological Society (APS) [22].

Some well-known fungicides used to control *P. infestans* and other oomycetes are shown in Table 5.1 in order of release, with additional information from *The UK Pesticide Guide 2023*, which also includes 14 approved mixtures [23]. In no way is it a complete list! I have also included the FRAC (Fungicides Resistance Action Committee) mode of action code (www.frac.info).

Copper salts were the dominant fungicides in the first era of chemical control (1885–1934). Out of the many copper compounds and mixtures which were tried, three became widely used, Bordeaux Mixture, copper oxychloride and cuprous oxide. They had a strictly protective action (protectants) which necessitated repeated applications at short intervals during epidemics. This is because it is difficult to get a complete

Table 5.1 History of fungicides used to control late blight of potatoes [21–23]

Name	Year released	Class [FRAC mode of action code]	UK approval for 2023
Bordeaux Mixture	1885	Copper	
Copper oxychloride	1900	Copper	Not on potatoes
Cuprous oxide	1932	Copper	
Zineb (Zn)	1943	Dithiocarbamate	
Maneb (Mn)	1950	Dithiocarbamate	
Fentin hydroxide	1954	Triphenyltin	
Fentin acetate	1956	Triphenyltin	
Mancozeb	1961	Dithiocarbamate [M3]	GB only*
Cymoxanil	1976	Cyanoacetamide-Oxime [27]	Yes
Metalaxyl	1977	Phenylamide [4]	In mixtures
Propamocarb hydrochloride	1978	Carbamate [28]	In mixtures
Dimethomorph	1988	Cinnamic acid amide (CAA) [40]	Yes
Fluazinam	1992	Dinitroaniline [29]	Yes
Mandipropamid	1999	Mandelic acid amide (CAA) [40]	Yes
Cyazofamid	2001	Cyanoimidazole sulfonamide [21]	Yes
Zoxamide	2001	Benzamide [22]	In mixtures
Benthiavalicarb-isopropyl	2003	Valinamide carbamate (CAA) [40]	Yes
Fluopicolide	2006	Benzamide/pyridine [43]	In mixtures
Amisulbrom	2007	Sulfonamide [21]	Yes
Ametoctradin	2009	Triazolopyrimidine [45]	In mixtures
Valifenalate	2009	Valinamide carbamate (CAA) [40]	In mixtures
Oxathiapiprolin (OXTP)	2017	Piperidinyl thiazole isooxazoline (OSBPI) [49]	Yes

*United Kingdom excluding Northern Ireland, but ended in 2024

FRAC Fungicide Resistance Action Committee, *CAA* carboxylic acid amides group of fungicides, *OSBPI* oxysterol binding protein inhibitors group of fungicides covering of the chemical on the plant, the deposits can be washed off by rain, and new growth is not protected. Nevertheless, they were effective in reducing infection pressure and resulted in significant yield increases. They are classed as residual contact fungicides because their copper ions act on fungal structures on the leaf surface (zoospore release and germination). Hence, they prevent infection but cannot eradicate an existing one. Common rates of use are 1–3 kg of metallic copper per hectare. Copper oxychloride is still in use today but is no longer approved for potatoes.

The second era of chemical control (1934–1975) began with the discovery of the fungicidal activity of dithiocarbamates in 1934. They were the first organic fungicides, but again, protectants with residual contact action. The ones introduced from 1943 to 1963 for the control of late blight were zineb, maneb, metiram, mancozeb (a mixture of maneb and zineb) and propineb. Common rates of application are 1.2–2 kg of active ingredient per hectare. There have been formulations and mixtures of dithiocarbamates with other fungicides.

The dithiocarbamates have better crop tolerance, ease of handling and sprayability than copper compounds and have been widely used. Again, they are strictly protective fungicides that affect zoospore release and germination on the leaf surface, but mancozeb does have multi-site activity. Nevertheless, they have to be applied at regular intervals. The actual fungicide is the dialkyl dithiocarbamidate (sulphur-containing) anion. It is toxic to plant cells but not taken up by them. Mancozeb was still approved for use in Great Britain in 2023, but this approval ended on 31 May 2024 because the GB Health and Safety Executive had concluded that mancozeb has endocrine-disrupting properties. Stocks must be used by 30 November 2025. The withdrawal was widely reported in the farming press and is a cause for concern to growers, as mancozeb is also used to control *Alternaria* [24].

The other group of effective protectants against late blight are the triphenyltin compounds. They combine high fungicidal activity with acceptable crop tolerance (they are phytotoxic) and show protective activity against tuber blight as well as foliar blight. Fentin hydroxide was introduced in 1954 and fentin acetate in 1956, with the former applied at 250–350 g active ingredient per hectare and the latter at 160–260 g. As well as affecting zoospore release and germination, they have an anti-sporulant effect. This reduces the risk of sporangia being washed down to the tubers and causing infection. Because of their phytotoxicity and effectiveness against tuber blight, they are used late in the season.

When I was assessing potato breeding material in yield trials in the 1990s, our typical spraying program for late blight control was four sprays of mancozeb followed by two sprays of fentin hydroxide at 2-week intervals from mid-June (plants meet in rows) to mid-August. The potato plants would then be burnt down (haulm destroyed by sulphuric acid or pulverized) at the end of August before harvest in mid-September. If farmers thought it was necessary to spray at weekly intervals, their crops could be sprayed as many as 12 times! In the humid highland tropical regions of the world where potatoes are grown all year round, even more spays may be required.

The period from 1960 to 1970 saw the rapid development of the crop protection industry, including an expansion of research and development and an interest in the biochemical mode of action of fungicides. Young potted plants were sprayed with test compounds, inoculated with pathogens, and disease symptoms observed, all under controlled environment conditions.

The introduction of cymoxanil (a cyanoacetamide-oxime) in 1976 opened the third era of chemical control of late blight. It was the first systemic fungicide with action against *P. infestans* but only had very short persistence in the plant. It was therefore recommended for use in combination with residual fungicides. Metalaxyl, a phenylamide, quickly followed in 1977 and proved the breakthrough as the first curative of late blight. By 1989, there were eight systemic compounds involving five chemical classes, but propamocarb and the phosphites were not effective enough under field conditions, leaving dimethomorph (a cinnamic acid derivative) as the newest class. Dimethomorph has anti-sporulant activity and is locally systemic, but little transport occurs from treated leaves. It is persistent for 7–10 days. Three other phenylamides (ofurace, benalaxyl and oxadixyl) were introduced from 1978 to 1983, but metalaxyl became the most widely used. It is quickly taken up by roots, stems and leaves and translocated from old to new tissue (acropetally). It inhibits mycelial growth of *P. infestans* and particularly sporulation by interfering with RNA synthesis, thus acting inside the plant tissue. Hence, a strong curative effect is observed on application to existing infections. Its effect on sporulation results in protection against tuber blight. Application rates of 200–250 g active ingredient per hectare give protection for 10–14 days, partly dependent on its persistence.

Although metalaxyl initially provided effective control of late blight, *P. infestans* developed resistance (insensitivity) to the fungicide. It became clear that the widespread use of a systemic fungicide puts a selection pressure on the pathogen to mutate and evolve new races that can grow on treated plants. This is particularly true of systemic fungicides with a specific or single mode of action. In contrast, resistance to protective fungicides was not expected to evolve because they have multi-site action against the pathogen. It therefore came as a surprise when insensitivity appeared in 2015 to fluazinam (e.g., in strain EU 37_A2), a protective fungicide with non-specific action which was introduced in 1992 [25]. Although still approved, its use has declined except in mixtures. *A Short History of Fungicides* also mentions two protectants, mandipropamid (with translaminar action against spore infection) and cyazofamid, released in 1999 and 2001, respectively. However, in 2022, there were reports that mandipropamid was insensitive to an aggressive blight strain (EU 43_A1) prevalent in Denmark and moving into the Netherlands and Belgium, and concerns about it spread to the United Kingdom in 2023 and its possible resistance to other fungicides in the CAA group (Table 5.1) [26]. Nevertheless, despite concerns, its use in 2023 was advocated in the UK farming press in what proved to be a blight year with repeated periods of warm (not hot) weather and high humidity in July combined with windy conditions that made spraying difficult. EU 43_A1 did increase in Germany, the Netherlands, and Belgium, and spread to France, Spain, and the Republic of Ireland, but not to the United Kingdom [27].

The period 2001–2009 saw six more fungicides which are currently approved: zoxamide (acts after spore germination to inhibit fungal penetration), benthiavalicarb-isopropyl, fluopicolide, amisulbrom, ametoctradin and valifenalate. They are protective with some curative effects, although valifenalate is classed as systemic.

The newest type of fungicide is oxathiapiprolin, which was introduced in 2017. It is the first member of a new class of piperidinyl thiazole isoxazoline fungicides, which are oxysterol binding protein inhibitors (OSBPI) with exceptional activity against plant diseases caused by oomycete pathogens. It provides an example of the modern approach to the discovery, synthesis, optimization, mode of action and biological efficacy of a new class of fungicide [28]. The researchers, based at a chemical company, purchased a diverse (bis-amide) organic compound library and found one of the compounds had modest fungicide activity. Using the techniques of organic chemistry, they systematically modified the compound until they had synthesized compound 20 (oxathiapiprolin) with a relative activity of 10,000 compared with compound 1. It was effective when applied at 12–30 g active ingredient per hectare. The researchers found that "preventatively, it inhibits zoospore release and stops zoospore and sporangia germination. Curatively, it stops mycelial growth within the host plant before visible lesions occur, offering protection at 1, 2 and 3 days. Post-infection, it stops mycelial growth and inhibits further lesion expansion, and as an antisporulant, it inhibits spore production and viability". The researchers also found that "it is rapidly absorbed into the epicuticular waxy layer of the plant, making it extremely resistant to wash-off. Once inside the plant, it shows translaminar and acropetal systemic movement, protecting treated leaves as they grow and expand and new leaves as they emerge and grow". Oxathiapiprolin was also shown to have an excellent toxicity profile, with low oral, neuro, developmental, dermal, aquatic, avian and bee toxicity. It acts via inhibition of a novel fungal target, an oxysterol binding protein, but the cellular role of this protein is not currently understood. As it has a single-site mode of action, use in mixtures is being advocated to prevent the spread of resistance, which was found in mainland Europe in 2022 and 2023 in strain EU 43_A1 and in 2023 in the Netherlands and northwest Germany in new strain EU 46_A1 [27]. Strain EU 43_A1 can be resistant to both CAA and OSBPI fungicides, to one but not the other, and to neither! [27]. As a result, FRAC have updated their guidelines for CAAs and OSBPIs [www.frac.info].

Experience has therefore shown that there is a need for both protectant and systemic fungicides, often as admixtures, sometimes with unexpected synergistic effects. Other materials are often incorporated with the fungicides to increase their wetting properties (spreaders) and their ability to adhere to leaves (stickers). During the last 20 years new products have become available and others have been withdrawn from use for a variety of reasons, usually associated with concerns about toxicity in the environment and human health. Growers in the United Kingdom, for example, need to check *The UK Pesticide Guide* for available products. They will try to use products belonging to different modes of action in their spraying regimes and will benefit from the monitoring which is now done of *P. infestans* populations for race composition and fungicide resistance. They will also benefit from information from the industry-led Fungicide Resistance Action Committee [FRAC]. The search for safer and more environmentally acceptable fungicides that give high levels of control will no doubt continue.

Experience has also shown the need to apply the first spray as a protectant before infection occurs. As a consequence, many countries have developed blight warning systems based on monitoring weather conditions favourable for late blight, such as those initially developed in the United Kingdom by Beaumont in the 1930s [29]. Beaumont conditions are periods of not less than 48 consecutive hours of relative humidity above 75% and temperatures above 10 °C. Improvements on this have developed over the years, such as the Smith Periods in 1956 (two consecutive days each with a minimum temperature of 10 °C and at least 11 h with a relative humidity above 90%) and, more recently, in 2017, the Hutton Criteria (two consecutive days each with a minimum temperature of 10 °C and at least 6 h with a relative humidity above 90%) [30].

5.7 Disease Resistance to 1968

Potato breeding in Europe and North America during the second half of the nineteenth century did not produce any cultivars which were highly resistant or immune to late blight, although sufficient resistance was supposedly built up in some cultivars to prevent further disasters. This is difficult to verify, and the situation could have been helped by weather less favourable to epidemics and the use of Bordeaux mixture on crops after 1885 [31]. The apparent breakthrough was the discovery of genes (*R*-genes) for high levels of resistance in the wild tuber-bearing species *S. demissum*, which is found in Mexico, where potatoes and *P. infestans* (both mating types) co-evolved. Breeding for resistance started in 1908 in Germany and 1909 in Great Britain [32]. I am going to tell the story of what happened in Great Britain through the work of three potato breeders. The global literature on the subject is too vast for me to do it justice here, but I like to think that my book on potato breeding published in 2021 contained a reasonable attempt [33]. Furthermore, what happened in Britain is typical of what happened globally.

A key figure in England in the early years of the twentieth century was Redcliffe N. Salaman (1874–1955), who trained as a doctor but contracted tuberculosis in 1903. On his recovery, with guidance from his friend William Bateson, and at the suggestion of his gardener, he turned to the scientific study of the potato [1, 34]. Bateson, while based at Cambridge University from 1900 to 1910, coined the word "genetics" for the study of heredity in 1906, then in 1910 became the first director of the John Innes Horticultural Institution at Merton, London, and in 1919 founded the Genetical Society of Great Britain. Salaman started his breeding and genetics experiments in 1906 in the garden of his home in Barley in Hertfordshire and published his seminal paper on potato genetics in the first issue of the *Journal of Genetics* [35]. He summarized the results of his work at Barley in 1926 in his book *Potato Varieties* [36]. His work then transferred to the new Potato Virus Research Institute in Cambridge, where he became its founding director, a position he held until his retirement in 1939. In 1906, he requested tubers of *S. maglia* from Kew Gardens [1]. The mislabelled tubers that he received were in fact *S. edinense*, a natural hybrid of *S. demissum* and *S. tuberosum* from the botanic garden in Edinburgh, which was interestingly recreated in 2018 by their staff [37]. At Barley, *S. edinense* proved highly but not completely resistant to late blight in the years 1906–1910. Salaman was able to self-pollinate *S. edinense* and in 1909 raised a family of 40 plants, 7 of which proved resistant to the bad blight attacks of that year and the next. One clone was allowed to remain in the kitchen garden at Barley for 17 consecutive years and remained completely resistant. In 1910, Salaman made crosses between the resistant clones and domestic cultivars. He also secured *S. demissum* itself, confirmed that it was resistant, and from 1911 hybridized it as a female parent with domestic cultivars. He crossed some of the resistant offspring with the resistant stocks he had previously established. By 1926, he had a number of clones combining blight resistance with other desirable traits. However, in 1932 some previously resistant material showed signs of being attacked by blight, an omen that was also experienced on similarly derived material in Scotland [34].

Experiments in crossing potatoes were also done by John H. Wilson of St Andrews University, Scotland [38, 39]. He bred two cultivars, Kate Kennedy and Bejant, from his first cross in 1901 with British Queen as the female parent and Myatt's Kidney as the pollen-producing male. After he died in January 1920, a collection of his breeding material was given to the newly founded (also in 1920) Scottish Plant Breeding Station (SPBS) at Craigs House in Edinburgh. The material included derivatives of Wilson's breeding experiments with blight resistance from *S. edinense* and *S. demissum*. Selections from this material initially remained free of blight but in 1932 succumbed to a new race of blight, whereas the original *S. demissum* source remained resistant.

Hence, in 1932, William Black, the first and longest-serving (1926–1968) potato breeder at SPBS, started a new breeding program to introduce new resistance factors from *S. demissum*. His first two cultivars with blight resistance were Craigs Bounty [(The Bishop × W800(2)) × (The Bishop × Laures)] and Craigs Snow White [Craigs Defiance × W800(2)], released in 1946 and 1948 from crosses made in 1936 and 1939, respectively [40]. The source of blight resistance [W800(2)] was a clone produced by Wilson from crosses involving both *S. demissum* and *S. edinense*. Craigs Snow White contains what became known as the *R1* (strictly *Rpi-R1*) gene for blight resistance from *S. demissum*, once appropriate genetic studies had been done. The story of its failure to provide durable resistance can be found in Van der Plank's 1968 book *Disease Resistance in Plants* [13]. In 1955, cultivars with the *R1* gene were being grown widely in West Germany, races virulent on *R1* were frequent, and the gene delayed blight epidemics by only 5 days, whereas the *R3R4* combination delayed them by 15 days. In contrast, there were no commercial cultivars of any importance with the *R1* gene in the Netherlands in 1956, virulent races were rare, and the gene delayed blight epidemics by about a month. However, the potential rate of increase of virulent races could be seen in the United States, where cultivar Kennebec was the first with the *R1* gene to be widely grown. When released in 1948, it either escaped blight altogether or had a few lesions too late in the season to do any harm. By 1954, it had increased in popularity, virulent races were abundant, and it was no longer very resistant. These results can be explained in terms of Curve C in Fig. 5.4 and the proportion of spores that can cause infection and hence are able to initiate an epidemic, again

Table 5.2 Genetics of host-pathogen interactions where the *R* gene encodes a receptor for the pathogen ligand (effector protein) specified by the *Avr* gene

		Pathogen	
		Avr	*avr*
Host	*R*	Resistance	Susceptibility
	r	Susceptibility	Susceptibility

as explained by Van der Plank in his 1968 book [13].

It is tempting to say that *P. infestans* had overcome the resistance gene, but recent understanding of the molecular basis of host-pathogen interactions allows a more informative account of what happens [41]. *P. infestans* produces and delivers effector molecules (proteins) into the cells of potatoes (from a haustorium) where they try to suppress host defence mechanisms. The potato responds by detecting the effector proteins (a kind of immune response) through intracellular receptors encoded by disease resistance genes, and this response triggers programmed cell death to stop the spread of the pathogen. Recognized effectors are referred to as avirulent (Avr) effectors because pathogen spread does not occur. If the pathogen can mutate so that it does not produce the effector detected by the host receptor, it will be able to spread and cause disease. In genetic terms, we say that *R* and *r* are the resistant and susceptible alleles (variants) of a gene in the host, and *Avr* and *avr* are the avirulent and virulent alleles of a gene in the pathogen (Table 5.2).

Thus, these interactions between *P. infestans* and potatoes are an example of Harold H. Flor's original 1942 gene-for-gene (concept) hypothesis for the genetics of resistance to the rust *Melampsora lini* in flax (*Linum usitatissimum*). Flor's work is considered a landmark in understanding the genetics of host resistance, where a dominant host resistance (*R*) gene interacts with a dominant pathogen avirulence (*Avr*) gene to give race-specific hypersensitive resistance [42]. Van der Plank called this type of resistance to infection "vertical" [13].

By 1947 four *R* genes for resistance to *P. infestans* had been recognized, with all four genes present in SPBS clone 2070(54). Similar work was being done in other countries, and in 1953, Black, Mastenbroek, Mills and Petersen established an international system for the nomenclature of races of *P. infestans* in which they were numbered to indicate their virulence towards the four *R* genes (*R1*, *R2*, *R3* and *R4*) [43]. Thus, Race 0 was avirulent on all of these genes, whereas Race 1, 2 was virulent on *R1*, *R2* and *R1R2*. By Black's retirement in 1968, 11 *R* genes had been discovered in *S. demissum*, but it was already clear from race surveys that they would not provide durable resistance, either singly or in combination, due to the evolution of new races of *P. infestans* [44, 45]. Cultivar Pentland Dell, used for making French fries, entered production in Britain in 1963 when race 4 was the prevalent race. It had three genes, each of which conferred resistance to race 4 but succumbed to blight in 1967. The "three" genes are now known to be *Rpi-R1*, *Rpi-R3a*, $Rpi\text{-}R3b^{G3111}$, $Rpi\text{-}R3b^{G1696/G3111}$ and $Rpi\text{-}abpt^{T86}$ [33]. Nevertheless, it went on to become one of SPBS's most successful cultivars and was still the 18th most widely grown cultivar in Britain in 2020, albeit under the protection of fungicides. Interestingly, Black's differentials (the potato clones which differentiate the *R* genes) BlR8 and BlR9 still provide resistance to many isolates of *P. infestans*, but we now know that they harbour more resistance genes than simply *R8* and *R9* (BlR9 contains *Rpi-R1*, $Rpi\text{-}R1^{\Delta T4109}$, *Rpi-R3a*, $Rpi\text{-}R3b^{G1696/G3111}$, *Rpi-R8*, *Rpi-R9a* and $Rpi\text{-}abpt^{T86}$) [33].

5.8 Field Resistance

The failure of the *R* genes from *S. demissum* to provide durable resistance prompted potato breeders to seek other types of resistance. As early as 1954, Black at SPBS started to breed for field resistance using races of *P. infestans* compatible with the *R* genes that were present in his germplasm. He defined field resistance as the degree of resistance exhibited by a plant to all races of *P. infestans* to which it is not hypersensitive [46]. He thought that field resistance would be more durable than hypersensitivity and provide more reliable protection against crop failure. Hence, he favoured using field resistance to supplement that provided by *R* genes and had made good progress by the time he retired in 1968, the year Van der Plank published his book. Arguments then intensified over the durability and effective use of the two types of resistance and have continued to the present day. Van der Plank argued that cultivars with field resistance were a historical fact. For example, he presented data from 117 potato fields in the Netherlands that compared the rate of disease progress on susceptible cultivar Bintje ($r = 0.42$ per day), intermediate cultivar Eigenheimer ($r = 0.21$ per day) and resistant cultivar Voran ($r = 0.16$ per day). He called the decrease in the rate of infection "horizontal" resistance and developed what proved to be a controversial but testable theory of vertical (hypersensitive) and horizontal (field) resistance.

Norman Simmonds, who was director of the Scottish Plant Breeding Station when I started work there in 1975, was enthusiastic about horizontal resistance, particularly for late blight of potatoes, but not everyone shared his views! After Black retired, selection continued for resistance, both in the tubers and the foliage, as one does not guarantee the other. Promising blight-resistant cultivars were Teena (released in 1986), Shelagh (1986), Torridon (1989), Brodick (1990), Stirling (1991) and Cramond (1992) [47]. They all had *S. demissum* in their pedigrees [40, 47]. However, none of them had sufficient table or processing quality for commercial success.

Fig. 5.5 Organic potatoes from Egypt, bought from a supermarket in Scotland on 7 May 2018, and marketed as a source of thiamine (vitamin B1) and potassium

This was eventually achieved in 2001 with the cultivar Lady Balfour, which is still one of the leading organic cultivars in Britain (Fig. 5.5). With good intentions we had named it after Lady Eve Balfour, an aristocratic Christian, one of the pioneers of the organic movement in England, founder and president of the Soil Association, and author of *The Living Soil*, which was published in 1943 [48]. There was just one problem with the name. Her Scottish uncle was Arthur (Lord) Balfour, Prime Minister of the United Kingdom from 1902 to 1905, who in 1916 became Foreign Secretary in Lloyd George's coalition government. It was in 1917 that he issued the Balfour Declaration on behalf of the cabinet, which supported a home for the Jewish people in Palestine; and sadly, as I write this book in October 2023, there is war in the Gaza Strip and a humanitarian disaster. So, Lady Balfour was not a good name for a cultivar that is grown in Egypt from Scottish seed potatoes for export back into British supermarkets.

On a lighter note, the oldest Chair of Genetics in the English-speaking world is the Arthur Balfour Professorship of Genetics, which was founded in 1912 at Cambridge University, but that's another story! I must complete the one on the cultivar. White and Shaw [49] assessed the resistance of Lady Balfour and other cultivars at a site conducive to late blight on the Isle of Anglesey, North Wales, from 2005 to 2008. In 2005 and 2006, the trials relied on natural infection, whereas in 2007 and 2008, they were inoculated with an isolate of genotype EU 13_A2, a new metalaxyl-resistant race of *P. infestans*, which came to prominence in 2007. The disease progress curves for Lady Balfour looked like those of C in Fig. 5.3 (i.e., vertical, not horizontal, resistance). In 2005, Lady Balfour delayed the epidemic by 27 days compared with the susceptible cultivar Bintje, whereas in 2008, the delay was down to 7 days. In other words, the resistance of Lady Balfour was now disappointing, as also found in trials in Scotland from 2008 to 2010 with EU 13_A2 (Race 1,2,3,4,5,6,7,10,11) [50]. The trials in Scotland also showed that the same disappointment occurred with cultivar Stirling, which shares a common parent with

Fig. 5.6 Late blight resistant cultivar Sarpo Mira (left, still flowering) and susceptible cultivar Vales Sovereign (right) in a garden in Currie, Edinburgh, Scotland, on 27 August 2023 (July and August had been warm, wet months)

Lady Balfour. This showed that supposed field resistance is not necessarily more durable than *R*-gene resistance. Nevertheless, Lady Balfour continued to do well under organic farming in Britain and in Egypt and is still commercially successful, despite its disappointing level of resistance to EU 13_A2. In contrast, White and Shaw [49] found that cultivar Sarpo Mira was completely resistant in 2005 and slowed the rate of infection (horizontal resistance) in 2008 so that the proportion of infected tissue did not exceed 0.40 after 55 days of the epidemic. Cultivar Sarpo Mira is grown by gardeners in Britain, including me (Fig. 5.6), for its blight resistance but has not proved popular with farmers and hence hasn't achieved truly commercial success.

So, are there two types of resistance? I don't know, but speaking as a plant breeder and geneticist, I am fairly sure that whenever a large genetic difference in resistance exists in breeding material, the gene (allele) giving the greater resistance will be selected (either in a laboratory or glasshouse test or in a field disease nursery). It may be a gene for hypersensitive resistance, or a defeated *Rpi* gene with a useful residual effect, or a gene for late maturity, or a gene making a large contribution to quantitative resistance (and hence to the rate of infection) [16]. It will manifest itself as resistance in the field but may or may not provide durable, broad-spectrum resistance. Thus, the *r* value for cultivar Stirling in the disease nursery in Ayrshire in 1998 (Fig. 5.3) was 0.16, but the gene on chromosome 4, which accounted for most of the quantitative field resistance, later proved susceptible to EU 13_A2. On the other hand, the field resistance of Sarpo Mira held up against EU 13_A2 and is probably due to a gene (*Rpi-Smira2*) of large effect on chromosome 9, which could in fact be *R8* [51]. In contrast to selecting such a gene, accumulating ones of small effect (to increase horizontal resistance) in the presence of one of large effect will always be difficult.

5.9 CIP's B Populations

Since its establishment in Lima, Peru, in 1971, the International Potato Centre (CIP) has made broad-based germplasm available to breeders in National Agricultural Research Systems (NARS) in developing countries, particularly germplasm with high and stable resistance to late blight. Best known are the supposedly *R*-gene (race-specific) free B populations with quantitative resistance to late blight, of which B3 has been the most widely used. It was set up in 1990 and included both Group Andigena and Group Tuberosum clones as well as complex hybrids of the latter with *S. demissum*, *S. acaule*, *S. bulbocastanum* and Group Phureja. Recurrent selection was practised for late blight resistance and other desirable traits under high endemic disease pressure in the Andean highlands, as well as selection for adaptation to other geographical areas [52, 53]. Good progress was reported over three cycles of recurrent selection in the period to 2008 [53–57].

5.10 GILB

Despite reports of good progress, in 1996 the world appeared to be struggling to control late blight, including the breeding of widely grown cultivars with high levels of durable resistance. As a consequence, on 17 March 1996, CIP launched a Global Initiative on Late Blight (GILB). It was a two-phase, 10-year program with the goal "to reduce the threat that late blight is posing to the potato crop especially in less developed countries, through stimulating collaborative and complementary research and technology transfer among developing and developed countries by: fostering…, enhancing…, assisting, promoting and catalysing" [58]. Crucially, GILB was neither a research nor funding organization. I attended the three international conferences that reported on progress: 16–19 March 1999 in Quito, Ecuador; 11–13 July 2002 in Hamburg, Germany; and 3–5 April 2008 in Beijing, China. Much interesting research was reported and ideas discussed, both at the conferences and since. We will see some outcomes in the next two sections: the discovery of new sources of resistance, a better understanding of effector proteins, and a better network for monitoring changes in *P. infestans* populations around the world. Here I can give two examples of breeding progress from the CIP-derived B3 population.

Bangladesh is in a subtropical monsoon region of the world (20°34'N to 26°38'N) where 80% of agriculture is smallholdings of less than 1 ha. It is one of the top 15 potato-producing countries in the world (Table 1.1). Within Bangladesh, potatoes are the third most important crop after paddy rice and wheat. They are grown across the country as a winter crop. Late blight is a regular occurrence, and farmers commonly spray susceptible cultivars at least seven times per season. Twelve clones from CIP's B3C1 and B3C2 populations (first and second cycles of recurrent selection) were introduced from Peru as disease-free in vitro plantlets in 2008 and selected down to two potential cultivars on the basis of yield and high and consistent resistance to late blight, as well as other important traits [59]. After extensive trials the two B3C1 clones were released as late-blight-resistant cultivars, BARI Alu-46 (CIP 393371.58 coded LB-7) in 2013, and BARI Alu-53 (CIP 393280.64 coded LB-6) in 2014. When harvested 90 days after planting, their yields in on-farm trials in three locations were 35.7 and 32.0 t/ha, respectively, compared with 24.8 t/ha for the blight-susceptible control Diamant. BARI Alu-46 had already been released in Peru (2007) as Chugmarina, in Kenya as Kenya MPYA and in Ethiopia as Belete.

My second example provides an example of participatory varietal selection in Peru where more than 50% of the area planted with potatoes in 2017 was occupied by three susceptible cultivars, Canchan, Yungay and UNICA, resulting in reduced yields [60–62]. Thirty-six clones from B3C3 (third cycle of recurrent selection) were evaluated in three highland areas in Peru (five locations) over the period 2016 and 2017 by farmers (participatory) as well as scientists from CIP [60]. The fives selected clones were evaluated again in 2017 and 2018 at four locations. Three clones (CIP308488.92, CIP308495.227 and CIP308478.59) were selected for their resistance and superior yield to local cultivars (INIA-302 Amarilis, INIA-303 Canchan and Yungay), when harvested 120 days after planting, as well as for good organoleptic quality and low glycoalkaloid content. They are undergoing further trials prior to cultivar registration [60]. In the meantime, three B3C2 (second cycle) blight resistant clones were released as cultivars in 2022 and officially registered in Peru in 2023 [61, 62]. They were CIP-Poderosa Crocante (CIP396026.101), CIP-Poderosa Pollera (CIP395123.6) and CIP-Poderosa Watia (CIP396034.103). They derive from crosses made back in 1995 and 1996. As well as high horizontal resistance to late blight, they have high tuber yield, low content of glycoalkaloids and excellent quality for French fries (high percentage of dry matter, low content of reducing sugars and low acrylamide on frying). They should help to meet the increasing demand in Peru for French fries, which accompany a traditional grilled chicken dish "Pollo a la brasa". Watia also has high baking quality [61, 62]. The durability of these new sources of resistance to late blight will only be apparent in the years to come.

5.11 Back to *R* Genes

By 2008, many wild relatives of the potato had been identified as sources of simply inherited resistance to late blight, and in particular to the new populations of *P. infestans* that were spreading around the world. After 2008, the search intensified (Fig. 5.7), and two reviews published in 2022 listed 61 known *Rpi* genes for resistance to *P. infestans* from 27 wild species [63, 64]. One of the reviews also listed 51 cultivars with *Rpi* genes [64]. Furthermore, from 2008, the discovery and increased understanding of pathogen effectors in *P. infestans* led pathologists to develop effector libraries to catalogue *Rpi* genes on the basis of which effectors they detected [65, 66]. Without going into detail, *Rpi* genes and putative effectors can be transiently co-expressed in leaves of *Nicotiana benthamiana* to see if a hypersensitive reaction occurs, having been delivered into the leaf tissue via *Agrobacterium tumefaciens* in a process called agroinfiltration. This research revealed that the number of functionally distinct *Rpi* genes is less than the apparent number present in diverse wild species. It also allowed useful predictions to be made about the durability of the resistance [41]. An effector that is shared between multiple races of *P. infestans* is more likely to be indispensable for the pathogen than an effector that is present in some but not all races (effector molecules are used by pathogens to suppress host defence mechanisms but are detected by the product of the corresponding *Rpi* gene). Thus, the most promising *Rpi* genes for durable, broad-spectrum resistance can be prioritized for use in breeding new cultivars, and different ones stacked together to make their use even more durable. This can be done by traditional potato breeding (called introgression breeding), and the process can be speeded up by using molecular markers that identify the resistance genes without the need to expose the potato germplasm to appropriate isolates of *P. infestans* in a special laboratory or glasshouse test or a field nursery. However, a faster and more precise method called cisgenesis is now available.

5.12 Cisgenesis

Cisgenesis uses the techniques of genetic engineering and genetic transformation to "cut and paste" a resistance gene from a wild species into a modern cultivar. I don't need to go into the technical details, but we do need to look at the scientific opinion (requested by the European Commission) of the European Food Safety Authority (EFSA) published in 2012 [67]. First, they produced a definition: "Cisgenesis is the genetic modification of a recipient organism with a gene from a crossable, sexually compatible, organism (same species or closely related spe-

Fig. 5.7 Screening the Commonwealth Potato Collection for new sources of resistance to late blight. (From Bradshaw JE (2021) Potato Breeding: Theory and Practice. Springer, Cham with permission; 4 July 2008)

cies). This gene includes its introns and is flanked by its native promoter and terminator in the normal sense orientation. Cisgenic plants can harbour one or more cisgenes, but they do not contain any parts of transgenes or inserted foreign sequences. To produce cisgenic plants, any suitable technique used for the production of transgenic organisms may be used. Genes must be isolated, cloned and transferred back into a recipient where stably integrated and expressed." Secondly, they expressed an opinion: "similar hazards can be associated with cisgenic and conventionally bred plants, while novel hazards (from new genes) can be associated with intragenic and transgenic plants". EFSA updated their opinion in 2022, but with only one minor change concerning where the cisgene is inserted in cisgenic plants [68]. Back in 2012 they had pointed out that cisgenesis (and intragenesis) makes use of the same transformation techniques as transgenesis and hence results in the same hazards from random insertion (as opposed to hazards from a novel gene), namely potential unintended modifications of the host genome. EFSA now points out that new genomic techniques (NGTs) allow (genome) site-directed integration of the added genetic material. Hence, on a case-by-case basis, a lesser amount of data might be needed for the risk assessment of cisgenic (or intragenic) plants obtained through NGTs. Therefore, in my opinion, cisgenic plants should be exempt from the regulations on genetically modified organisms, but in 2023, there was still a global debate and disagreements [69].

There are currently two good examples of three *Rpi* genes being stacked. The first comes from the DuRPh project done at Wageningen University in the Netherlands from 2006 to 2015 [70]. It provided the first scientific report on the production and functional evaluation of cisgenic *Rpi*-gene stacking in different potato cultivars. The three genes were *Rpi-vnt1.1* (from *S. venturii* from Argentina), *Rpi-sto1* (from *S. stoloniferum* from Mexico) and *Rpi-chc1* (from *S. chacoense* from Bolivia/Argentina/Paraguay/Uruguay). All three genes had broad-spectrum resistance in their own right, but the three together should prove more durable. In field trials in Wageningen in 2014 and 2015, potatoes carrying the genes did not become infected with *P. infestans*, unlike the susceptible cultivar Desiree. Even with 2–3 years of field trials, it takes a total of just 3–4 years to produce cisgenic late blight-resistant potato cultivars, which can be released for seed tuber multiplication, and this is less than conventional introgression breeding. The second example comes from highland potato production in Uganda, where high humidity and low temperatures of 14–25 °C during the two rainy seasons (October to January and March to June) favour the spread of *P. infestans*, which can cause significant yield loss (15%-30%) for smallholder farmers [71]. The three resistance genes were *Rpi-blb1* and *Rpi-blb2* (from *S. bulbocastanum* from Mexico and Guatemala) and *Rpi-vnt1.1* (from *S. venturii* from Argentina). Resistance to natural infection by *P. infestans* was assessed in a field trial at 2200 masl in southwestern Uganda in each of three seasons (2015–2016). No blight lesions were seen on the plants with the resistance genes, whereas Desiree and local cultivar Victoria were dead some 20 days after the disease had started. In the third season, 110 days after planting, the resistant plants of Desiree and Victoria had total yields of 29 and 45 t/ha, respectively, whereas the susceptible equivalents without the *Rpi* genes had yields of less than 10 t/ha and did not produce any marketable tubers. The researchers concluded that their late-blight-resistant potato cultivars, which are the farmers' preferred cultivars, could be rapidly adopted and bring significant income to smallholder farmers in Sub-Saharan Africa.

These two examples, along with the new fungicides considered earlier, bring us up to date with the control of late blight. Pathologists are currently working on other ways to attack *P. infestans* which may bear fruit in the future (e.g., using gene editing to knock out susceptibility genes in potatoes that unintentionally aid *P. infestans* or seeking genes for extracellular rather than intracellular recognition of *P. infestans* effectors). Furthermore, the search goes on for new types of fungicides with novel chemistry. In the meantime, worried farmers seek help and advice on blight control from their agronomists and advisers.

5.13 The Success of *P. infestans*

P. infestans isolates can be characterized, discriminated and monitored using a combination of nuclear DNA markers (simple sequence repeats, SSRs) and maternally inherited (non-recombining) mitochondrial DNA markers [72]. A molecular marker is also available for determining mating type (A1 or A2) [73]. Furthermore, isolates can be assessed for their resistance to fungicides and their non-compatibility with *Rpi* genes. Together all of this information is valuable in attempts to control late blight through resistance genes and fungicides. Here, however, I want to finish with a brief history of the evolution and spread of populations of *P. infestans*. In other words, an account of the success of *P. infestans*! The key point to remember is that sexual reproduction generates genetically variable populations from which many different clonal lineages can be propagated by asexual reproduction. These clonal lineages accumulate mutations over time.

There are two main hypotheses about the origin and migration of *P. infestans*, one based on a Peruvian origin and the other on a Mexican origin. Recent (2024) evidence for a Peruvian origin is based on an analysis of a dataset of 1706 microsatellite genotypes of *P. infestans*, isolated between 1994 and 2019, from North and South America, Europe and Asia. A Peruvian origin of *P. infestans* was considered most probable on the basis of a posterior Bayesian probability, but this is not proof. There followed an initial migration to Colombia and Mexico, a later migration from Mexico to the United States, and from there to Europe and Asia, with no return to northern South America. Following the migration to Colombia, there were also migrations within Colombia [74]. The evidence for a Mexican origin comes from an earlier (2014) analysis of isolates of the pathogen [75]. The conclusion was that *P. infestans* originated in central Mexico rather than the

Andes, so that the populations found in the Andes (Peru and Colombia) today are descendants of the Mexican populations and not a return from Europe. Hence, the first hosts of *P. infestans* were the potato wild relatives that are found in Mexico. The Mexican populations were sexual (both mating types), whereas the South American populations were clonal (one mating type), and part of the argument is over whether or not sexual reproduction preceded asexual reproduction [74, 75]. However, the very latest genome divergence analyses favour a South American origin [76]. The researchers concluded that all seven 1c clade *Phytophthora* species, including *P. infestans* and its clonal lineage which caused the famine, had an Andean origin and that modern globalization has resulted in admixture of populations from both Mexico and the Andes.

Wherever the origin of *P. infestans*, a (meta-) population of *P. infestans* became established at the periphery of its Mexican centre of diversity, or possibly in North America, sometime before the nineteenth-century pandemic. A comparison of 11 herbarium and 15 modern strains of *P. infestans* revealed that the pandemic was caused by a single clonal lineage, FAM-1 (or Herb-1), with the 1a mitochondrial haplotype, from the metapopulation that persisted globally into the twentieth century [77]. During the twentieth century, from the 1930s to the 1950s, another clonal lineage, US-1, with the 1b mitochondrial haplotype, spread from the metapopulation, replacing FAM-1, as it came to dominate the global *P. infestans* population until the 1980s. Some of the history of these migrations has been revealed by studying blight outbreak samples from mycological herbaria [78]. A total of 137 historic samples of *Solanum* species were genotyped with microsatellites, of which 112 were *S. tuberosum* from Africa (13), Asia (23), Europe (24), North America (43), Oceania (2) and South America (7). The earliest FAM-1 genotype was from France in 1845 and the latest from Malaysia in 1987, although the latter was the only post-1950s sample of FAM-1. The researchers proposed tentative migration routes and times of FAM-1 migration from Europe to Africa, Asia and Oceania over the period 1870–1940. The earliest US-1 genotype was from the United States in 1931 and the latest from India in 1991, the last collection date. US-1 was first found in Bolivia in 1944, China in 1952 and Cameroon and Nigeria in 1953.

From the end of the 1970s, new populations, comprising both the A1 and A2 mating types, spread from Mexico to the rest of the world [79, 80]. New clonal lineages emerged from these genetically diverse global populations and became locally dominant, such as US-8 and then US-22 and US-23 in North America and EU 13_A2 in Europe [72]. Sometimes these changes occurred too quickly for the timescale on which breeders operate to produce new cultivars and chemical companies operate to produce new fungicides. One example will suffice. In Britain, there was a dramatic change over the period 2003–2007: genotype EU 13_A2 (A2 mating type) was absent in 2003 but comprised 71.4% of blight samples (1023) in 2007, whereas genotypes EU 2_A1, EU 7_A1 and EU 8_A1 (A1 mating type) comprised 67.5% in 2003 but only 7.3% in 2007 [81]. As already mentioned, cultivar Lady Balfour proved disappointingly susceptible to genotype EU 13_A2, but this did not affect its commercial success. Since then, there have been further changes in the *P. infestans* population in Britain so that in 2021 (and still in 2023, a blight year) EU 36_A2 (an aggressive strain), and EU 6_A1 were the dominant genotypes [82].

Recently (2024), important results have been secured from 29 herbarium specimens from 1845 to 1954 that contain DNA from both host (potato) and pathogen (*P. infestans*) [83]. The *P. infestans* genomes comprised both diploids and triploids. Sixteen FAM-1 genotypes were definitely diploid, whereas three of the more recent US-1 genotypes were definitely triploid, an interesting evolutionary trend which we need not discuss here. As expected, most potatoes had a Chilean origin, as shown by the potato plastid trnVUAC/ndhC spacer marker. A technique called baited enrichment sequencing was used to increase the proportion of host *R*-genes and pathogen (RXLR) effectors in the extracted DNA for comparison with the whole genomes of both the *Solanum* host and *P. infestans*. Ten RXLR-R gene pairs were observed, of which the most interesting was *Avr1* and *R1*, the first resistance gene from *S. demissum* to be used by breeders. None of the 29 potatoes sampled in the research had a complete functional *R1* gene, whereas all of the *P. infestans* sampled exhibited the resistance-breaking, virulent allele of *Avr1*. The researchers concluded that at the time of the Irish potato famine, the pathogen already possessed the ability to overcome the first resistance gene that breeders would deploy later in the twentieth century, presumably because the pathogen had already been exposed to this resistance gene in wild hosts. We really do need to know more about the 500 predicted putative effectors in the *P. infestans* genome.

5.14 Still a Disaster Waiting to Happen?

Back in 1822 in England, Joseph Sabine, the secretary of the Royal Horticultural Society, was advocating the breeding of new potato varieties with superior flavour and productiveness, but less subject to injury by changes of weather during the growing season, and with good keeping quality after harvest [84]. As the potato was the most important human food after wheat and rice, he also gave the following warning: "A general failure of the year's crop, whenever it shall have become the chief or sole support of a country, must inevitably lead to all the misery of famine, more dreadful in proportion to the numbers exposed to its ravages". The Irish potato famine occurred 24 years later.

The front cover of *New Scientist* on 2 August 2008 had a picture of potatoes under the heading "Super Crop: How the humble potato could feed the world" and a cover story "Let them eat spuds" [85]. The article ended with a warning: "As the humble potato spreads around the world, and more and more people depend on it for sustenance, the need to win the battle against disease becomes more urgent. Blight is a disaster waiting to happen, and this time we have no alternative but to fight back".

In other words, *P. infestans* is endemic in most potato-growing countries and has the potential for exponential production of pathogenic spores, but one interesting fact remains. In Mexico, where both mating types and the greatest diversity of *P. infestans* occur, the wild species *S. demissum* has successfully resisted its attacks. In Scotland, back in 1932, when breeding material succumbed to late blight, the original accession of *S. demissum* remained resistant. Much more recently, in the summers of 2020 and 2021, 94 *S. demissum* accessions from the US Potato Genebank (USPG) were grown from true potato seed and assessed for blight resistance under natural infection in the Toluca Valley, Mexico [86]. In both seasons, the climatic conditions (temperature and rainfall) were typical and ideal for both potatoes and late blight. Eighty-three (88%) accessions were resistant in both seasons, whereas cultivar Agatha was highly susceptible. The unavoidable conclusion is that Nature has done a better job of utilizing and deploying resistance genes than modern breeders have achieved over the last 100 years. Is it simply a matter of the diversity of *S. demissum* and its resistance genes in Mexico compared with a few resistance genes deployed in cultivars grown globally in monocultures?

References

1. Reader J (2008) Propitious esculent. William Heinemann, London
2. Young A (1892) Arthur Young's tour in Ireland (1776–1779). George Bell & Sons, London (published by Legare Street Press: support.creativemedia.io)
3. Zadoks JC (2008) The potato murrain on the European continent and the revolutions of 1848. Potato Res 51:5–45
4. Nally DP (2011) Human encumbrances. University of Notre Dame Press, Notre Dame
5. FAO statistics on 24 April 2023 (http://faostat.fao.org)
6. Worldometer. https://www.helgilibrary.com/indicators/potato-consumption-per-capita/ on 24 April 2023
7. Hunter J (2019) Insurrection: Scotland's famine winter. Birlinn Limited, Edinburgh

8. Berkeley MJ (1846) Observations, botanical and physiological, on the potato murrain. J Hort Soc Lond 1:9–34
9. de Bary A (1861) Die gegenwärtig herrschende Kartoffelkrankheit, ihre Ursache und ihre Verhütung. Arthur Felix, Leipzig
10. de Bary A (1876) Researches into the nature of the potato fungus *Phytophthora infestans*. J R Agric Soc Engl 2nd Ser 12:239–269
11. Ingram DS, Williams PH (eds) (1991) Advances in plant pathology volume 7 *Phytophthora infestans*, the cause of late blight of potato. Academic, London
12. Fry W (2008) *Phytophthora infestans*—the plant (and *R* gene) destroyer. Mol Plant Pathol 9:385–402
13. Van der Plank JE (1968) Disease resistance in plants. Academic, New York
14. Wheeler BEJ (1975) The control of plant diseases. Oxford University Press, London
15. Wheeler BEJ (1976) Diseases in crops. Edward Arnold, London
16. Bradshaw JE, Pande B, Bryan GJ, Hackett CA, McLaren K, Stewart HE, Waugh R (2004) Interval mapping of quantitative trait loci for resistance to late blight [*Phytophthora infestans* (Mont.) de Bary], height and maturity in a tetraploid population of potato (*Solanum tuberosum* subsp. *tuberosum*). Genetics 168:983–995
17. Cruickshank G, Stewart HE, Wastie RL (1982) An illustrated assessment key for foliage blight of potatoes. Potato Res 25:213–214
18. Wilson A (1993) The story of the potato through illustrated varieties. Balding and Mansell, England
19. Moggridge M (1846) Letter in the Gardeners' Chronicle on Saturday 26 September 1846
20. Evans C, Miskell L (2020) Swansea copper: a global history. John Hopkins University Press, Baltimore
21. Schwinn FJ, Margot P (1991) Control with chemicals. Adv Plant Pathol 7:225–265
22. Morton V, Staub T (2008) A short history of fungicides. Online, APS*net* Features. https://doi.org/10.1094/APSnetFeature-2008-0308
23. Lainsbury MA (ed) (2023) The UK pesticide guide 2023. BCPC, Cambridge
24. HSE confirms end date for mancozeb approval. The Vegetable Farmer June 2024:6
25. Schepers HTAM, Kessel GJT, Lucca F, Förch MG, van den Bosch GBM, Topper CG, Evenhuis A (2018) Reduced efficacy of fluazinam against *Phytophthora infestans* in The Netherlands. Eur J Plant Pathol 151:947–960. https://doi.org/10.1007/s10658-018-1430-y
26. Anderson E (2023) EU_43 could be coming to a place near you. The Vegetable Farmer July 2023:22–25
27. Cooke D (2024) Confronting potato late blight in 2024. The Vegetable Farmer April 2024:28–29
28. Pasteris RJ, Hanagan MA, Bisaha JJ, Finkelstein BL, Hoffman LE, Gregory V, Andreassi JL, Sweigard JA, Klyashchitsky BA, Henry YT, Berger RA (2016) Discovery of oxathiapiprolin, a new oomycete fungicide that targets an oxysterol binding protein. Bioorg Med Chem 24(3):354–361
29. Beaumont A (1947) The dependence on the weather of the dates of outbreak of potato blight epidemics. Trans Br Mycol Soc 31:45–53
30. Hutton Criteria (2017). https://www.hutton.ac.uk/news/hutton-criteria-potato-late-blight-risk-analysis-unveiled-ahdb-conference
31. Forbes GA (2012) Using host resistance to manage potato late blight with particular reference to developing countries. Potato Res 55:205–216
32. Muller KO, Black W (1951) Potato breeding for resistance to blight and virus diseases during the last hundred years. Z Pflanzenzuchtung 31:305–318
33. Bradshaw JE (2021) Potato breeding: theory and practice. Springer, Cham
34. Weintraub B (2019) Redcliffe Nathan Salaman (1874–1955) and the first potato plant with "genuine resistance" to late blight. Isr Chem Chem Eng 5:28–34
35. Salaman RN (1910) The inheritance of colour and other characters in the potato. J Genet 1:7–46
36. Salaman RN (1926) Potato varieties. Cambridge University Press, Cambridge
37. Sánchez-Ganfornina D, Tapson J, Sutherland L, Coleman M (2020) Recreation of the Edinburgh potato *Solanum* × *edinense* Berthault: a historically and economically important potato hybrid. Sibbaldia Int J Bot Garden Hortic 19:116–130. https://doi.org/10.24823/Sibbaldia.2020.302
38. Wilson JH (1907) Experiments in crossing potatoes. Trans High Ag Soc 1907:74–92
39. Wilson JH (1916) Further experiments in crossing potatoes. Trans High Ag Soc 1916:33–55
40. Bradshaw JE (2009) Potato breeding at the Scottish Plant Breeding Station and the Scottish Crop Research Institute: 1920–2008. Potato Res 52:141–172
41. Jones JDG, Witek K, Verweij W, Jupe F, Cooke D, Dorling S, Tomlinson L, Smoker M, Perkins S, Foster S (2014) Elevating crop disease resistance with cloned genes. Philos Trans R Soc B 369:20130087. https://doi.org/10.1098/rstb.2013.0087
42. Flor HH (1942) Inheritance of pathogenicity in *Melampsora lini*. Phytopathology 32:653–659
43. Black W, Mastenbroek C, Mills WR, Peterson LC (1953) A proposal for an international nomenclature of races of *Phytophthora infestans* and of genes controlling immunity in *Solanum demissum* derivatives. Euphytica 2:173–179
44. Malcolmson JF, Black W (1966) New R genes in *Solanum demissum* Lindl. and their complementary races of *Phytophthora infestans* (Mont.) de Bary. Euphytica 15:199–203
45. Malcolmson JF (1969) Races of *Phytophthora infestans* occurring in Great Britain. Trans Br Mycol Soc 53:417–423
46. Black W (1970) The nature and inheritance of field resistance to late blight (*Phytophthora infestans*) in potatoes. Am Potato J 47:279–288

47. Wastie RL (1991) Breeding for resistance. Adv Plant Path 7:193–224
48. Mann CC (2018) The wizard and the prophet. Picador, London
49. White S, Shaw D (2009) The usefulness of late-blight resistant Sárpo cultivars-a case study. Acta Hortic 834:161–166
50. Lees AK, Stewart JA, Lynott JS, Carnegie SF, Campbell H, Roberts AMI (2012) The effect of a dominant *Phytophthora infestans* Genotype (*13_A2*) in Great Britain on host resistance to foliar late blight in commercial potato cultivars. Potato Res 55:125–134
51. Jo KR (2013) Unveiling and deploying durability of late blight resistance in potato from natural stacking to cisgenic stacking. PhD Thesis, Wageningen University, Wageningen, The Netherlands
52. Trognitz BR, Bonierbale M, Landeo JA, Forbes G, Bradshaw JE, Mackay GR, Waugh R, Huarte MA, Colon L (2001) Improving potato resistance to disease under the global initiative on late blight. In: Cooper HD, Spillane C, Hodgkin T (eds) Broadening the genetic base of crop production. CAB International, Wallingford, pp 385–398
53. Landeo JA (2002) Durable resistance: quantitative/qualitative resistance. In: Lizarraga C (ed) Proceedings of the global initiative on late blight conference, 11–13 July 2002, Hamburg, Germany, pp 29–36
54. Mulema JMK, Olanya OM, Adipala E, Wagoire W (2004) Stability of late blight resistance in population B potato clones. Potato Res 47:11–24
55. Landeo JA, Gastelo M, Díaz L et al (2007) Phenotypic stability for horizontal resistance to late blight in potato. African Potato Association Conference Proceedings. vol 7, Alex, Egypt, pp 58–62
56. Mulema JMK, Adipala E, Olanya OM, Wagoire W (2008) Yield stability analysis of late blight resistant potato selections. Expl Agric 44:145–155
57. Lindqvist-Kreuze H, Gastelo M, Perez W, Forbes GA, de Koeyer D, Bonierbale M (2014) Phenotypic stability and genome-wide association study of late blight resistance in potato genotypes adapted to the tropical highlands. Phytopathology 104(6):624–663
58. Collins W (1999) Overview of GILB and objectives of meeting. In: Crissman L, Lizarraga C (eds) Late blight: a threat to global food security volume 1, proceedings of the global initiative on late blight conference, March 16–19, 1999, Quito, Ecuador. CIP, Lima
59. Mahmud AA, Bonierbale M, Kadian MS, Hossain M, Kundu BC, Rahaman S, Alam J, Hasan M, Rahman M, Kawochar A, Uddin N (2022) Selection of late blight resistant varieties in Bangladesh from CIP population B3 potato clones. Potato J 49(1) https://epubs.icar.org.in/index.php/PotatoJ/article/view/124847
60. Gastelo M, Burgos G, Bastos C, Perez JM, Otiniano R, Cabrera H, Ccanto R, de Scurrah MM, Felde TZ (2021) Identification of elite potato clones with resistance to late blight through participatory varietal selection in Peru. Potato Res 64(4):611–634
61. Gastelo M, Pérez W, Eyzaguirre R, Quispe K, Sanabria K, Bastos C, Otiniano R, Pérez JM, Mendoza A, Unda T, Andrade-Piedra J (2023) CIP-PODEROSA CROCANTE, CIP-PODEROSA POLLERA, and CIP-PODEROSA WATIA: new potato varieties for family farming with resistance to late blight and high quality for the frying industry. Am J Potato Res 100:288–303
62. Gastelo M, Pérez W, Eyzaguirre R, Quispe K, Sanabria K, Bastos C, Otiniano R, Pérez JM, Mendoza A, Unda T, Andrade J (2024) New potato varieties resistant to late blight and with high quality for French fries generated in Peru. Potato Res 67:1435–1468
63. Blossei J, Gäbelein R, Hammann T, Uptmoor R (2022) Late blight resistance in wild potato species—resources for future potato (*Solanum tuberosum*) breeding. Plant Breed 141(3):314–331. https://doi.org/10.1111/pbr.13023
64. Paluchowska P, Śliwka J, Yin Z (2022) Late blight resistance genes in potato breeding. Planta 255:127. https://doi.org/10.1007/s00425-022-03910-6
65. Vleeshouwers VG, Rietman H, Krenek P, Champouret N, Young C, Oh SK, Wang M, Bouwmeester K, Vosman B, Visser RG, Jacobsen E, Govers F, Kamoun S, van der Vossen EAG (2008) Effector genomics accelerates discovery and functional profiling of potato disease resistance and *Phytophthora infestans* avirulence genes. PLoS One 3:e2875
66. Vleeshouwers VGAA, Raffaele S, Vossen JH, Champouret N, Oliva R, Segretin ME, Rietman H, Cano LM, Lokossou AA, Kessel GJT, Pel M, Kamoun S (2011) Understanding and exploiting late blight resistance in the age of effectors. Annu Rev Plant Physiol Plant Mol Biol 49:507–531
67. EFSA (2012) EFSA Panel on Genetically Modified Organisms (GMO). Scientific opinion. Addressing the safety assessment of plants developed through cisgenesis and intragenesis. EFSA J 10(2):2561 and online www.efsa.europa.eu/efsajournal)
68. EFSA (2022) EFSA Panel on Genetically Modified Organisms: Mullins E, Bresson J-L, Dalmay T, Dewhurst IC, Epstein MM, Firbank LG, Guerche P, Hejatko J, Moreno FJ, Naegeli H, Nogue F, Sanchez Serrano JJ, Savoini G, Veromann E, Veronesi F, Casacuberta J, FernandezDumont A, Gennaro A, Lenzi, P, Lewandowska A, Munoz Guajardo IP, Papadopoulou N and Rostoks N. Updated scientific opinion on plants developed through cisgenesis and intragenesis. EFSA J 20(10):7621 and online https://doi.org/10.2903/j.efsa.2022.7621
69. Chaurasia A, Kole C (2023) Cisgenic crops: safety, legal and social issues. Springer, Cham
70. Haverkort AJ, Boonekamp PM, Hutten R, Jacobsen E, Lotz LAP, Kessel GJT, Vossen JH, Visser RGF (2016) Durable late blight resistance in potato through dynamic varieties obtained by cisgenesis: scientific

and societal advances in the DuRPh project. Potato Res 59:35–66
71. Ghislain M, Byarugaba A, Magembe E, Njoroge A, Rivera C, Roman ML, Tovar JC, Gamboa S, Forbes G, Kreuze J, Barekye A, Kiggundu A (2018) Stacking three late blight resistance genes from wild species directly into African highland potato varieties confers complete field resistance to local blight races. Plant Biotech J 17(6):1119–1129. https://doi.org/10.1111/pbi.13042
72. Martin FN, Zhang Y, Cooke DEL, Coffey MD, Grünwald NJ, Fry WE (2019) Insights into evolving global populations of *Phytophthora infestans* via new complementary mtDNA haplotype markers and nuclear SSRs. PLoS One 14(1):1–24. https://doi.org/10.1371/journal.pone.0208606
73. Judelson HS, Spilman LJ, Shattock RC (1995) Genetic mapping and non-Mendelian segregation of mating type loci in the Oomycete *Phytophthora infestans*. Genetics 141:503–512
74. Patarroyo C, Lucca F, Dupas S, Restrepo S (2024) Reconstructing the global migration history of *Phytophthora infestans* towards Colombia. Phytopathology Online: 18 Jun 2024. https://doi.org/10.1094/PHYTO-05-24-0163-R
75. Goss EM, Tabima JF, Cooke DEL, Restrepo S, Fry WE, Forbes GA, Fieland VJ, Cardenas M, Grünwald NJ (2014) The Irish potato famine pathogen *Phytophthora infestans* originated in Central Mexico rather than the Andes. PNAS 111:8791–8796
76. Coomber AL, Saville AC, Carbone I, Martin M, Bieker VC, Ristaino JB (2025) A pangenome analysis reveals the center of origin and evolutionary history of *Phytophthora infestans* and 1c clade species. PLoS One 20(1):e0314509. https://doi.org/10.1371/journal.pone.0314509
77. Yoshida K, Schuenemann VJ, Cano LM, Pais M, Mishra B, Sharma R, Lanz C, Martin FN, Kamoun S, Krause J, Thines M, Weigel D, Burbano HA (2013) The rise and fall of the *Phytophthora infestans* lineage that triggered the Irish potato famine. eLife 2013:2. https://doi.org/10.7554/eLife.00731
78. Saville AC, Ristaino JB (2021) Global historic pandemics caused by the FAM-1 genotype of *Phytophthora infestans* on six continents. Nat Sci Rep 11:12335. https://doi.org/10.1038/s41598-021-90937-6
79. Hohl HR, Iselin K (1984) Strains of *Phytophthora infestans* with A2 mating type behaviour. Trans Br Mycol Soc 83:529–530
80. Goodwin SB, Drenth A (1997) Origin of the A2 mating type of *Phytophthora infestans* outside Mexico. Phytopathology 87:992–999
81. Cooke DEL, Lees AK, Shaw DS, Taylor MC, Prentice MWC, Bradshaw NJ, Bain RA (2008) The status of GB blight populations and the threat of oospores. Proc Crop Prot North Br 2008:217–222
82. Potato Yearbook & Buyers Guide 2023. Act Publishing, Maidstone
83. Coomber A, Saville A, Ristaino JB (2024) Evolution of *Phytophthora infestans* on its potato host since the Irish potato famine. Nat Commun 15:6488. https://doi.org/10.1038/s41467-024-50749-4
84. Sabine J (1822) On the native country of the wild potato, with an account of its culture in the garden of the Horticultural Society; and observations on the importance of obtaining improved varieties of the cultivated plant. Trans Hortic Soc Lond 5:249–259
85. MacKenzie D (2008) Let them eat spuds. NewScientist 2 August 2008:30–33
86. Díaz-García G, Enciso-Maldonado GA, Díaz-García LA, Legaria-Solano JP, Bamberg J, Lozoya-Saldaña H (2024) Field screening of *Solanum demissum* confirms its late blight resistance in the Toluca Valley, Mexico. Am J Potato Res. https://doi.org/10.1007/s12230-024-09945-7

6 Seed Certification, True Potato Seed and Disease-Free Planting Material

As potato cultivation spread around the world, so did the many pests and pathogens that can attack both growing plants and stored tubers. Some of the potatoes that were taken from one country to another carried pests and diseases within their tubers, on the surface of their tubers, and in the soil attached to their tubers. Ships provided a global network for their spread. Saving tubers to plant the next crop exacerbated the problem. Some pests and diseases became major problems globally, others were of more local concern, and yet others of only minor significance; this could change with global warming. Disease-free planting material is essential if potato crops are going to deliver the yield and quality of tubers required of a food security crop. Statutory seed certification schemes for the multiplication of seed tubers and propagation through true potato seed (TPS) are two important ways of achieving this objective. Resistance genes are available and could be more widely deployed for vector-borne viruses and soil-borne nematodes and wart disease (caused by a fungus). However, it is also clear that wherever potatoes are grown and tubers stored, they will be under attack from various pests and diseases.

6.1 Spread of Pests and Diseases

As potato cultivation spread around the world, so did the many pests and pathogens (diseases) that can attack potatoes, reducing the yield and quality of their tubers. This happened because not all of the potatoes that were taken from one country to another were clean and healthy. Some carried pests and diseases within their tubers, on the surface of their tubers, and in the soil attached to their tubers. Ships provided a global network for the spread of potato pathogens. Saving some tubers to plant the next crop exacerbated the problem. Some pests and diseases became major problems globally, others were of more local concern, and yet others of only minor significance; but this could change with global warming. The importance of planting healthy tubers into soil free of pests and diseases became apparent. Furthermore, practical control measures emerged from increased knowledge of the life cycles of the various pests and pathogens. Today statutory seed certification schemes for the multiplication of seed tubers and propagation through true potato seed (TPS) are two important ways of providing farmers with healthy planting material. They will be considered in this chapter, along with vector-borne viruses, and soil-borne nematodes and wart disease (caused by a fungus), for all of which resistance genes are available and could be more widely deployed. Detailed information on the many other major pests (over 18) and diseases (over 39) of potatoes can be found in appropriate books and reviews, such as references [1–3]. The pests include moths, beetles, aphids, flies and nematodes, and the diseases include fungal, bacterial and viral ones.

J. E. Bradshaw, *Can Potatoes Feed the World?*, Sustainable Development Goals Series,
https://doi.org/10.1007/978-3-031-92890-1_6

6.2 Infectious Juice

A severe condition of potatoes known as "the curl" was first recorded in Great Britain in Lancashire in the 1760s, where it threatened potato cultivation in the 1770s [4]. It was soon realized that "the curl" was more of a problem in some areas than others, such as Scotland. A trade in seed tubers from these areas developed (seed tuber is often confusingly abbreviated to seed). Although the resulting crops were free of "the curl" in their first year, it appeared and worsened in subsequent generations of farm-saved seed tubers. There was also a more general degeneration of potato stocks over generations, as recognized by Knight in 1810 [5]. Knight advocated the production of new cultivars from true potato seed as a solution to the problem of degeneration (true seed is botanical seed). It is reasonable to assume that the degeneration was the result of infection with diseases transmissible by tubers but of a milder nature than those causing the severe symptoms of "the curl" [4]. However, it was not until 1921 that Salaman [6] recognized that "the curl" and degeneration were the result of viral infections, although the causative agents were unknown. This lack of knowledge is clear from reading books such as *Potato Varieties* by Redcliffe Salaman, published in 1926 [7], and *The Potato: Its History, Varieties, Culture and Diseases* by Thomas McIntosh, published in 1927 [8]. They present Leaf Roll, Mosaic, Crinkle and Streak as diseases that can be transmitted by means of the juice (sap) of affected plants; juice that is still infective after passage through filters with pores small enough to keep back known micro-organisms. However, filters with very minute pores could produce a non-infective filtrate. It was concluded that the juice contained infectious particles, and these were called viruses. It was known that viral diseases persist in vegetative propagation, causing secondary infection, unless the primary infection of foliage occurred too late in the season for the virus to reach the tubers. It was also known that viruses could move through a graft, whereas transmission through true seed was rare. Vectors such as aphids could transmit viruses from plant to plant. The symptoms of viral diseases varied with cultivar, environment and age of plant. The viruses reduced the yield of tubers and produced a dwarfed condition of the haulm. Three ways were recognized to control these diseases: the production of virus-free seed tubers by selecting disease-free plants, maintaining virus-free stocks by using isolated sites where there were virtually no transmitting insects combined with roguing out any diseased plants, and growing tolerant (resistant) cultivars.

6.3 Plant Viruses

The use of the word virus came from work published in 1898 by the Dutch microbiologist and botanist Martinus Willem Beijerinck [9], who was one of the founders of virology and environmental microbiology. He used filtration experiments on infected sap to demonstrate that tobacco mosaic disease was caused by an infectious agent smaller than a bacterium (<1/5000 mm) and hence too small to see with a light microscope. Furthermore, although he could not culture the filterable infectious agent, he concluded that it could replicate and multiply in living plants. He used the word virus to describe the "contagium vivum fluidum". Later, Stanley in 1935 [10], isolated and characterized tobacco mosaic virus (TMV) as a crystallizable protein; Bawden and Pirie in 1937 [11], reported that TMV comprised 95% protein and 5% nucleic acid; and Kausche, Pfankuch and Ruska in 1939 [12], secured the first electron microscope picture of the rod-shaped particles of TMV. Subsequently, work published in the 1950s showed that the nucleic acid (DNA or RNA) is the genetic information (infective agent) which is enclosed (encapsulated) within a protective coat of identical protein sub-units [13–15]. One paper published in 1955 became a classic in genetics. Fraenkel-Conrat and Williams achieved the natural co-aggregation of inactive TMV protein subunits and inactive but freshly prepared viral RNA to give reconstituted nucleoprotein rods which were infective. Biological activity was therefore the result of physicochemical relationships [16].

In summary, disease-causing plant viruses are nucleoprotein particles that can multiply in plant cells and move from cell to cell (systemic) but are dependent on the host cell's metabolism for their multiplication, including the host's ribosomes for protein production.

6.4 Coronaviruses

As a result of the spread of the coronavirus SARS-CoV-2 from China at the end of 2019, we have all experienced the COVID-19 pandemic, and we are all experts on the detection and diagnosis of viruses! We know COVID-19 has characteristic symptoms but that they vary between individuals, as does the severity of the disease (symptomless carriers can transmit the disease). We became accustomed to taking throat and nasal swabs for diagnostic tests to check if we were infected with the virus: either a rapid lateral flow (antigen) self-test, which gives a result in 15–30 min, or a more sensitive laboratory PCR test. The former looks for the protein on the outside of the virus that elicits an immune response (i.e., it is an antigen); the latter looks for the presence of the viral RNA (nucleic acid) using what is called the reverse transcription polymerase chain reaction (RT-PCR). The reverse transcription converts RNA to DNA. The polymerase chain reaction amplifies the DNA. We also learnt that it is possible to find out if someone has been infected with the virus by looking for the corresponding antibodies (the body's response to combat infection) in their blood. One such serology test is the Enzyme-Linked ImmunoSorbent Assay (ELISA) test. Finally, we know that viral RNA can be sequenced, and hence new mutations (variants/strains) of the virus can be detected as they arise and the virus evolves as it spreads. I am not going to attempt more detailed accounts of these and other techniques used by virologists. I simply want to make the point that similar methods are used by plant virologists and can be found in textbooks of plant virology [17, 18]. I want to move on to what we know about potato viruses.

6.5 Potato Viruses

In a review in 2018, Jeffries and Lacomme [19] listed 54 viruses that can infect tuber-bearing *Solanum* species, of which 42 had single-stranded RNA as their nucleic acid. Infection can cause a wide range of symptoms in the foliage (leaf rolling, mosaic, chlorosis, necrosis, ringspot and stunting) and in the tubers (external necrosis, internal necrosis, growth cracks and misshapen), affecting yield and tuber quality, and hence marketability. Infection can be primary (current season) or secondary (tuber-borne). Out of the 54 viruses, 14 are transmitted by aphids, 8 by whiteflies, 7 by thrips, 6 by nematodes, 4 by beetles, 4 by fungi and 4 by leafhoppers. Five do not require a vector and can be spread by contact or mechanical means. Furthermore, 2 are transmitted through true potato seed (TPS). That accounts for the 54, although 5 can be transmitted through TPS as well as by a vector. Some viruses are common in potatoes and have a worldwide distribution, whereas others are either rare in potatoes or confined to particular geographic areas. Some are dependent on potatoes for their survival, whereas others are not, having alternative hosts [20].

Table 6.1 lists the viruses that became of particular concern to farmers and hence of interest to potato breeders and geneticists in many countries. They are all single-stranded RNA viruses with a wide distribution, and all except TRV are dependent on potatoes for their spread and survival. The three that have attracted the most attention worldwide are PVY, PVX and PLRV. Also included is a viroid (infectious RNA without a coat protein), which was known to spread through contact between plants or contaminated machinery. It is not endemic in the United Kingdom and hence is treated as a quarantine disease. In 1970 it was shown to be transmitted through true potato seed (TPS) [21]. This caused problems for the Commonwealth Potato Collection, which had been transferred to Scotland (a high-grade seed area) as botanical seed in 1967 without passing through quarantine. The whole collection had to be screened and declared free of potato spindle tuber viroid [22].

Table 6.1 Some of the viruses of interest to potato geneticists and breeders

Virus	Acronym	Genus	Vector	Persistence in Vector
Potato virus Y	PVY	*Potyvirus*	Aphid	Nonpersistent
Potato virus A	PVA	*Potyvirus*	Aphid	Nonpersistent
Potato virus V	PVV	*Potyvirus*	Aphid	Nonpersistent
Potato virus X	PVX	*Potexvirus*	Contact	–
Potato virus M	PVM	*Carlavirus*	Aphid	Nonpersistent
Potato virus S	PVS	*Carlavirus*	Aphid	Nonpersistent
Potato leafroll virus	PLRV	*Polerovirus*	Aphid	Persistent
Tobacco rattle virus	TRV	*Tobravirus*	Nematode	Persistent
Potato mop-top virus	PMTV	*Pomovirus*	Fungus	Persistent
Potato spindle tuber viroid	PSTVd	Viroid	TPS	–

Six of the viruses in Table 6.1 are aphid–transmitted, and of these, five are nonpersistent in the aphid. Monitoring aphids in crops using yellow water traps allows identification of those present, and sharing of data across whole regions allows maps of aphid populations to be constructed. The monitoring allows decisions to be made about insecticide use, where appropriate, and desiccation timings to destroy the haulm and prevent/minimize virus spread [23].

PVY is considered the most economically damaging virus in cultivated potatoes [19]. It is a *Potyvirus* which gives severe mosaic symptoms and is transmitted by about 65 aphid species, most of which do not colonize potato plants. The aphids acquire the virus in seconds to minutes on probing infected plants and may immediately infect other plants. They usually lose their infectiousness within an hour. These attributes make chemical control with insecticides ineffective because they do not kill the aphids quickly enough. However, mineral oils may provide significant control by repelling or reducing the probing of aphids, provided coverage is maintained as the plant grows without causing phytotoxic effects from over-application [19]. A number of PVY strains have been characterized based on differential ability to trigger a hypersensitive response to specific resistance genes, symptoms on tobacco (vein necrosis or mosaic), development of potato tuber necrotic ringspot disease (PTNRD) on tubers, serology (ELISA) and genome phylogeny (molecular methods: RT-PCR and sequencing) [19]. Genes with extreme resistance to all of the currently known strains have been found and utilized by breeders. They suppress virus multiplication in the initially infected cells so that no visible signs of infection can be observed apart from possible tiny, necrotic pinpoint lesions, and no virus is detected in the inoculated tissue. The *R*-genes produce R proteins that stop viral proteins from suppressing RNA silencing, the most important basal antiviral defence system in plants, and thus elicit a quick and powerful defence response. The genes Ry_{adg}, Ry_{sto}, $Ry\text{-}f_{sto}$ and Ry_{chc} were derived from *S. tuberosum* Group Andigena, *S. stoloniferum* and *S. chacoense*, respectively [24–27]. Although PVY is naturally transmitted by aphids, in breeding for resistance, large numbers of young seedlings in seed pans at the cotyledon stage can be inoculated with PVY using infective plant extracts mixed with an abrasive in a spray gun. Susceptible seedlings showing mosaic symptoms can be discarded from about 2 weeks after inoculation [28]. For more detailed studies, young plants can be tested in a glasshouse by spray inoculation with infected tobacco sap containing celite powder as an abrasive. Susceptible seedlings develop mottle symptoms within a few weeks. Finally, resistance can be confirmed in field exposure trials. Major genes for resistance to PVA, PVV, PVM and PVS have also been used by breeders [2].

PLRV is the only potato virus that is aphid-transmitted in a persistent manner. It is a *Polerovirus* that causes economic damage to potatoes throughout the world [19], and as implied in its name, leaf rolling is a typical symptom. In the initial acquisition phase, aphids feed on an infected plant's phloem for minutes, but days are required for maximum transmission efficiency. Acquisition is followed by a latent period of about 24–72 h during which the virus moves through the aphid to the salivary gland. After this, the virus can be transmitted to healthy plants throughout the aphid's life, including moults. The relatively long acquisition phase and latent period mean that colonizing aphids are responsible for the spread of PLRV, the most important of which is *Myzus persicae*. It can be controlled by aphicides, but some clones of *M. persicae* are resistant to a number of major insecticide classes, for example, MACE (modified acetyl cholinesterase) clones that arrived and spread in the United Kingdom from the year 2000 [19]. Breeding for resistance to PLRV has often involved selection for field resistance to infection by the vector [28]. It can be assessed in field exposure trials where the virus is spread from infector plants to exposed plants by aphids, predominantly *M. persicae*. The frequency of secondary infection in tuber progeny plants is assessed by symptoms and ELISA in comparison with standard cultivars. Resistance to infection has been successfully combined with simply inherited resistance to accumulation so that the virus reaches only a relatively low concentration in an infected plant [2, 29–31]. Virus titre can be measured in tissue from glasshouse- or field-grown plants using quantitative ELISA, including the leaves of plants from daughter tubers. Resistance to accumulation may be due to restricted virus movement as well as restricted virus multiplication.

PVX is one of the few viruses that can be transmitted by contact under natural conditions in the field when infective sap contaminates wounds in healthy plants. This may be through abrasions caused by leaf contact during wind or rain, or during tuber storage when sprouts are developing. PVX can also be spread on clothing when walking through a crop containing infected plants and on farm machinery working such a crop. It is a *Potexvirus* with a worldwide distribution whose symptoms tend to be mild mosaic, but mixed infections with other viruses such as PVY or PVA are very damaging [19]. Different strains of the virus exist so that in breeding for resistance, genes with extreme resistance to virtually all currently known strains are preferred. Two such genes are Rx_{adg} (= *Rx1*) and Rx_{acl} (= *Rx2*), the former from an Andigena accession in the Commonwealth Potato Collection (CPC 1673) and the latter from the wild species *S. acaule* [2]. Not surprisingly for a contact-transmitted virus, young seedlings or young plants can be tested in a glasshouse by spray inoculation with infected sap containing an abrasive such as celite powder or carborundum.

TRV is a *Tobravirus* with two RNA genomes. It is transmitted by free-living nematodes in the genera *Paratrichodorus* and *Trichodorus*. These nematodes prefer light sandy or peaty soils and cool climates such as northern Europe [19]. The virus causes necrotic symptoms in the tuber flesh known as spraing or corky ringspot. As few as 2–5% of spraing-affected tubers may lead to rejection of consignments of potatoes by processors and supermarkets. TRV resistance can be assessed in field trials on land infected with Trichodorid nematodes by scoring tubers for frequency and severity of spraing symptoms. However, more reliable results can be achieved in a glasshouse test in which tubers are grown in pots of tested field soil containing viruliferous nematodes, and daughter tubers are scored for spraing symptoms [32]. Symptomless tubers need to be tested for infection to distinguish desirable extreme resistance from undesirable tolerance, which allows virus-free sites to become infected if the vector nematodes are present. Some well-known processing cultivars have resistance, including Markies and Shepody for French fries and Hermes, Lady Rosetta, Record and Saturna for crisping (chipping).

PMTV is a *Pomovirus* which also causes spraing symptoms in tubers and which is now widely distributed globally. It is transmitted by *Spongospora subterranean*, the soil-borne protozoan which causes powdery scab [19], but cultivars with resistance to *S. subterranean* are apparently not protected against PMTV infection. It can be disseminated to new fields in infected seed tubers or in PMTV-containing resting spores of *S. subterranea* adhering to the surface of the seed tubers. Treatment of soil with chemicals does not kill the resting spores and hence is ineffective as well as environmentally damaging. Cultivars with resistance to PMTV are highly desirable but have proved elusive [33].

So far, we have been considering the control of viruses in potato crops (ware) being grown for consumers such as supermarkets and processors. The growers' job is much easier if the tubers they plant are free of viruses and other tuber-borne diseases. Based on experience around the world in the twentieth century, this is best achieved through statutory seed certification schemes in areas where potatoes are grown only for seed. Such areas will usually be geographically and climatically less favourable to the vectors of viruses which cause systemic infection [20]. However, low-income countries, such as Bangladesh, have informal seed systems based on farm-saved tubers with very low use of certified seed, often below 10% [34].

6.6 Seed Systems and Certification Schemes

6.6.1 The United Nations

Although seed certification schemes predate the formation of the United Nations (UN), today the UN provides a global framework in which they operate. The United Nations Economic Commission for Europe (UNECE) was set up in 1947 by its Economic and Social Council (ECOSOC), which had been established by the UN Charter in 1945 (https://ecosoc.un.org/en/about-us). ECOSOC is at the heart of the UN system to advance sustainable development. Although UNECE's main aim is to promote pan-European economic integration, all interested UN member states may participate in its work. Interestingly, UNECE has a standard for seed potatoes that "sets common terminology and minimum commercial quality requirements for the certification of high-quality seed intended for marketing internationally" (https://unece.org/trade/wp7/SeedPotatoes-Standards). The latest version is the 2021 edition. It provides an international frame of reference for all aspects of seed-potato certification, namely: "(a) varietal identity and purity; (b) genealogy and traceability; (c) diseases and pests; (d) external quality; (e) sizing of tubers; and (f) labelling". It also lists the Designated Authorities for Seed Potato Certification in 49 countries, the information having been updated on 27 June 2023. Of the top 15 potato-producing countries over the period 2016–2020, India, the United States, Germany, France, Poland, the Netherlands, the United Kingdom, Canada and Egypt are in the list, whereas China, Russia, Ukraine, Bangladesh, Belarus and Peru are absent.

As one of the main aims of seed certification is healthy planting material, an even wider context is the International Plant Protection Convention (IPPC), which is "the intergovernmental treaty that aims to protect the world's plants, agricultural products and natural resources from plant pests" (https://www.ippc.int/en/about/overview/). It has been ratified by 185 contracting parties. The IPPC develops, adopts and promotes the application of International Standards for Phytosanitary Measures (ISPMs) to safeguard global food security, facilitate safe trade and protect the environment. Whilst ware potatoes tend to be consumed locally because of their bulk, there is an international trade in seed potatoes.

6.6.2 Scotland

Scotland is ideal for seed potato production as a result of its northern climate, with cold winters, cool summers and windy conditions, all of which aid good control of aphids and viruses. Scotland also has moderate rainfall and fertile soils which help good potato growth. This means that Scotland is also suitable for growing ware potatoes, which creates problems in areas where both types of crops are grown.

Seed potato production in Scotland is associated with East Craigs on the western side of Edinburgh, not far from the airport (http://www.sasa.gov.uk). It provides a good example of similar schemes around the world. The Board of Agriculture purchased the 111-acre farm in 1920, and in 1923 approval was given for the building of laboratory facilities. It was also the first home of the Scottish Plant Breeding Station (SPBS) from 1920, with the first director of SPBS living in Craigs House. Initially East Craigs was concerned with the certification and purity of new cultivars, but from 1929, their work included developing healthy virus-free potatoes. Today East Craigs is a modern housing estate in which Craigs House, a listed building, has been converted to three flats (Fig. 6.1).

The Certifying Authority (SASA: Science and Advice for Scottish Agriculture) for the Seed Potato Classification Scheme (SPCS) in Scotland is now based a few miles away at Gogarbank Farm and has modern laboratory facilities (Fig. 6.1). It administers the scheme on behalf of the Scottish Government, which controls all stages of seed potato production in Scotland, in particular, the production of the initial propagating material, the length of the multiplication chain, and the application of strict tolerances for diseases, particularly those caused by viruses. Scotland produces and markets only pre-basic (PB) and basic seed potatoes. Seed potatoes entering Scotland from the European Union (EU) must be of Community Grade, and those from outside the EU must go through quarantine. The aim of quarantine is to prevent entry to a country of a non-indigenous pest or disease or at least to isolate and eradicate any outbreak before the problem spreads. SASA also holds in vitro pathogen-free nuclear stocks of cultivars under multiplication and also fingerprints cultivars with molecular markers for unique identification (http://www.sasa.gov.uk).

Details of the SPCS can be found in the document "Scottish Seed Potatoes" produced by SASA [35]. A brief summary of the scheme is as follows [2]. Seed production starts from pathogen-free microplants produced in sterile laboratory conditions at SASA or under licence from SASA. The microplants are therefore the nuclear stock of the cultivar from which all seed crops are derived. Pathogen testing is done on rooted plantlets from single-node cuttings from tuber sprouts of tubers supplied by the breeder of the new cultivar. If necessary, virus elimination can be done by thermotherapy (heat treatment at 36–40 °C for 3 weeks) or chemotherapy (e.g., Virozole) prior to meristem-tip culture in liquid nutrient medium for plant regeneration. Once the resulting, or original, plantlets have been confirmed pathogen-free, they can be rapidly multiplied by in vitro nodal cuttings under artificial, aseptic conditions in the laboratory. Once sufficient cuttings have been produced, growth regulators are used to produce fully rooted in vitro plantlets (transplants).

The transplants are then used by licensed commercial growers to produce minitubers [Pre-Basic Tissue Culture (PBTC)] in a greenhouse or screenhouse with a high health status. The transplants can be grown in soil or in aeroponic, hydroponic or nutrient film cultures. Soil-grown transplants usually produce two to five minitubers per plant, whereas up to 40 can be achieved from frequent harvests in culture. The aeroponic

Fig. 6.1 Craigs House at East Craigs (left) and SASA at Gogarbank Farm (right), both near Edinburgh Airport

system is popular in many countries, such the one shown in Fig. 6.2, which is in China. The "below ground" plant parts are suspended in air and intermittently misted with nutrient solution.

Returning to Scotland, the minitubers are grown by officially approved commercial growers in the field to produce pre-basic (field-grown) seed tubers (PB1), which can then be multiplied for up to three more generations as pre-basic (PB2–PB4) (Fig. 6.3) and a further three generations as grades of basic seed (S, SE and E). Statutory seed multiplication can be done in other suitable areas in the United Kingdom, such as parts of Yorkshire (Fig. 6.3).

The land used for seed crops in Scotland must be free from wart disease (*Synchytrium endobioticum*) and also tested and confirmed free from cyst nematodes (*Globodera rostochiensis* and *G. pallida*). The interval between potato crops in the rotation must be 7 years for pre-basic seed crops and 5 years for basic seed crops. Seed crops are officially inspected twice during the growing season and must be 99.99% pure and true to type and have very low levels of PVY, PVA, PLRV and blackleg disease (*Pectobacterium* and *Dickeya* spp.). Any diseased plants should have been removed (rogued out), ideally before aphids colonize the crop. Harvested seed tubers must also be inspected for diseases and disorders and meet the required standards before they can receive the official label required for marketing and the phytosanitary certificate if required for export overseas.

More ware than seed crops are now grown in Scotland because they are profitable, and this can cause problems where they occur in the same area. According to the front page of *The Scottish Farmer* for 3 August 2024, the year had seen 64.4 hectares of virus-infected ware crops in Scotland destroyed by order of SASA to protect seed crops from an increasing prevalence of PLRV. This was double the previous year (ware crops on a seed farm must be grown from PB, S, SE or E stocks) [36].

6.6.3 India

In India, summer crops from the northern hills around Shimla (2276 masl) were able to provide good quality seed potatoes for the sub-tropical plains where most of the ware production took place, but the hill area could not satisfy the increasing demand [37]. A solution was found following the establishment of the ICAR-Central Potato Research Institute (ICAR-CPRI) in 1949. Surveys revealed a remarkable consistency in the build-up of the aphid population in the northwestern and central Indo-Gangetic plains during the winter growing season. The aphid population remained very low during October to December, giving a window of opportunity, through an earlier planting date and earlier haulm destruction, for seed potato production. This proved very successful from its introduction in 1970 to the present day. However, there is now concern about whitefly (*Bremisia tabaci*) transmitted viruses (e.g., tomato leaf curl

Fig. 6.2 Aeroponic culture of minitubers in Chengdu, China. (From Bradshaw JE (2021) Potato Breeding: Theory and Practice. Springer, Cham with permission; 8 April 2008)

Fig. 6.3 Pre-basic seed production in a field in northeast Scotland (PB1 and PB2; from Bradshaw JE (2021) Potato Breeding: Theory and Practice. Springer, Cham with permission; left, 14 July 2011) and in the Yorkshire Wolds in England (PB3 and PB4) (right, 26 June 2024)

New Delhi virus, ToLCNDV) becoming a major problem and threatening the disease-free seed production in these areas [38]. Today ICAR-CPRI, like SASA, oversees the production of certified seed tubers of both old and new cultivars which are being commercialized [39]. Annual demand is high, at 5.25–6.30 million tonnes, so rapid hi-tech seed systems are preferred over conventional, slower field multiplications, and private enterprises are becoming involved [40]. As in the SASA system, the process starts with disease-free microplants which are used to produce minitubers, with aeroponic minitubers gaining in popularity. The minitubers are planted in the field (or nethouse) the next year (Generation-1, pre-basic) and the produce is planted in the field the following year (Generation-2) to provide breeder (basic) seed tubers at harvest. Further multiplication is over two (Generation-3 and Generation-4) or three generations (Foundation seed-1, Foundation seed-2 and Certified Seed). More information about potato seed certification in India can be found in an article written in 2019 by Singh et al. [39].

6.6.4 China

Seed potato production in China takes place mainly in the north at high latitudes and the southwest at high altitudes.

Figure 6.4, for example, shows seed production in the Inner Mongolia Autonomous Region (41°N, 112°E), a high, dry, cool and windy plateau around 1400 masl where the natural vegetation is grassland. The potato is a summer crop that requires irrigation for high yields and quality, as the annual rainfall is less than 300 mm per annum. Virus-free seed potato production started in China in the 1970s but, by 2018, only accounted for 35% of the potatoes planted by farmers. Most farmers still used seed that they had produced themselves for the next season, seed that was not disease-free [41]. However, virus-free seed potato production is increasing as large enterprises become involved with their own complete systems which are similar to the one described for Scotland [41].

6.6.5 Uganda

Uganda is typical of many Sub-Saharan African (SSA) countries when it comes to seed potatoes. Smallholder farmers commonly use their own recycled seed potatoes or ones from the informal sector of family, neighbours and the local market. These are often highly degenerated with viruses. Research has shown that there is some benefit in more positive selection of seed tubers from healthy ware plants to reduce levels in the next ware crop. In positive selection, the healthiest looking plants are pegged and selected just before full flowering to identify plants whose tubers will serve as seed for the next generation, provided

Fig. 6.4 Seed potato production in Inner Mongolia Autonomous Region, 41°N, 112°E (Zhengfeng Seedpotato Co., Ltd.) (6 August 2010)

they are still without symptoms 2 weeks later and their tubers look healthy after storage. In contrast, traditional farmer selection just involves visual selection of tubers from the bulk of the potato harvest [42, 43]. Nevertheless, farmers in Uganda would benefit from new cultivars with resistance to viruses and other diseases, as well as from disease-free planting material. This need was recognized by the International Potato Centre (CIP) on its establishment in Lima, Peru, in 1971. CIP has therefore made broad-based germplasm, from its collection of South American landraces, available to national programmes in more than 20 countries in Africa, Asia and Latin America for breeding and selection, including germplasm with resistance to viruses, particularly PVY, PVX and PLRV [44]. Furthermore, following the world food crisis in 2008, CIP partnered National Agricultural Research Institutions (NARIs) in SSA to tackle the shortage of high-quality seed potatoes through a "3 Seed Potato Generation Revolution (3G Revolution)" involving public and private stakeholders, which was implemented in Kenya, Uganda and Rwanda from 2008 to 2011 [45]. A large number of minitubers were produced from in vitro plantlets using aeroponics, followed by two field generations of further multiplications. As a result, the production of healthier planting material is increasing [45].

6.6.6 Russia and Ukraine

Since 1991, Russia has made the transition from a government-regulated economy to a "free market" one. The resulting private sector for potatoes comprises small family farms of less than two hectares, which in total account for 80% of production. They have difficulty in accessing/affording certified seed tubers and, as a consequence, plant farm-saved ones with their accompanying problems. In contrast, certified seed (often imported) is used when much larger Russian agricultural organizations collaborate with foreign companies who want to grow cultivars suitable for processing [46]. The same is true for Ukraine, where over 80% of production is for domestic consumption and comes from non-professional growers in small-scale backyard plots using home-saved seed for many years. In contrast, just 50,000 hectares are in full commercial production. The Continental Farmers Group was set up in 2006 and is now the largest commercial producer with over 2000 hectares of potatoes, albeit on state-rented land [Dr Keith Dawson, personal communication]. They dealt with soil compaction and a huge weed problem, imported high-quality seed tubers of new cultivars and used GPS-controlled planters instead of horse-drawn manual planting. As a consequence, the fertile, stone-free black soils of western Ukraine delivered three times the previous yields. The Group also invested in temperature-controlled stores for potatoes destined for a crisping company. A more recent venture is the Central Plains Group, who are growing 1000 hectares of potatoes and have a new starch production factory for biodegradable plastic feedstock.

A detailed review of seed potato production, certification, and trade can be found in a recent (2023) book chapter by Hingrat and Quéré [47], but it is clear from my brief account that many countries have and will benefit from planting a greater quantity of healthy seed tubers. More generally, the quality of a seed tuber will also depend on its physiological age (developmental state) and size (initial energy source of sprouting tuber), with cultivar (genetic makeup, including disease and pest resistance) a factor.

6.7 Seed Storage

Once harvested, seed tubers usually need to be stored for several months before planting; for example, about 6 months for Scottish seed potatoes. They are stored in the dark at low temperature (2–4 °C) to prolong dormancy, suppress sprouting and prevent losses of water and storage carbohydrates and proteins [48]. However, in many countries cold stores are not available, and diffuse light storage (DLS) at ambient temperatures is used [49]. In DLS, some level of cooling is combined with the sprout-inhibiting and age-delaying effect of light [48, 50]. Seed tubers have been successfully stored for up to 8–9 months in small rustic ventilated stores in tropical regions with mean temperatures well above 20 °C [49]. Stores need to be monitored for disease and pest problems that can lead to serious tuber losses from rotting. Some of these problems, together with control measures, will be briefly dealt with later in this chapter.

6.8 True Potato Seed (TPS)

The production of healthy planting material can be difficult in some parts of the world. Knight in England in 1810 advocated breeding new cultivars in order to deal with the degeneration of seed stocks during generations of reproduction from tubers. It worked because there are very few diseases that are transmitted through the true potato seed (TPS) produced by sexual reproduction. But what about the multiplication of cultivars from TPS? It has not proved feasible to produce sufficiently uniform tetraploid cultivars in this way but is now possible in diploids with the advent of genetically uniform F_1 hybrid cultivars [51–53]. These are like diploid F_1 hybrid tomatoes which gardeners can either grow themselves from botanical seed or buy young plants from a nursery that has raised them in this way. The botanical seed can be produced by hand-pollinations in areas with abundant cheap labour, such as East Asia, and in the tropics and subtropics, altitudes above 1500 m during summer are conducive to the flowering and fruiting of potatoes [54]. China and India are both interested in the propagation of potatoes by TPS as well as being the two largest producers of tomatoes [55].

The early stages of a conventional breeding program can be used to illustrate propagation from true potato seed (Fig. 6.5). The plants grown from true seed, either in a glasshouse (top left) or after transplanting to a field (bottom left), produce tubers (seedling tubers) (top right). These can either be used for consumption or as seed tubers to plant the next crop (bottom right), although in the future, pelleted and primed seeds may be direct drilled by farmers. When potato transplants are grown as a ware crop, a whole new cultivation system must be developed, as is happening in East Africa [52].

Fig. 6.5 TPS. Young plants grown from true seed (top left) and their seedling tubers (top right) in the glasshouse at the Scottish Crop Research Institute (SCRI) on 29 April and 22 August 2008, respectively. Young plants grown from true seed and transplanted to the field (bottom left) in India (13 December 2008). Spaced plants (bottom right) grown from one tuber per pot (top right) at SCRI's seed site north of Dundee, Scotland (24 June 2008)

In contrast to transplanting young plants, seedling tubers are the kind of minitubers that farmers are used to handling. Potato transplants grown in dense nurseries at 100 plants per square metre (10 cm spacing) produce about 500 small but usable minitubers with an average size of 10–15 g that can be chitted before planting [54]. Hence, 100 square metres can produce the 50,000 minitubers required to plant 1 ha of potatoes, and these will weigh 750 kg (50,000 × 15 g) compared with 2500 kg of normal-sized seed tubers (50,000 × 50 g [50]). More importantly, the 10,000 transplants can be produced from 10,000 seeds, assuming 100% germination, that would weigh in total around 7.5 g and hence be easy to transport over long distances. The potato transplants could also be grown in a glasshouse or in an aeroponic system to produce minitubers.

A potato transplant grown from a true potato seed produces a different kind of plant than a seedling tuber. The former produces a single-stemmed plant, whereas the latter can produce multiple main stems. The single stem and multiple stems differ in branch production. The consequences of these differences for crop yield and tuber-size distribution at different stem densities per unit area of land are currently being researched [56]. One result has already confirmed that seedling tubers have a higher fresh yield per unit area of land as a result of larger tubers more than compensating for fewer tubers; effects which outweighed those due to stem density. More results are anticipated, as it is important to get the agronomy right for growing potatoes from TPS.

Today all types of potato cultivars can be grown from minitubers, with a diploid F_1 hybrid starting multiplication from TPS, whereas a clonally propagated cultivar would be multiplied from rooted in vitro plantlets. However, production of minitubers from TPS should prove cheaper and more flexible, with multiplication starting where and when required from relatively little TPS by weight. Furthermore, fewer rounds of multiplication should be able to satisfy the commercial demand for planting material, the ideal being just one round [57]. Currently diploid F_1 hybrid cultivars are lower yielding than clonally propagated tetraploid cultivars [58]; but this may change in the future with the former grown more widely than in areas where clean planting material is the main concern of farmers. The advantage of TPS in the latter situation can be seen in work reported in 2016 from the tropical highlands of East Africa (1680 masl) [59]. Hybrid seedlings were raised from botanical seeds and 6 weeks later transplanted into irrigated ridges in a field. Tubers of the 10 hybrids were harvested 4 months later. They yielded between 8 and 29 t ha^{-1} compared with average yields of potatoes in East Africa of 5–14 t ha^{-1}. In contrast, the best hybrid had yielded 26 t ha^{-1} in a trial in the Netherlands the previous year, where tetraploid cultivars yield between 35 and 40 t ha^{-1}. It should also be relatively easy to introduce major genes for disease and pest resistance into diploid F_1 hybrid cultivars such as those found in landraces and wild relatives for resistance to late blight (previous chapter) and viruses, wart disease and nematodes (this chapter). These genes should help maintain the disease-free nature of the starting material during the growing seasons.

6.9 Wart Disease

Seed crops in Scotland must be grown on land that is free from wart disease caused by the soil-borne fungus *Synchytrium endobioticum*. The disease has not been found in Scotland for over 30 years [35] but was a serious problem at the beginning of the twentieth century, causing disfiguring blemishes on potato tubers. It is assumed that *S. endobioticum* spread at the end of the nineteenth century from its centre of origin in the Andes, first to Europe and North America, and subsequently across the globe [60]. Its persistent resting spores can remain infectious for more than 20 years, and chemical control has not proved possible. Nevertheless, wart was largely eliminated from Britain during the first half of the twentieth century by coupling the breeding of cultivars with immunity/resistance to pathotype 1 (the only pathotype found in Britain) with scheduling of land, which prohibited the cultivation of susceptible cultivars on land known to be infested [61]. The resistance occurred in some of the cultivars being grown at the time. The only complication was the fact that wart was a quarantine disease. This meant that testing had to be done by the Quarantine Authority in Scotland and was restricted to potential cultivars in an advanced stage of development. This was not a serious problem in practice, and until 1990, all cultivars bred at the Scottish Plant Breeding Station (SPBS) were resistant to wart. After that date, wart was no longer considered a problem. However, in other European countries, wart is still a problem, and since 1941, other pathotypes of the fungus have been found and are evolving. Hence, there is still a need for research to ensure a continuation of effective and sustainable management of the disease, including quarantine measures and resistant cultivars [60]. Two distinct dominant genes have been found in diploid germplasm and mapped to chromosome 11, *Sen1* for resistance to pathotype 1 and *Sen2* for resistance to eight pathotypes that are common in Europe, including pathotype 1 [62, 63]. Three more genes were subsequently identified: *Sen3* for resistance to pathotypes 1, 2, 6 and 18 also mapped to chromosome 11 and is the main source of resistance in German and Polish cultivars; *Sen4* for resistance to pathotypes 2, 6 and 18 mapped to chromosome 12 and is the main source of resistance found in Dutch starch potatoes; and *Sen5*, also for resistance to pathotypes 2, 6 and 18, mapped to chromosome 5. Molecular markers have been developed for *Sen1*, *Sen3*, *Sen4* and *Sen5*, thus making screening for resistance much easier in a breeding program as there are no quarantine issues, whereas *Sen2* is still to be introgressed into tetraploid potatoes [64]. It is hoped that these resistances will prove durable, as wart is a slowly spreading soil-borne disease, in contrast to a pathogen like *Phytophthora infestans*, the cause of late blight.

6.10 Nematodes

Nematodes are small, slender, non-segmented worms of 0.1–2.5 mm in length which occur in the soil in great numbers [65, 66]. The potato cyst nematodes (PCN), *Globodera pallida* (pale/

white potato cyst nematode) (Fig. 6.6) and *G. rostochiensis* (golden potato cyst nematode), co-evolved with wild potatoes in the Andes of South America. They are a good example of how we need to solve a problem that we created by helping them to spread through shorter rotations, poor farm hygiene and failure to monitor their buildup.

Since the 1840s, they have spread to many parts of the world through the global trade in seed potatoes. They travel through infested tubers and soil, and if they avoid quarantine measures, can establish damaging infestations. Their second-stage juveniles hatch from viable eggs in response to chemicals exuded from the roots of growing potato plants. The juveniles penetrate and feed on the roots and thus reduce tuber yield, size and quality. The nematode life cycle is completed with the formation of males and females, the latter each producing 200–600 eggs, which, following fertilization, are protected within a cyst that can survive for 20 years or longer in the soil. A small proportion of eggs hatch each season, and in the absence of any host will die, resulting in a decline in the nematode population.

Infestations can result in quarantine restrictions, loss of land for seed production and increased rotation periods, but can be managed with chemical control. In the United Kingdom in 2023, fluopyram (a fungicide) and fosthiazate (an organophosphate contact nematicide) were approved for use, with the latter licensed up to 2029, but 2027 in the European Union and up for review there in October 2024 [67]. The likely loss of fosthiazate (Nemathorin 10G) would also make control of wireworms and free-living nematodes harder. Nevertheless, there is political pressure in many countries to reduce the use of nematicides because they are highly toxic pesticides that raise both environmental and health concerns. Field applications can leach through the soil into groundwater, and run-off can contaminate watercourses. Nematicide residues can get into the harvested potatoes, particularly with carbamates such as carbofuran that have both a contact and systemic mode of action [65]. So, carbofuran is now banned in the European Union and other countries. Hence, biological methods, trap crops and resistant cultivars are more attractive control measures. Liquid bio-formulations, for example, of *Pseudomonas fluorescens* and *Purpureocillium lilacinum* applied as a seed treatment and as a soil drenching, respectively, are as effective as the nematicide carbofuran [68]. Trap crops and resistant varieties cause PCNs to hatch but allow little or no reproduction, thus preventing multiplication. Breeding has already delivered resistant cultivars, and more can be expected in the future, but so far greater success has been achieved with *G. rostochiensis* than with *G. pallida*.

The cyst nematodes started to become a serious problem in Europe in the early 1950s. In 1952 Ellenby [69] screened the Commonwealth Potato Collection and found the first source of resistance to be successfully used in an Andigena accession (CPC 1673). It was a simply inherited

Fig. 6.6 The white potato cyst nematode (*Globodera pallida*) on potato roots. (Source: SCRI)

dominant major gene [70] that was named *H1*, from the old name for the nematode, *Heterodera*. It proved effective against what became known as pathotypes Ro1 and Ro4 of *G. rostochiensis.* It was quickly backcrossed into the European cultivated potato, with cultivar Maris Piper released from the Plant Breeding Institute, Cambridge, in 1966 and Pentland Javelin from the Scottish Plant Breeding Station, Edinburgh, in 1968 [2]. Maris Piper is still the most widely grown cultivar in Britain. The *H1* gene has been used extensively in potato breeding worldwide. It provides an extremely high level of resistance, which is effective in reducing the size of PCN populations. However, as it inhibits PCN juveniles in potato roots at various stages of their development prior to cyst formation, tolerance of root damage is also desirable, as found in the Irish-bred cultivar Cara, but requires expensive yield trials on infested soils, with and without nematicide control for its detection [71]. In contrast, breeders can assess a large number of potato clones for resistance using rapid and reliable laboratory or glasshouse tests; for example, a tuber in a canister in sterilized soil inoculated with a known number of eggs [72]. Molecular markers are also becoming available.

The *H1* resistance has proved durable in the United Kingdom where Ro1 appears to be the only pathotype of *G. rostochiensis*, but other pathotypes occur elsewhere. The PCN populations found in the United Kingdom that overcame the *H1* gene proved to be what we now know as pathotypes Pa1 and Pa2/3 of *G. pallida*, of which Pa2/3 became the main problem in the United Kingdom and elsewhere in Europe. In a survey in 2002 [73], PCN was detected in 64% of 484 potato fields in England and Wales, with 66% of these populations being wholly *G. pallida.* It appears that growing cultivars with the *H1* gene to control *G. rostochiensis* helped the spread of *G. pallida.* Quantitative resistance to *G. pallida* was found in another Andigena accession in the CPC (CPC 2802) [2] and quantitative resistance to both *G. rostochiensis* and *G. pallida* in a diploid wild species from South America, *S. vernei* (CPC 2487 and CPC 2488) [74]. As the resistance was provided by more than one gene, it proved difficult to breed commercially successful cultivars with high levels of resistance. Cultivars Lady Balfour from the Scottish Crop Research Institute, Sante from the Netherlands, and Nadine from Caithness Breeders, for example, were classified as partially resistant to *G. pallida* in the United Kingdom [75]. As a consequence, *G. pallida* continued to spread in Britain and elsewhere. Today one can find articles in the Scottish Farming Press stressing the urgent need for cultivars with both resistance and tolerance of root damage, of which a few are now available [76–78]. There does in fact appear to be good prospects for combining a few major genes and quantitative trait alleles (QTLs) to give a high level of resistance to a broad spectrum of pathotypes of *G. pallida* and *G. rostochiensis* [2]. One of the first examples is a new cultivar Buster, from the Teagasc program in Ireland, which combines resistances from Andigena [*H1* and *Gpa2* from CPC 1673 and *Gpa4* from CPC 2802] and *S. vernei* (*Grp1*, *Gpa5* and *Gpa6*) [2, 79]. Furthermore, they should remain effective for a long period of time provided their use is combined with appropriate crop rotations and farm hygiene. However, as populations of *G. pallida* are genetically variable, they may eventually evolve virulence against the resistance genes [80]. A detailed account of breeding for resistance can be found in a recent (2024) review by Spychalla and De Jong where they make the following very important point. In practice it is simple to produce potatoes with a resistance gene but much harder to develop clones that are commercially acceptable, let alone successful [81].

Finally, I am going to briefly mention the root-knot nematodes (RKN, *Meloidogyne* spp.), which are both similar to and different from the cyst nematodes in their life cycle. The economic impact of the different species depends on the climatic region: the Southern root-knot nematode (SRKN) group is largely confined to the tropics (*M. arenaria*, *M. incognita* and *M. javanica*), whereas the Columbian root-knot nematode (CRKN, *M. chitwoodi*), the False Columbian root-knot nematode (*M. fallax*) and the Northern root-knot nematode (NRKN, *M. hapla*) are adapted to cooler climates [65]. Current control of these nematodes relies heavily on synthetic nematicides, but as with the cyst nematodes, there is political pressure to reduce their use. Hence, resistant cultivars are an attractive proposition, with most progress made in the United States with *M. chitwoodi* in finding and utilizing resistance genes, although resistant cultivars are still in the pipeline [2]. It remains to be seen if climate change alters the global distribution of these nematodes and, more generally, of other pests and diseases. Currently there is concern about "tropical" root-knot nematodes moving further north in Europe [82].

6.11 Other Pests and Diseases

What about other pests and diseases of potatoes? I am going to finish with brief comments on the currently important ones and indicate appropriate control measures. More details can be found in my book on potato breeding and the original references therein [2], and photographs can be found in books on potato diseases, pests and defects [83, 84]. A few additional references are given in the text below.

6.11.1 Pests

The Colorado potato beetle (CPB, *Leptinotarsa decemlineata* Say) is not endemic in the United Kingdom and hence is a quarantine pest. When the beetle is found, the Animal and Plant Health Agency issues statutory notices to ensure containment and eradication, as happened in 2023 in Kent (a single colony of larvae) and Hampshire (a single live adult) and previously in the hot summer of 1976 [85]. Once established, adults and larvae can defoliate a whole field of potatoes, and hence the beetle has been described as a "ten-striped spearman" in the era of biotechnological warfare [86]. It is a major pest in northern latitudes, particularly in North America (from where it spread), Europe, Russia and Central Asia. It is difficult to control, as strategies rely on pesticides, but the beetle has developed resistance to most of them. No commercial cultivars with strong resistance had been produced by 2015 [87]. This was despite glandular trichomes and foliar leptine glycoalkaloids being known resistance factors present in wild relatives and transgenic resistance being provided by insecticidal proteins encoded by *cry* genes from the soil bacterium *Bacillus thuringiensis* (Bt).

Potato tuber moth (PTM, *Phthorimaea operculella* Zeller) is the most serious insect pest of potatoes worldwide, particularly in warm tropical and subtropical climates. Its larvae mine foliage, stems and tubers in the field and storage. Control measures are the use of healthy seed, deep planting, proper ridging, removal of host weeds and potato volunteers, irrigation, harvesting soon after maturity, use of insecticides both in the field and in storage, biological control agents, host resistance due to antibiosis (resistance to feeding when the moth has no choice) and antixenosis (the tuber is unattractive when the moth is given choice) and transgenic resistance provided by *cry* genes.

6.11.2 Diseases

Soil-borne Verticillium wilt (*Verticillium dahliae* Kelb.) and air-borne early blight (*Alternaria solani* Sor.) are warm-climate fungal diseases which occur especially where crops are irrigated. Verticillium wilt has recently been spreading in China with a relatively new species, *V. nonalfalfae*, of increasing importance. *Verticillium* species infect the vascular tissue of potatoes through their roots, disrupting the transport of water and nutrition. The microsclerotia (resting bodies) produced by the fungus can survive in the soil for a long time. Control by soil fumigation is effective but expensive with a negative environmental impact. Early blight results in necrotic lesions on the leaves that spread and coalesce, resulting in defoliation. The disease can be controlled with fungicides, but this can be costly and ineffective. Improved resistance to both diseases is desirable and feasible, as screening methods have been developed and differences in cultivar resistance/susceptibility have been found, although true resistance needs to be distinguished from late maturity.

Powdery scab (*Spongospora subterranea* (Wallroth) Lagerheim f. sp. *subterranea* (Sss)), like wart, is another persistent soil-borne, disfiguring, blemish-forming disease of tubers in temperate regions, where the pathogen thrives in cool, damp, soil conditions and can also infect the roots. It is also the vector of potato mop-top virus (PMTV). The pathogen is now considered a protozoan belonging to the Order Plasmodiophorida [88]. As the disease can be transmitted on infected and infested seed tubers, seed certification systems usually include low tolerance levels for powdery scab. Improved resistance, as part of integrated disease management, is feasible as screening methods have been developed and differences in cultivar resistance/susceptibility have been found. More information on disease management can be found in a recent (2024) review by Strydom et al. [88].

Dry rot [(*Fusarium coeruleum* (Libert.) Sacc.), *F. sambucinum* (Fuckel) (synonymous with *F. sulphureum* and *Gibberella cyanogena*), *F. avenaceum* Sacc. (Corda ex. Fr.) and other species] and gangrene (*Phoma foveata* Foister) are fungal diseases which cause spoilage of tubers in storage. Whilst dry rot species occur worldwide, gangrene is associated with northern Europe. Dry rot can be both soil- and tuber-borne, with inoculum transmitted by infected or contaminated seed tubers. The disease can be controlled by good agronomic practice to minimize damage, the use of fungicides (but resistance to thiabendazole does occur), in-store ventilation and host resistance, although resistance to *F. coeruleum*, for example, is different and separate from resistance to *F. sulphureum*. Gangrene is tuber-borne, and improved resistance is relatively easy to achieve if the disease is considered important enough for this to be a breeding objective. Heritable variation and screening methods are available.

Black scurf is a soil- and tuber-borne skin blemish disease caused by the fungus *Rhizoctonia solani* (Kühn.), which also causes stem canker. Skinspot (*Polyscytalum pustulans* (Owen and Wakef.) M.B. Ellis), silver scurf (*Helminthosporium solani* Dur. and Mont.), and black dot (*Colletotrichum coccodes* (Wallr.) Hughes) are three tuber-blemishing fungal diseases. All three pathogens cause a deterioration of tubers in store, and all are mainly or partly tuber-borne as well as soil-borne. They can be controlled with fungicides such as fluxapyroxad [89]. Although new cultivars are tested for resistance/susceptibility in Britain [75], resistance is not a high-priority breeding objective, and the same is true in many other countries. Nevertheless, it is always desirable to avoid growing extremely susceptible cultivars to these and other currently minor disease problems.

Common scab (*Streptomyces scabies* (Thaxt.) Waksman and Henrici and other pathogenic *Streptomyces* species) is a global actinomycete disease that causes scab-like (surface/raised/pitted) lesions on tubers. Infection is aided by the production of a phytotoxin, thaxtomin A. The bacterium is dispersed by spores and survives on seed, in soil and in soil water. Tubers are most susceptible to infection up until 3–4 weeks after initiation and with soil moisture below field capacity. Hence, irrigation is one control strategy but may encourage powdery scab. Host resistance is therefore considered one of the best options and has been a high priority among North American potato breeders. Ironically in the United Kingdom, in the 1970s, susceptible cultivar Maris Piper replaced the resistant cultivar Pentland Crown as the most widely grown cultivar, a position it still held in 2023.

Bacterial wilt of plants and brown rot of tubers are caused by *Ralstonia solanacearum* Smith, a soil-borne bacterium that enters via the roots, invades xylem vessels and spreads rapidly to the aerial parts of the plant through the vascular system, resulting in wilting. Classification of strains of the bacterium is complicated. The bacterium can spread through contaminated soil and debris, weed hosts (e.g., *Solanum dulcamara* on riverbanks) and irrigation water, and over longer distances by latently infected seed tubers. In some countries the disease can be as serious as late blight [90, 91]. As chemical controls are not available, breeding for resistance has been a high priority, but with disappointing results to date.

Blackleg of potato plants is a tuber-borne disease caused by a number of pectolytic bacteria species, most commonly *Pectobacterium atrosepticum* (synonym *Erwinia carotovora* subsp. *atroseptica*). The disease results in brown-black discolouration of the stem and stunting and wilting, as well as some initial non-emergence of seed tubers. It is increasing in incidence and severity in northwestern Europe and North America. Other bacterial species such as *Pectobacterium carotovorum* (synonym *E. carotovora* subsp. *carotovora*) and *Dickeya solani* (previously classified as *E. chrysanthemi*) can also cause field symptoms but are more commonly associated with soft rots of tubers in storage. Which *Pectobacterium* and *Dickeya* pathogens dominate depends on the location and prevailing environmental conditions. Blackleg and soft rot can be controlled by management methods (seed free of pathogen, avoiding prolonged periods of high soil moisture, proper store ventilation) and resistant cultivars, but not by chemical means. However, despite much research and breeding effort, there is not an obvious increase in the number of new cultivars with higher levels of resistance.

Clavibater michiganensis subsp. *sepedonicus* (Spieck and Kotth.) Davis et al. is the cause of potato ring rot, a recurring seed-tuber-transmitted disease problem in temperate regions, despite many countries treating it as a quarantine disease and having a zero-tolerance policy for the import of infected seed potatoes, which can be symptomless. The disease causes the lower leaves to wilt and discolouration of the vascular ring tissue of tubers. Little breeding work appears to have been done, although *S. acaule* has been reported as a source of extreme resistance (immunity to infection) [92, 93].

Zebra chip disease was first observed in Mexico in 1994 and then spread to the southern parts of the United States, other countries in Central America, and New Zealand [94]. The name comes from the characteristic dark stripes which appear when infected tubers are fried. The causal agent of the disease is believed to be the unculturable bacterium *Candidatus* Liberibacter solanacearum (Lso), which resides in the plant phloem and is primarily vectored by the potato psyllid (lice) *Bactericera cockerelli*. The most common method of control is with pesticides, but insecticide resistance has been reported.

6.12 Conclusions

The emphasis in this chapter has been on the importance of disease-free planting materials if potato crops are going to deliver the yield and quality of tubers required of a food security crop. However, it is also clear that wherever potatoes are grown and tubers stored, they will be under attack from various pests and diseases. Farmers must therefore take steps to control them, or they will end up at best with lower-yielding and poorer-quality crops or at worst with crop failures. Farm hygiene and management are important, including sensible crop rotations, removal of weed hosts and volunteer potato plants from previous crops, as well as disinfecting farm machinery and stores. Growing resistant cultivars is highly desirable, but these are not always available for the target market and end use. Hence, there will continue to be a need for the thoughtful rather than indiscriminate use of chemical control of pests and diseases. Finally, it is important to appreciate that we are in a dynamic situation: pests and pathogens can evolve to overcome host resistance and to develop resistance to pesticides and fungicides, as well as adapting to climate change. New attacks must be anticipated.

References

1. der Zaag V (ed) (1994) Potato diseases. NIVAA, Den Haag
2. Bradshaw JE (2021) Potato breeding: theory and practice. Springer, Cham
3. Kroschel J, Mujica N, Okonya J, Alyokhin A (2020) Insect pests affecting potatoes in tropical, subtropical, and temperate regions. In: Campos H, Ortiz O (eds) The potato crop: its agricultural, nutritional and social contribution to humankind. Springer, ebook, pp 251–306
4. Glendinning DR (1983) Potato introductions and breeding up to the early 20th century. New Phytol 94:479–505
5. Knight TA (1810) On potatoes (*Solanum tuberosum*). Trans Hortic Soc Lond 1:187–193

6. Salaman RN (1921) Degeneration of potatoes. In: Report on the international potato conference. Royal Horticultural Society, London, pp 79–91
7. Salaman RN (1926) Potato varieties. Cambridge University Press, Cambridge
8. McIntosh TP (1927) The potato: its history, varieties, culture and diseases. Oliver and Boyd, Edinburgh
9. Beijerinck MW (1898) Over een contagium vivum fluidum als oorzaak van de vlekziekte der tabaksbladen. Versl Gewone Vergad Wis Natuurk Afd K Akad Wet Amsterdam 7:229–235
10. Stanley WM (1935) Isolation of a crystalline protein possessing the properties of tobacco mosaic virus. Science 81:644–645
11. Bawden FC, Pirie NW (1937) The isolation and some properties of liquid crystalline substances from solanaceous plants infected with three strains of tobacco mosaic virus. Proc R Soc Lond Ser B Biol Sci 123:274–320
12. Kausche GA, Pfankuch E, Ruska H (1939) Die sichtbormachung von pflanzlichem virus im ubermikroskop. Naturwissen 27:292–299
13. Markham R (1953) Nucleic acids in virus multiplication. Soc Gen Micro Symp 2:85–98
14. Gierer A, Schramm G (1956) Infectivity of ribonucleic acid from tobacco mosaic virus. Nature 177:702
15. Crick FHC, Watson JD (1956) Structure of small viruses. Nature 177:473–475
16. Fraenkel-Conrat H, Williams RC (1955) Reconstitution of active tobacco mosaic virus from its inactive protein and nucleic acid components. Proc Natl Acad Sci 41:690–698
17. Walkey DGA (1985) Applied plant virology. Heinemann, London
18. Hull R (2013) Plant virology, 5th edn. Academic, London
19. Jeffries C, Lacomme C (2018) Viruses affecting potatoes. In: Wale S (ed) Achieving sustainable cultivation of potatoes. Volume 2: Production, storage and crop protection. Burleigh Dodds Science Publishing, Cambridge, pp 209–241
20. Jeffries C, Barker H, Khurana SMP (2006) Viruses and viroids. In: Gopal J, Khurana SMP (eds) Handbook of potato production, improvement, and postharvest management. Food Products Press, New York, pp 387–448
21. Singh RP (1970) Seed transmission of potato spindle tuber virus in tomato and potato. Am Potato J 47:225–227
22. The Commonwealth Potato Collection Inventory for 1994. Scottish Crop Research Institute, Dundee
23. Fox A (2024) Potato virus. The Vegetable Farmer September 2024 p 30. ACT Publishing, Maidstone
24. Hämäläinen JH, Watanabe KN, Valkonen JPT, Arihara A, Plaisted RL, Pehu E, Miller L, Slack SA (1997) Mapping and marker-assisted selection for a gene for extreme resistance to potato virus Y. Theor Appl Genet 94:192–197
25. Song Y-S, Hepting L, Schweizer G, Hartl L, Wenzel G, Schwarzfischer A (2005) Mapping of extreme resistance to PVY (*Rysto*) on chromosome XII using anther-culture-derived primary dihaploid potato lines. Theor Appl Genet 111:879–887
26. Flis B, Hennig J, Strzelczyk-Zyta D, Gebhardt C, Marczewski W (2005) The *Ry-fsto* gene from *Solanum stoloniferum* for extreme resistance to *Potato virus Y* maps to potato chromosome XII and is diagnosed by PCR marker GP122718 in PVY resistant potato cultivars. Mol Breed 15:95–101
27. Sato M, Nishikawa K, Komura K, Hosaka K (2006) Potato virus Y resistance gene, *Rychc*, mapped to the distal end of potato chromosome 9. Euphytica 149:367–372
28. Solomon-Blackburn RM, Barker H (2001) Breeding virus resistant potatoes (*Solanum tuberosum*): a review of traditional and molecular approaches. Heredity 86:17–35
29. Solomon-Blackburn RM, Barker H (1993) Resistance to potato leafroll luteovirus can be greatly improved by combining two independent types of heritable resistance. Ann Appl Biol 122:329–336
30. Solomon-Blackburn RM, Nikan J, Barker H (2008) Mechanism of strong resistance to Potato *leafroll virus* infection in a clone of potato (*Solanum tuberosum*). Ann Appl Biol 152:339–347
31. Mihovilovich E, Alarcón L, Pérez AL, Alvarado J, Arellano C, Bonierbale M (2007) High levels of heritable resistance to Potato leafroll virus (PLRV) in *Solanum tuberosum* subsp. *andigena*. Crop Sci 47:1091–1103. https://doi.org/10.2135/cropsci2006.06.0369
32. Dale MFB, Solomon RM (1988) A glasshouse test to assess the sensitivity of cultivars to tobacco rattle virus. Ann Appl Biol 112:225–229
33. Valkonen JPT (2015) Elucidation of virus-host interactions to enhance resistance breeding for control of virus diseases in potato. Breed Sci 65:69–76. https://doi.org/10.1270/jsbbs.65.69
34. Forbes GA, Charkowski A, Andrade-Piedra J, Parker ML, Schulte-Geldermann E (2020) Potato seed systems. In: Campos H, Ortiz O (eds) The potato crop: its agricultural, nutritional and social contribution to humankind. Springer, ebook, pp 431–447
35. SASA (2017) Scottish seed potatoes, 14th edn. SASA © Crown Copyright
36. Sleigh J (2024) £1.5 m potato crop loss. The Scottish Farmer August 3, 2024, front page
37. Singh BP, Venkatasalam EP, Singh RK (2020) Seed potato production in India: then and now. In: Souvenir, global potato conclave 2020, Mahatma Mandir, Gandhinagar, Gujarat, India 28–31 January 2020, pp 270–275
38. Subramanian S (2020) Virus-vectors relationship with special reference to vectors of potato. In: Souvenir, global potato conclave 2020, Mahatma Mandir, Gandhinagar, Gujarat, India 28–31 January 2020, pp 232–235

39. Singh RK, Buckseth T, Sharma AK, Tiwari JK, Chakrabarti SK (2019) Potato seed certification in India. Indian Hortic, November–December 2019:54–58
40. Buckseth T, Singh RK, Sharma AK, Tiwari JK, Chakrabarti SK (2019) Hi-tech seed potato production system. Indian Hortic, November–December 2019:18–21
41. Jin L (2018) Potato production and breeding in China. In: Wang-Pruski G (ed) Achieving sustainable cultivation of potatoes. Volume 1: Breeding improved varieties. Burleigh Dodds, Cambridge, pp 259–270
42. Priegnitz U, Lommen WJM, van der Vlugt RAA, Struik PC (2019) Impact of positive selection on incidence of different viruses during multiple generations of potato seed tubers in Uganda. Potato Res 62:1–30. https://doi.org/10.1007/s11540-018-9394-z
43. Priegnitz U, Lommen WJM, van der Vlugt RAA, Struik PC (2020) Potato yield and yield components as affected by positive selection during several generations of seed multiplication in southwestern Uganda. Potato Res 63:507–543. https://doi.org/10.1007/s11540-020-09455-z
44. Bonierbale MW, Amoros WR, Salas E, de Jong W (2020) Potato breeding. In: Campos H, Ortiz O (eds) The potato crop: its agricultural, nutritional and social contribution to humankind. Springer, ebook, pp 163–218
45. Muthoni J, Shimelis H (2023) An overview of potato production in Africa. In: Caliskan ME, Bakhsh A, Jabran K (eds) Potato production worldwide. Academic, London, pp 435–456
46. Osipov VS, Zeldner AG (2023) Potato production in Russia and Ukraine. In: Caliskan ME, Bakhsh A, Jabran K (eds) Potato production worldwide. Academic, London, pp 341–363
47. Hingrat YL, Quéré B (2023) Seed potato production, certification, and trade. In: Caliskan ME, Bakhsh A, Jabran K (eds) Potato production worldwide. Academic, London, pp 241–260
48. Gottschalk K, Ezekiel R (2006) Storage. In: Gopal J, Khurana SMP (eds) Handbook of potato production, improvement and postharvest management. The Haworth Press, Binghamton, pp 489–522
49. Johansen TJ, Molmann JAB (2018) Seed potato performance after storage in light at elevated temperatures. Potato Res 61:133–145
50. Struik PC (2018) Understanding ageing processes in seed potatoes. In: Wang-Pruski G (ed) Achieving sustainable cultivation of potatoes. Volume 1: Breeding improved varieties. Burleigh Dodds, Cambridge, pp 33–55
51. Lindhout P, Meijer D, Schotte T, Hutten RCB, Visser RGF, van Eck HJ (2011) Towards F1 hybrid seed potato breeding. Potato Res 54:301–312
52. Lindhout P, de Vries M, ter Maat M, Ying S, Viquez-Zamora M, van Heusden S (2018) Hybrid potato breeding for improved varieties. In: Wang-Pruski G (ed) Achieving sustainable cultivation of potatoes. Volume 1: Breeding improved varieties. Burleigh Dodds, Cambridge, pp 99–122
53. Lindhout P (2020) Diploid breeding: a paradigm shift in potato. In: Souvenir, global potato conclave 2020, Mahatma Mandir, Gandhinagar, Gujarat, India, 28–31 January 2020, pp 146–148
54. Gopal J, Ortiz R (2006) True potato seed. In: Gopal J, Khurana SMP (eds) Handbook of potato production, improvement and postharvest management. The Haworth Press, Binghamton, New York, USA, pp 557–585
55. Bradshaw JE (2022) Diploid F_1 hybrid potatoes for propagation from botanical seed (TPS): comparisons with theory and other crops. Plants 11:1121. https://doi.org/10.3390/plants11091121
56. Gu J, Evers JB, Driever SM, Shan K, Struik PC (2024) Branching response to stem density and its impact on yield in hybrid potato grown from true seeds and seedling tubers. Field Crops Res 317:109548
57. van Dijk LCM, Lommen WJM, de Vries ME, Kacheyo OC, Struik PC (2021) Hilling of transplanted seedlings from novel hybrid true potato seeds does not enhance tuber yield but can affect tuber size distribution. Potato Res 64:353–374
58. Stockem J, de Vries M, van Nieuwenhuizen E, Lindhout P, Struik PC (2020) Contribution and stability of yield components of diploid hybrid potatoes. Potato Res 63:345–366
59. De Vries M, ter Maat M, Lindhout P (2016) The potential of hybrid potato for East-Africa. Open Agric 1:151–156. https://doi.org/10.1515/opag-2016-0020
60. Obidiegwu JE, Flath K, Gebhardt C (2014) Managing potato wart: a review of present research status and future perspective. Theor Appl Genet 127(4):763–780. https://doi.org/10.1007/s00122-014-2268-0
61. Simmonds NW (1969) Prospects of potato improvement. Scottish Plant Breeding Station Forty-Eighth Annual Report 1968–69, pp. 18–38
62. Hehl R, Faurie E, Hesselbach J, Salamani F, Whitham S, Baker B, Gebhardt C (1999) TMV resistance gene *N* homologues are linked to *Synchytrium endobioticum* resistance in potato. Theor Appl Genet 98:379–386
63. Plich J, Przetakiewicz J, Śliwka J, Flis B, Wasilewicz-Flis I, Zimnoch-Guzowska E (2018) Novel gene *Sen2* conferring broad-spectrum resistance to *Synchytrium endobioticum* mapped to potato chromosome XI. Theor Appl Genet 131(11):2321–2331
64. Marhadour S, Prodhomme C (2023) Recent trends in genetics studies and molecular breeding of potato. In: Caliskan ME, Bakhsh A, Jabran K (eds) Potato production worldwide. Academic, London, pp 273–301
65. Davie K, Pickup J (2018) Nematode pests of potatoes. In: Wale S (ed) Achieving sustainable cultivation of potatoes. Volume 2: Production, storage and

crop protection. Burleigh Dodds Science, Cambridge, pp 263–283
66. Perry RN, Moens M, Jones JT (2018) Cyst nematodes. CABI, Wallingford
67. Lainsbury MA (ed) (2023) The UK pesticide guide 2023. BCPC, Cambridge
68. Seenivasan N (2020) Management of potato cyst nematodes using liquid bio-formulations of *Pseudomonas fuorescens*, *Purpureocillium lilacinum* and *Trichoderma viride*. Potato Res 63(4):479–496
69. Ellenby C (1952) Resistance to the potato root eelworm, *Heterodera rostochiensis* Wollenweber. Nature 170:1016
70. Toxopeus HJ, Huijsman CA (1953) Breeding for resistance to potato root eelworm. Euphytica 2:180–186
71. Dale MFB, Phillips MS, Ayres RM, Hancock M, Holliday M, Mackay GR, Tones SJ (1988) The assessment of the tolerance of partially resistant potato clones to damage by the potato cyst nematode *Globodera pallida* at different sites and in different years. Ann Appl Biol 113:79–88
72. Phillips MS, Forrest JMS, Wilson LA (1980) Screening for resistance to potato cyst nematodes using closed containers. Ann Appl Biol 96:317–322. https://doi.org/10.1111/j.1744-7348.1980.tb04782.x
73. Minnis ST, Haydock PPJ, Ibrahim SK, Grove IG (2002) Potato cyst nematodes in England and Wales-occurrence and distribution. Ann Appl Biol 140:187–195
74. Ellenby C (1948) Resistance to the potato root eelworm. Nature 162:704
75. NIAB TAG (2019) Potatoes 2019. NIAB, Cambridge
76. The Vegetable Farmer November 2022 p 27. ACT Publishing, Maidstone
77. The Vegetable Farmer February 2023 p 26. ACT Publishing, Maidstone
78. The Vegetable Farmer August 2023 p 21. ACT Publishing, Maidstone
79. The Vegetable Farmer September 2023 p 27. ACT Publishing, Maidstone
80. Fournet S, Eoche-Bosy D, Kerlan M-C, Grenier E, Montarry J (2018) Phenotypic and genomic modifications associated with *Globodera pallida* adaptation to potato resistances. Potato Res 61:65–71
81. Spychalla P, De Jong WS (2024) Breeding for potato cyst nematode resistance in *Solanum tuberosum*. Crop Sci 64:1167–1182. https://doi.org/10.1002/csc2.21244
82. The Vegetable Farmer February 2024 p 10. ACT Publishing, Maidstone
83. Van der Zaag DE (ed) (1996) Potato diseases. The NIVAA, Holland
84. BPC (2001) BPC handbook—pests & diseases. British Potato Council, Oxford
85. The Vegetable Farmer August 2023 p 6. ACT Publishing, Maidstone
86. Cingel A, Savić J, Lazarević J, Ćosić T, Raspor M, Smigocki A (2016) Extraordinary adaptive plasticity of Colorado potato beetle: "ten-striped spearman" in the era of biotechnological warfare. Int J Mol Sci 17(9):1538. https://doi.org/10.3390/ijms17091538
87. Maharijaya A, Vosman B (2015) Managing the Colorado potato beetle; the need for resistance breeding. Euphytica 204(3):487–501
88. Strydom RF, Wilson CR, Tegg RS, Balendres MA, van der Waals JE (2024) Advances in *Spongospora subterranae*: current knowledge, management strategies and research gaps. Potato Res 67:1497–1537
89. The Vegetable Farmer Supplier Profiles 2024: page S8. ACT Publishing, Maidstone
90. Fuglie KO (2007) Priorities for potato research in developing countries: results of a survey. Am J Potato Res 84(353–365):353
91. Guchi E (2015) Disease management practice on potato (*Solanum tuberosum* L.) in Ethiopia. World J Agric Res 3(1):34–42
92. Jansky S (2000) Breeding for disease resistance in potato. Plant Breed Rev 19:69–155
93. Zimnoch-Guzowska E, Lojkowska E, Perombelon M (2005) Resistance to bacterial pathogens. In: Razdan MK, Mattoo AK (eds) Genetic improvement of solanaceous crops. Volume I: Potato. Science Publishers, Enfield, pp 339–395
94. Sjolund MJ, Kelly R, Saddler GS, Kenyon DM (2018) Bacterial diseases affecting potatoes. In: Wale S (ed) Achieving sustainable cultivation of potatoes. Volume 2: Production, storage and crop protection. Burleigh Dodds Science Publishing, Cambridge, pp 189–208

7 Farming, Potential Yields and Increased Production

Since the 1960s, the world has seen big regional changes in the areas devoted to potato cultivation and in production. Nevertheless, yields per hectare have increased in all regions to give a global average of 21.41 t/ha for the period 2016–2020. Increased production in the future is likely to come from further yield increases by narrowing the gaps between actual and potential yields, with a current global average ratio of around 50% for actual divided by potential yields. Although modern industrial farming has given relatively large short-term gains in yields to over 40 t/ha in a number of countries, its sustainability is being questioned together with its impact on the environment, biodiversity and climate change. There is increasing debate about the importance of regenerative (restorative) agriculture, with its emphasis on crop rotations to maintain soil health, including the more restrictive requirements of organic farming. It is difficult and costly, but necessary, to do research to compare farming systems that provides important answers that are both statistically significant and biologically meaningful. Longer-term research may also result in new potato cultivars with increased resource use efficiency through the better capture and utilization of light, carbon dioxide, water and nutrients.

7.1 Production Statistics

Potatoes were confined to South America before 1562 [1], yet 404 years later, their cultivation was global, with an area of 21.5 million hectares, of which just under 1.0 million hectares was still in South America, a truly remarkable expansion. Table 7.1 shows potato production in 1966 and 50 years later in 2016 for different regions of the world, as well as for a few individual countries [2]. Although the global figures show an increase in production over the fifty-year period, as a result of an increase in yield per hectare, they are misleading. There have been increases in yield per hectare in all regions and countries, but also big changes in production and areas. Production in Europe declined from 224 to 109 million tonnes as the area fell from 16.24 to 4.88 million hectares. Poland mirrored these dramatic changes, whereas there was a slight drop in production in the United Kingdom, despite the area halving, and an increase in production in the Netherlands, where there was a slight increase in area. In complete contrast, Asia saw a massive increase in production from 30 to 175 million tonnes, with the area increasing from 3.16 to 9.18 million hectares, as dramatic changes took place

J. E. Bradshaw, *Can Potatoes Feed the World?*, Sustainable Development Goals Series,
https://doi.org/10.1007/978-3-031-92890-1_7

Table 7.1 Potato production in 1966 and 2016 (http://faostat.fao.org)

Region	Production million tonnes (t)		Area million hectares (ha)		Yield t/ha	
	1966	2016	1966	2016	1966	2016
Africa	2.29	23.05	0.27	1.66	8.35	13.87
North America	16.43	25.77	0.72	0.56	22.75	46.20
United States	13.94	20.43	0.59	0.42	23.54	48.64
Central America	0.41	2.54	0.05	0.10	8.08	26.70
South America	7.05	17.91	0.97	0.99	7.27	18.09
Asia	30.30	174.60	3.16	9.18	9.58	19.03
China	18.02	84.99	2.00	4.81	9.00	17.69
India	4.08	43.42	0.48	2.12	8.51	20.51
Europe	224.41	109.12	16.24	4.88	13.82	22.35
Netherlands	4.12	6.53	0.13	0.16	31.56	42.00
Poland	45.80	8.87	2.71	0.31	16.89	26.63
United Kingdom	6.90	5.40	0.28	0.14	24.37	38.81
Oceania[a]	0.93	1.61	0.05	0.04	17.72	42.35
World	281.96	354.83	21.49	17.41	13.12	20.38

[a]Australia and New Zealand

in China and India. By 2016, total production in Africa was similar to North America, but yield per hectare was low in Africa (13.87) compared with North America (46.20).

So, what is going to happen in the future? If potatoes are going to help feed the growing world population, there will need to be an increase in production, particularly in Africa, Asia and Latin America. So, will this be achieved by bringing more land into cultivation, or using more of the currently cultivated land for potatoes rather than other crops, or increasing production per unit of land, or by all of these means? If additional good agricultural land is used for potatoes, both production and yields will increase, whereas if marginal land is used, production will increase but yields per hectare will fall.

It is instructive to look at the Fenlands of Eastern England, an area north of Cambridge where I was born and brought up. Fens are a type of peatland fed by mineral-rich groundwater and river water as well as rainfall. Peat results from the partial decay of plant material under waterlogged conditions. In England and Wales, it is defined as any soil with an organic matter content exceeding 20% and a depth of 40 cm or more [3]. The story of the Fenlands epitomizes the history and challenges of increasing potato production. Every country into which potatoes have been introduced and production increased, either by bringing more land into cultivation or by increasing yield per hectare, will have its "Fenlands Story".

7.2 Bringing Land into Cultivation

Today Ely Cathedral towers over flat, rich farmland, which was still wet fens when its Norman crossing-tower collapsed in 1322 (Fig. 7.1). It was rebuilt from 1322 to 1328, larger than before, with an octagonal shape, which between 1328 and 1342, was crowned by an octagonal lantern of timber, the wonder of the building we see today [4]. Attempts at reclamation of the land by the Romans, and then by monastic communities, had only limited success because it was not sufficient to cut water channels in the peat and to build dykes.

The "Great Drainage" of the Fens was started in 1626 by King Charles I, who employed the skills of the great Dutch engineer Cornelius Vermuyden. The first area to be drained was the "Great Level" between Ely and Peterborough, to the northwest. We can get a reminder of the for-

Fig. 7.1 Ely Cathedral (2 February 2016) and rich farmland to the west (14 July 2014). (From Bradshaw JE (2016) Plant Breeding: Past, Present and Future. Springer, Cham with permission)

Fig. 7.2 Windmill for pumping water (Wicken Fen) and sluice for regulating flow of water (8 July 2013). (From Bradshaw JE (2016) Plant Breeding: Past, Present and Future. Springer, Cham with permission)

mer landscape and what happened to it by visiting the nature reserve at Wicken Fen, close to Ely (Fig. 7.2). Water had to be pumped from the low levels into higher drainage channels that flowed northeastwards to the sea. Two massive parallel channels, 22 miles long, were dug from the village of Earith in Cambridgeshire to Denver in Norfolk, and from there to the Wash (North Sea). The first channel (Old Bedford River) was completed in 1637 and the second (New Bedford River) in 1651. Initially the pumping power came from windmills, later from steam, oil and electricity. For example, the drainage of Whittlesey Mere near Peterborough had to await the invention of Appold's Centrifugal Steam Pump in 1851. Sluices, such as the one at Denver, were also needed to control the flow of rainwater from the higher ground in the centre of England to the Wash and also to prevent inundations from the sea, such as the floods of 1953.

The drainage was deeply resented by the local Fennish people who saw their way of life being destroyed. Initially they supported the sympathetic Oliver Cromwell and his Parliamentarians in the English Civil Wars (1642–1651) that resulted in the King being tried and executed in 1649 and a period of 11 years without a monarch. However, in 1646, the Earl of Bedford and other landowners wanted new laws to facilitate the draining of the Great level and, as a consequence, supported Cromwell and got their way when, in May 1649, the Act for the Draining of the Great Level was passed. Furthermore, parliamentary soldiers were deployed to oversee and protect the necessary labour provided by prisoners of war and

Fig. 7.3 Welney Wetland Centre in Norfolk, England, 29 January 2016

immigrants [5]. The draining of the Great Level was effectively complete by 1653. When the monarchy (King Charles II) was restored in 1660, it was by "Consent of Parliament" and not by "Divine Right".

Eventually, by about 1900, 7000 km^2 of wet fens in Eastern England had been brought into cultivation in the form of privately owned and managed fields by drainage, clearance and enclosure [6]. A consequence of this "Agricultural Improvement" was the loss of fens rich in fish, fowl and deer; the loss of peat for fuel and sedge for thatching and kindling; and the loss of the livelihoods of local people who were driven into cities. Hence, whilst some people view the drainage of the fens as a technological triumph, others see it as an ecological disaster [6]. I think that it was a technological triumph, but I realize that we need wetlands to hold water to prevent flooding in winter and to provide irrigation in summer. In other words, we need to align modern intensive farming with more sustainable flood defences, an adequate water supply, nature conservation and a vibrant local economy. Welney Wetland Centre in Norfolk, for example, is a water holding area (Ouse washes) between the Old and New Bedford rivers, created by Vermuyden as part of his Great Drainage of 1626–1653 (Fig. 7.3). In modern terminology, it has provided an ecosystem service for 400 years!

It can be seen from the Ramsar Convention on Wetlands 50th Anniversary Special Report of 2021 that globally, the different types of wetlands (around 5% of Earth's surface) provide a range of ecosystem services. As well as essential services such as the provision of adequate quantities of good quality water, they can impact food security and climate change mitigation [7]. Likewise, globally it is recognized that other kinds of habitats, such as rain forests and grasslands, are equally important in maintaining the world's major ecosystems. Under these circumstances I think that there will be political pressure to increase food production per unit area of land rather than by bringing more land into cultivation. So, let us continue the story of the Fens with improvements in farming practice, including potato cultivation.

7.3 Improvements in Farming Practice

Welney is in Norfolk, a county associated with improvements in farming practice, in particular the Norfolk four-course rotation of crops. It was during the eighteenth century that Charles Townsend (2nd Viscount Townsend) (1674–1738) of Raynham Hall and Thomas Coke (1st Earl of Leicester) (1697–1759) of Holkham Hall, both in Norfolk, convinced British farmers of the value of a rotation of wheat, turnips, barley and clover [8]. The turnips were drilled in rows wide enough apart for weeds to be removed by a horse-hoe, both the drill and hoe being early eighteenth-century innovations of Jethro Tull (1674–1741). The turnips were either grazed in situ by folded (enclosed) sheep or lifted to fatten cattle in winter fold yards, something that had not previously been possible. The clover was a nitrogen-fixing forage crop grown on land previously left fallow. Furthermore, the clover left nitrogen in the soil for the subsequent cereal crops. Both the turnips and clover allowed livestock numbers to be increased, with a resultant increase in manure production. These improvements in agriculture provided increased quantities of food for an expanding population in Britain and, more generally, throughout Europe during the nineteenth century. The twentieth century saw the advent of new farming methods and new rotations, whereas the twenty-first century has seen debates over the sustainability of such farming practices. In 2018, Thomas Coke, the 8th Earl of Leicester, made Jake Fiennes the Director of Conservation for the Holkham Estate on the North Norfolk Coast, Holkham Hall being the Cokes' family home built between 1734 and 1764 by the 1st Earl of Leicester. In his 2022 book *Land Healer* [9], Jake Fiennes puts forward practical solutions to producing food without destroying nature, based on his work at Holkham to bring back wetlands, hedgerows, birds and butterflies to productive farms; but let us return to the Fens.

The potato was introduced into the Fens of Cambridgeshire and Lincolnshire in 1758 and was being widely grown by 1811. A history can be found in Sly's 2010 book *Soil in Their Souls* [10]. Newly drained fens and marshes were free of soil-borne diseases with silt and peat soils of excellent structure. The potato proved a good crop in the rotation, followed by wheat. Furthermore, the low rainfall of about 560 mm per year meant that crops were less prone to late blight than those in the wetter parts of the country. According to Sly, the potato created more wealth than any other crop grown in the Fens, despite being labour intensive, and in 2010, the Fens accounted for 24% of the English crop. Potato markets throughout Britain opened up with the advent of the railways, starting with the Peterborough to London line in 1863. By the 1870s, almost every small town in the Fens was part of the national rail network. The two world wars, particularly the second, saw an increase in the potato area as old pasture was turned into arable land. The crop was still largely planted and harvested by hand, with labour provided by the Women's Land Army and Prisoners of War. The nation had to be fed at any cost and rotations became short in the Fens, with potatoes commonly grown every third year (e.g., potatoes, sugar beet and wheat; in contrast, the 7-year rotation at Great Farm, Holkham, in 2018 was potatoes, winter wheat, sugar beet, winter barley, parsnips, maize and sugar beet [9]). The soil was farmed to its limits, and almost all farms grew potatoes. As a consequence, cyst nematodes built up in the soil and were a problem in the Fens by the 1950s. The demand for more food from British agriculture continued into the 1960s, the start of the era of intensive petrochemically driven arable farming, an aspect of modern industrial farming.

Today the Fenlands of Eastern England (Fig. 7.4) account for around half of the most productive agricultural land in England, but the drainage resulted in the physical shrinkage of the peat soils from water loss, wind and water erosion, and new challenges for lowland peat farming. These are discussed in Issue 53 of Landmark, *Meeting the challenges of lowland peat farming*, published in Autumn 2023 by NIAB in Cambridge, England [11]. High greenhouse gas (GHG) emissions occur in drained peatlands from the decomposition of the organic soils, which release stored

Fig. 7.4 Potato crop between Peterborough and Ely, Eastern England, 14 July 2014

carbon dioxide (CO_2) into the atmosphere and contribute to global warming. These drained lowland peatlands account for about 3% of the total GHG emissions from all sources in England. One solution to this problem is to rewet lowland peatlands by raising the water table (not flooding) so that the carbon in the soil is not exposed to oxidation, but this will affect land usage and water management and the crops that can be grown (if soils become too wet, they start to emit methane, CH_4, another greenhouse gas). So, complete rewetting may not be a sustainable solution, but partial rewetting would mean these areas are unlikely to maintain the current level of crop production, which will need to be increased elsewhere on mineral soils to maintain food security [3, 12]. As a consequence, there may be a decrease in UK potato yields, which had been steadily rising until 2008 and then plateaued. Put another way, whereas past changes in the farm and food systems resulted in higher yields, current changes may result in similar or slightly lower yields but sustainable over a longer period of time and with lower greenhouse gas emissions.

7.4 UK Potato Yields

Potato yields in Great Britain increased from 14.8 t/ha in the period 1885–1889, via 17.1 t/ha in the period 1935–1939 [13], to 22.9 t/ha in the period 1961–1965 (Fig. 7.5). Then there was a faster rate of increase to 42.4 t/ha for 2006–2010, before a drop back to 38.6 t/ha for 2016–2020 [2]. One could argue that yields reached a plateau of 40.6 t/ha for the period 1991–2020. Interestingly, we saw in Table 7.1 that the area of potatoes halved from 1966 to 2016.

A number of factors have been involved in these yield increases [10]. Modern chemical fertilizers were already available by 1900 but not widely used until the 1960s. Healthy seed tubers were guaranteed from the 1920s through cultivar registration and seed certification. Modern chemical fungicides, herbicides and insecticides became available from the 1950s. Although copper compounds for the control of late blight had been available since 1885, it was the period from 1960 to 1970 that saw the rapid development of the crop protection industry. The first

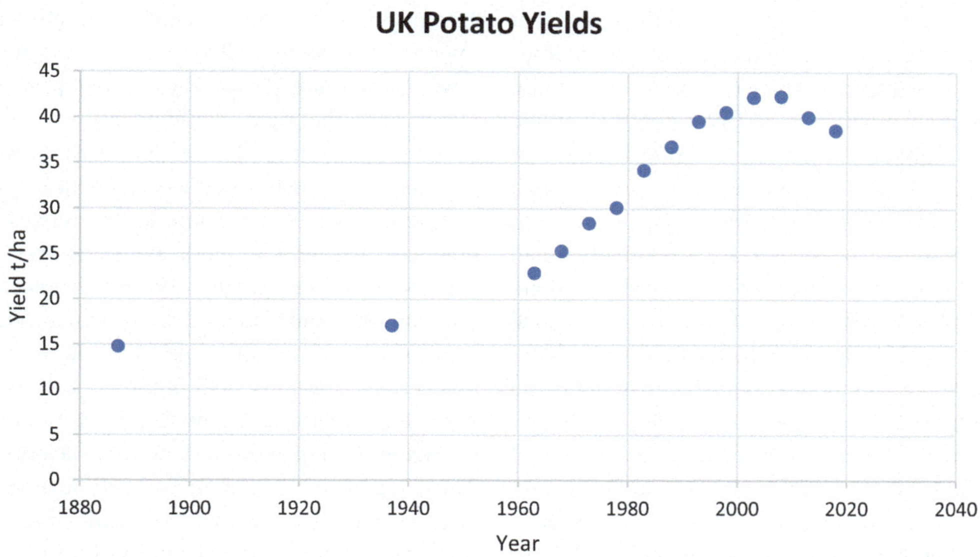

Fig. 7.5 UK potato yields in t/ha from 1887 to 2020, where data are 5-year averages [2, 13]

systemic fungicide with action against *P. infestans* was cymoxanil in 1976, followed in 1977 by metalaxyl, which is considered the first effective curative of late blight. Mechanization, such as tractors, planters, harvesters and graders, and larger fields, had been trends since the 1960s, which either directly or indirectly contributed to the end of potato growing on heavy, difficult-to-work soils. Post-harvest losses have been reduced by the use of controlled environment stores since the 1960s. The 1960s also saw the birth of road haulage for the transport of potatoes and the demise of the railways in many rural areas. Extensive irrigation, compared with other crops, has been used since the 1980s to ensure high yields of high-quality potatoes. The industrial production of crisps since the 1920s and French fries since the 1950s encouraged modern farming practices together with cultivars suitable for processing. The English company Walkers Crisps was founded in Leicester in 1948 and opened the world's largest crisp-producing factory in the city in 1982. The company was bought by PepsiCo in 1989. The Canadian company McCain Foods Limited was founded in 1957 and opened a French fry factory in Scarborough, Yorkshire, in 1968 (they now also have one in Whittlesey in the Fens). Finally, since the 1980s, supermarkets have dominated the retail market. Today it can be argued that a few processing companies and a few supermarkets control most of the supply chain, from breeding new cultivars through the production of seed tubers and ware potatoes for packing or processing to the supermarket shelves.

Globally, modern industrial farming has increased potato yields to over 40 t/ha in countries like Britain with a temperate climate and a long summer growing season, whereas the global average for the period 2016–2020 was 21.41 t/ha. In other words, this kind of farming has given relatively large short-term gains in production, but its sustainability has been questioned, as seen, for example, in the National Food Strategy for England, which was proposed in July 2021 (nationalfoodstrategy.org). So, what is the way forward in England and globally?

7.5 Organic Farming and Regenerative Agriculture

Since the foundation of the Soil Association in 1946, there has been an interest in organic farming in the United Kingdom. According to the UK Government website (GOV.UK), this is "A method of farming that requires farmers to operate to a system based on ecological principles, which impose strict limitations on the inputs that can be used to minimise damage to the environment and wildlife. Emphasis is placed on natural methods of production and pest control". In the United Kingdom in 2023, 498,000 hectares of land (2.8% of total farmed area) were farmed organically, of which 10,300 was used for vegetables, including potatoes (i.e., a relatively small area of potatoes). Hence, in the United Kingdom, there is a small group of enthusiastic potato growers and consumers who regard organic farming as better and more sustainable than industrial farming, but they account for only a small proportion of total production and usage. Hence, it may not be possible to economically justify breeding cultivars specifically for organic farming, rather than finding ones such as Lady Balfour that were bred for conventional farming but yield well under organic growing conditions. In other words, they can cope with competition from weeds and the nutrient availability from the soil, particularly nitrogen [14]. This raises the question of whether or not organic farming systems in general can produce the increased yields, which we think will be required globally for our major food crops by 2050. In 2016, Reganold and Wachter concluded that organic farming currently results in lower yields (8–25%) than conventional agriculture but is more environmentally friendly, provides equally or more nutritious food with less or no pesticide residues, and can be more profitable [15]. These conclusions are in line with what appears to be a general consensus on the topic. I will therefore give just one recent potato example from Brazil [16]. A set of 14 advanced clones and 3 control cultivars was evaluated on a farm in Brasília (15°55‘S, 48°08‘W, 999 masl) in the winter season of 2020, under a conventional and an organic farming system. The winter season was June to September, with 112 days from planting to harvest. Total and marketable (>45 mm diameter) yields were higher under the conventional system at 40.17 versus 37.00 t/ha and 30.10 versus 25.87 t/ha, respectively.

It is not yet clear which farming systems will prove most appropriate for food production in 2050, a mere 25 years away. There is, however, currently in the United Kingdom and elsewhere interest in what is called regenerative (restorative) agriculture, based on guiding principles rather than the rigid requirements of organic farming [17]. One UK company that provides agronomy advice has defined regenerative agriculture as "A system of farming principles and practices that aims to work alongside nature to increase biodiversity, improve soils, protect the environment, and enhance ecosystem services" [https://www.agrovista.co.uk]. The claim is that regenerative farming will improve soil health with time, whereas sustainable farming simply maintains it, and industrial farming destroys it. Thus, regenerative agriculture will reverse the errors created by previous nonsustainable practices. The emphasis in regenerative farming is on reduced or zero tillage (i.e., not disturbing the soil), keeping the soil covered with cash and cover crop residues, using cover crops to keep living roots in the soil between cash crops, increasing plant diversity through longer and more diverse crop rotations and multi-species cover crops, and integrating livestock into the system (use of animals for grazing and for fertilizing soil). It is recognized that with an increasing global population, regenerative farming techniques must not reduce yields and ideally should increase them, which is important for potatoes. Hence, in 2023, in the United Kingdom, a consortium of commercial and academic organizations launched a 3-year project, "Transformative Reduced Inputs in Potatoes (TRIP)", to investigate if potatoes can be grown regeneratively on a commercial basis. It is too soon for measurable outcomes that will allow fair judgements to be made on the claims of the advocates of regenerative farming. Indeed, it is difficult and costly, but necessary, to do research into farming systems that provides important answers that are both statistically significant and biologically meaningful.

Two global players in the potato processing industry have set ambitious global targets for regenerative farming. PepsiCo, the owner of Walkers Crisps in the United Kingdom, has announced a 2030 goal for regenerative farming practices across the 2.8 million hectares they farm for a range of crops, including potatoes [https://www.pepsico.co.uk/sustainability/sustainable-food-system/agriculture]. McCain, a producer of French fries, with factories in the United Kingdom, says that they are also committed to implementing regenerative agricultural practices across 100% of their potato acreage worldwide by the end of 2030 [https://www.mccain.com/media/4036/mccain-foods-regenag-framework.pdf]. In their regenerative agriculture framework, McCain defines regenerative agriculture "As an ecosystem-based approach to farming that aims to improve farmer resilience, yield, and quality by improving soil health, enhancing biodiversity, and reducing the impact of synthetic inputs".

Interestingly, in their regenerative agriculture framework, they give slightly different versions of a regenerative potato grower for the northern and southern hemispheres to reflect the differences in climate and soil types. These frameworks are seen as a guide for growers in progressing through four stages (Onboarding, Beginner, Master and Expert) towards a more regenerative model. The growers will receive financial rewards for their participation. The final stage of the journey will see "increased soil protection, minimized tillage, increase in soil organic matter and 30% of the crop nutrient needs provided from organic sources (plants/manure)". Some of the changes in farming practice will apply specifically to the potato crop; others to the farm and rotation cycle as a whole.

There is interest in zero tillage and paddy straw mulching in the coastal saline soils of the Ganges Delta [18]. Long-duration rice is grown during the wet kharif season, but the land is often left fallow during the dry rabi season because of soil salinity and lack of good quality irrigation water. However, researchers showed in two years of farm trials that there is scope for potato cultivation using a combination of zero tillage for early planting and paddy straw mulching to conserve residual soil moisture from the kharif season and to reduce the development of soil salinity.

Endorsement of the COP28 UAE Declaration on Sustainable Agriculture, Resilient Food Systems and Climate Action included an Action Agenda on Regenerative Landscapes. "This will see leading food and agriculture organizations join forces to scale regenerative agriculture, transitioning 160 million hectares to regenerative agriculture by 2030, accompanied by USD$2.2 billion in future investment, and engaging 3.6 million farmers world-wide." [https://www.cop28.com/en/news/2023/12/COP28-UAE-Presidency-puts-food-systems-transformation]. Time will tell what impact these changes have on global potato yields and the extent to which they are adopted in various countries. In the meantime, the thoughts of some English farmers about the appeal of regenerative farming, the need to make a living from farming, and what it means to be a farmer in today's England, can be found in Sarah Langford's 2022 book *Rooted: How Regenerative Farming Can Change the World* [19].

7.6 Weed Control

Whatever farming systems are adopted in the future, weed control will be an issue. An example from Egypt reveals the yield losses that can result from competition with weeds and also three methods of control: physical removal, use of chemical herbicides and suppression with mulches [20]. The investigators found that the fresh-weight yield of cultivar Spunta increased from 17.9 t/ha in unweeded plots to 37.6 t/ha in ones where weeds had been removed by hoeing twice (average of autumn trials in 2015/16 and 2016/17). Although application of the herbicide metribuzin achieved better control than hoeing (98% efficacy compared with 83%), yield was slightly lower at 33.8 t/ha. Interestingly, slightly higher yields were achieved with mulches: 38.3 t/ha with biodegradable ones (94% efficacy), 38.1 t/ha with infra-red transmitting ones (88%), 40.9 t/ha with black plastic (98%) and 40.5 t/ha with blue plastic (84%). The investigators concluded that black and blue plastic mulches were economically viable alternatives to hoeing and synthetic herbicides, especially for organic farms and newly reclaimed areas of land. Control of weeds is also important because they can be hosts to pests and diseases of potatoes.

7.7 New Cultivars and Improved Husbandry

I now want to ask if any of the yield increase in the United Kingdom and elsewhere was due to new, higher-yielding cultivars or was it all due to improved husbandry? The first point to make is that there is surprisingly very little data available to determine the relative importance of new cultivars and better husbandry. My conclusions for the United Kingdom are based primarily on two papers published by Norman Simmonds after he moved from the Scottish Plant Breeding Station to the Edinburgh School of Agriculture [21, 22]. He compared the years 1964 and 1976 using UK potato yields drawn from agriculture rather than trials. In 1964, King Edward (released in 1902) and Majestic (1911) were still two of the most important cultivars, having seen the United Kingdom through two world wars. However, in 1976 they had largely been replaced by three new cultivars, Pentland Crown (1958), Maris Piper (1966) and Desiree (1961). Simmonds concluded that 4.0 t/ha of the 8.9 t/ha increase in yield came from better husbandry and management, 5.5 t/ha from the change in cultivars and that the difference of 0.6 t/ha was a small but negative genotype × environment interaction. Simmonds also concluded that for the 6 years (1971–1976), Pentland Crown out yielded Majestic by 10.1 t/ha. In 1970, Pentland Crown had replaced Majestic as the most widely grown cultivar in Great Britain (GB). As well as high yield, Pentland Crown had good resistance to the viruses PVY and PLRV and to common scab. Maris Piper had resistance to the golden potato cyst nematode (*Globodera rostochiensis*) and soon became the most widely grown cultivar, a position it still held in 2023. Hence, for growers of Maris Piper, newer cultivars haven't contributed to higher yields. However, two other cultivars appeared in the GB top ten in 1970: Pentland Dell at number four and Record at number five. Pentland Dell, released in 1960, proved suitable for French fries as a result of its long oval shape and high dry matter content. Record, released in 1934, suddenly gained importance because it proved suitable for crisps (chips) as a result of its round shape, high dry-matter content and resistance to spraing. In 1990, Record was the number two cultivar and Pentland Dell number seven. Since then, new cultivars have appeared for processing, such as Hermes, Lady Rosetta and Brooke for crisping; Innovator, Markies, King Russet, Elland and Daisy for French fries; and Atlantic and Royal for both crisping and French fries [23]. King Russet (Fig. 7.6) has resistance to potato cyst nematodes (Ro1, Pa2 and Pa3 according to the McCain leaflet). Although high yielding, these processing cultivars appear not to have increased the fresh-weight yield of the GB/UK potato crop. Since 2013, each year, about a third of the UK potato crop has been processed [24].

Interestingly in Japan, yields of the summer crops in Hokkaido decreased from 35 to 32 t/ha over the period 1993–2013 as a result of an

Fig. 7.6 Cultivar King Russet growing in the Vale of Pickering, Yorkshire, on 26 June 2024, and destined for the McCain processing factory at Scarborough

increase in higher dry-matter cultivars for the processing and starch industries [25].

There was another consequence of the change of cultivars in the United Kingdom worth mentioning. Cultivars Maris Piper and Desiree flower profusely and produce very many berries, and cultivars Pentland Dell and Record also produce many berries. Research done in Scotland and published in 1983 demonstrated the potential of these cultivars to add large quantities of true potato seed (TPS) to the soil and create serious arable weed problems in subsequent years [26], including germination in the next potato crop in a 7-year rotation. Furthermore, tubers from TPS seedlings could add to the traditional groundkeeper problem from tubers not lifted from the soil during harvest of a potato crop.

Turning to potato yields, it is instructive to look at the NIAB (National Institute of Agricultural Botany, Cambridge) trials for the period 1969–1978 as they overlap with the analyses of Simmonds. NIAB produced its first Recommended List of Potato Varieties in 1954. Ten years later they undertook the statutory tests and trials of varieties required by the 1964 Plant Varieties and Seeds Act, which established Plant Breeders' Rights in the United Kingdom (i.e., royalty income to breeders for their new varieties), as well as an official index of varieties and statutory performance trials of new varieties before they could be marketed [27]. In 1984 an analysis was published of yield data from the Official UK Variety Testing System for 1969–1978. It included 178 potato trials, in which 36 varieties (cultivars) had been assessed at 20 test centres (locations) over 10 sowing years. As the data were not a complete set (all varieties assessed at all centres in all years), the components of variance (sources of variation as a percentage of total) were estimated using a technique known as residual maximum likelihood (REML) [28, 29]. The mean fresh-weight yield over all trials was 45.2 t/ha, which was higher than farmers were achieving during this period (Fig. 7.5). The macro-environmental variation (centres 6.1%, years 24.6% and centres × years 42.0%) was large compared to the variation between varieties (72.7% compared with 5.4%), with the centres × years interaction accounting for 42% of the variance. This confirms what we observe in Britain with our weather: not only does it vary from year to year and from location to location, but it is also unpredictable from year to year for any given location, as are disease and pest problems. The varieties × environments interaction (varieties × centres 1.7%, varieties × years 1.5%, and varieties × centres × years 8.3%) was also larger than the variation between varieties (11.5% compared with 5.4%). This means that the differences between varieties vary with the macro-environment. The variation between plots within a trial is micro-environmental variation, and this accounted for 10.4% of the variance. These components of variation allow an important calculation to be made. If the official system uses 2 years of trials with 2 replicates at each of 12 centres (which was typical), a new variety needs to be almost 10% higher-yielding than the best control to be declared higher yielding, based on what is called the critical percentage difference; in other words, quite a big difference. These analyses seem to confirm that much higher yielding cultivars were not appearing in official trials during a period when farmers were benefiting from the earlier introductions of Pentland Crown, Maris Piper and Desiree. Interestingly, a study in Angus, Scotland, published in 1985, found that the yields of four crops in a local competition, estimated from test digs, ranged from 64 to 80 t/ha and were 85, 94, 101 and 102% of the estimated potential yields. At the time, United Kingdom average yields were around 35 t/ha. Hence, lack of progress in producing potentially higher-yielding cultivars did not seem to be holding back the best farmers.

My own research in Dundee, Scotland (56.46°N, 2.97°W), in the 1990s revealed the existence of genetic variation for yield but a failure to translate it into higher yielding cultivars. Six high-yielding parents, including Maris Piper and Desiree, were crossed in all combinations to generate genetic variation both between and within the resulting 15 full-sib families. These were assessed in local trials in 1993 and 1994 with encouraging results. Seventeen out of 218 clones outyielded the best of the six parents [30,

31]. Furthermore, if the 17 clones were inter-crossed to produce a new population of clones, and the process repeated for four more generations, the final best 17 clones were predicted to be 40% higher yielding than the best original parent. However, when the first set of 17 clones were assessed as potential cultivars, none was good enough for other traits to be selected as a new cultivar, and they were not inter-crossed to produce a new population. Instead, the breeding concentrated on a new program started in 1991 to combine improvements in yield with better processing quality and disease and pest resistance [32]. Nevertheless, the message is clear: in potatoes, it is difficult to combine higher yields with all of the other characteristics that go to make up a successful cultivar.

In the United States, national potato yields increased from 5.6 t/ha in 1920 through 33.6 t/ha in 1989 [33] to the figure of 48.6 t/ha shown in Table 7.1 for 2016. These increases have been attributed to improved cultural practices, use of fertilizers and irrigation, pest management and a shift in production to the western United States (Idaho, Washington). New cultivars came from public sector breeding efforts, but did they contribute to the yield increases? A much-quoted result is the lack of improvement in yield found when cultivars released over the period 1861–1991 were compared for three years in Michigan in a high-yielding environment (50 t/ha) [34]. This is contrasted with the dramatic and steady increase in maize yields in the United States from 1.88 t/ha in 1927 to 8.78 t/ha in 1995, as open-pollinated cultivars were replaced by hybrids, first double-cross and then single-cross ones. About half of the yield increase in maize was attributed to improvements in agronomy and half to new cultivars [35]. There was, however, significant progress from 1960 to 1998 in developing potato cultivars with good processing quality [36]. Nevertheless, one of the most popular cultivars still being grown for processing was Russet Burbank, which had been released back in 1914. A maximum measured yield of 124 t/ha was reported for this cultivar in 1987 in the Columbia Basin, Washington, United States [37]. This yield was fairly close to the predicted yields of 140 and 160 t/ha based on intercepted radiation and length of growing season, respectively (the former is more accurate). Again, the best farmers did not appear to be held back by a lack of new, higher-yielding cultivars. Indeed, 124 t/ha is nearly 25 t/ha of dry matter!

So, on the basis of very limited data from Europe and North America, a consensus emerged that potato breeding during the twentieth century contributed little to the yield increases seen in potato cultivation. However, this conclusion may not be true for developing countries. In India, for example, half of the increase from 6.6 t/ha in 1949 to 23.97 t/ha in 2018 has been attributed to new, high-yielding cultivars with better resistance or tolerance to various biotic and abiotic stresses [38]. Although I have not seen the experimental evidence for this claim, it is not unreasonable given that the expansion in area, which also occurred, was to a new set of environmental conditions in the plains of India, namely, those of the lowland subtropics in winter. Furthermore, an assessment of the economic impact of CGIAR-related crop technologies on agricultural productivity in developing countries from 1961 to 2020 estimated potato yield increases of around 50% from new cultivars [39].

7.8 Photosynthesis and Mineral Nutrition

It is now time to look at increasing potato yields through yield potential. Potatoes, like all plants, have fairly simple needs, but they must all be met to achieve the maximum possible yield. They require water (H_2O) from the soil and carbon dioxide (CO_2) from the air for photosynthesis in which the sun's light energy is converted into chemical energy. This is used to synthesize sugars and other carbohydrates made of carbon (C), oxygen (O) and hydrogen (H) (i.e., to fix carbon). The CO_2 provides the C and O, and the H_2O provides the H with the release of O into the air. Potatoes also require 14 mineral elements in the form of ions in aqueous solution, which they need to extract from the soil unless being grown hydroponically. These are divided into macro-

and micro-elements, with the former at tissue concentrations greater than 1 mg g^{-1} dry matter (DM). The six macro-elements are nitrogen (N), phosphorus (P), potassium (K), calcium (Ca), magnesium (Mg) and sulphur (S), and the eight micro-elements are chlorine (Cl), boron (B), iron (Fe), zinc (Zn), manganese (Mn), copper (Cu), nickel (Ni) and molybdenum (Mo) [40]. The main driving force used to move water and minerals from the soil to all parts of the plant is the transpiration of water from the plant's leaves. This evaporation has a cooling effect and complements the convection of heat to keep plants cool [41]. The water retained in the plant maintains the turgor of its living cells, which is necessary for their functioning, and prevents wilting of tissues and organs. It is also the medium in which cellular metabolism (chemical reactions) takes place and comprises about 80% by weight of a potato plant with tubers. None of these requirements is obvious from simply looking at growing plants, so it is worth briefly mentioning how and when some of our knowledge was gained.

Our understanding of photosynthesis developed over a period of 300 years. A very readable account can be found in the book by John King, *Reaching for the Sun* [41]. Five key steps were as follows. (1) The Belgian physician Jan Baptista van Helmont (1580–1644) grew a willow sapling in a weighed amount of soil in a pot to which he added only water over a five-year period. He found that the willow's large increase in mass was not to any extent at the expense of the soil and wrongly concluded that virtually all of it came from the water. These results were published in 1648, after his death. (2) Experiments done by the English chemist and clergyman Joseph Priestly (1733–1804) in the 1770s included two with sprigs of mint. When he placed the mint in an air-tight glass vessel with a burning candle, the latter went out but could be re-lit some days later. The initial oxygen in the container was used up by the candle but was then replenished by the plant (the air was restored). More dramatically, he showed that a mouse would die in the closed vessel but could be kept alive by oxygen from the mint. (3) Then in 1779 the Dutch physician Jan Ingenhousz (1730–1799) discovered that plants placed in a transparent container and submerged in water give off bubbles of oxygen from the undersides of their leaves in the presence of sunlight but that this stops in the dark. He also discovered that plants produce carbon dioxide in the dark (i.e., they respire) and realized that more oxygen is given off in the light than carbon dioxide in the dark. Hence, some of the mass of plants comes from the air. (4) The work of Swiss scientist Nicholas Theodore de Saussure (1767–1845), published in 1804, completed the basic overall chemical equation of photosynthesis. He studied the increase in the dry weight of plants grown with their roots in water and their shoots in an atmosphere of air with added carbon dioxide. He found that the increase in dry weight could not be explained solely by the assimilation of available carbon dioxide. The difference had come from the water. He also demonstrated that plants obtain their carbon from the carbon dioxide in the air and not from humus in the soil. He also showed that plants require mineral nutrients, which they take up from the soil. (5) Finally, in work done in Cambridge, England, and published in 1939, the British plant biochemist Robert (Robin) Hill (1899–1991) used isolated leaf chloroplasts (the sites of photosynthesis) to demonstrate that the light reactions of photosynthesis don't involve carbon dioxide. The pigment chlorophyll absorbs light energy, and through electron transfers, water is split to provide energy and hydrogen as a source of reducing power with the release of oxygen. Then carbon dioxide is fixed in the dark reactions of photosynthesis, as subsequently elucidated between 1947 and 1958 by the American biochemist Melvin Calvin (1911–1997), along with Andrew Benson and James Bassham. The dark reactions are known as the Calvin cycle. Plant physiologists have been able to study how the rate and efficiency of photosynthesis are determined by temperature and carbon dioxide concentration as well as being dependent on the light intensity. Furthermore, in 2020 an EU-funded project, Photoboost (www.photoboost.org), started to explore the possibility of improving the photosynthetic performance and productivity of potatoes and rice. It is an interesting project but too early to draw conclusions about its success. These must await the results of extensive field trials. Plant physiologists have also been able to make predictions about potential yields in the absence of other limiting factors such as mineral requirements and abiotic stresses such as heat, drought and salinity.

The scientific study of plant nutrition developed from the seventeenth century onwards. Again, John King's book *Reaching for the Sun* [41] provides a readable account. Progress was made possible by advances in chemistry that allowed analyses of plants, soils and fertilizers. It was also helped by the use of hydroponics from the late nineteenth century, in which plants were grown directly in aqueous solutions of mineral nutrients. We have already met Jan Baptista van Helmont and Nicholas Theodore de Saussure in connection with photosynthesis. Three more commonly mentioned names from the middle of the nineteenth century are the German agricultural scientist Carl Sprengel (1787–1859), the French chemist Jean-Baptiste Boussingault (1801–1887) and the German chemist Justus von Liebig (1803–1873). Sprengel was the first to formulate the "Law of the Minimum", which was popularized from 1840 by von Liebig. This states that plant growth is determined by the scarcest resource available, not by the total resources. In other words, growth is limited by the essential mineral that is in the shortest supply. Boussingault established the first research station for agricultural experiments, in Alsace, France, in 1836. Then, in 1842, in England, John Benett Lawes (1814–1900) discovered and patented a method of producing superphosphate fertilizer from calcium phosphate (phosphate-rock) and sulphuric acid. Today two-thirds of the global production of phosphate rock comes from China, the United States and Morocco, with world reserves predicted to last for another 100 years [42]. In 1842 Lawes also opened the first factory for the manufacture of artificial fertilizers. A year later he hired a partner, the English chemist Joseph Henry Gilbert (1817–1901), who had studied under von Liebig. They initiated a series of long-term field experiments at the Lawes' family estate, Rothamsted Manor in Harpenden, England. The year 1843 is taken as the foundation of the Rothamsted Experimental Station, which continues today as Rothamsted Research, as do some of the long-term field experiments. Over the period 1843–1847, Lawes and Gilbert showed that nitrogen, in addition to that present in manure, was key to increasing crop yields. However, the breakthrough in producing synthetic nitrogen fertilizer came in Germany. In 1909 the German chemist Fritz Haber found a way of efficiently producing ammonia (NH_3) from hydrogen (H_2) and atmospheric nitrogen (N_2) at high temperatures (300–400 °C) and pressures (35 MPa) in the presence of a metal catalyst. Then, in 1913, the German industrial chemist Carl Bosch set up the first commercial plant for the manufacture of ammonia by the Haber-Bosch process. A range of nitrogenous fertilizers can be produced from ammonia. Today there are concerns about the cost and sustainability of using fossil fuels to produce nitrogen fertilizers and the environmental impacts of their production and use. Consequently, there is increased emphasis on crop rotations which include legumes and other plants that can enter a symbiosis with atmospheric nitrogen-fixing bacteria.

7.9 Fertilizer Requirements

Some of the late 19th and early twentieth century field experiments on fertilizer application were done at the Rothamsted Experimental Station. One was on the manurial response of different potato cultivars. Yields were recorded for three replicates of all combinations of 12 cultivars (varieties) and 6 manurial treatments, which included potassium fertilizers. The data were analysed and published by Fisher and Mackenzie in 1923 [43]. The yields of the 12 varieties, averaged over the manurial treatments, ranged from 4.02 to 9.75 kg/plot, and those of the 6 manurial treatments, averaged over the varieties, ranged from 2.03 to 9.13 kg/plot. Through a novel analysis of variance, the authors were able to demonstrate that the differences between varieties and the differences between manurial treatments were statistically significant but that the varieties by manurial treatments interactions could have arisen by chance. Potatoes went on to feature in Fisher's famous 1935 book *The Design of Experiments* [44]; for example, he provides potato data from an experiment at Ely (Fig. 7.1) in 1932 as an exercise for readers on the appropriate analysis of variance for a 6 × 6 Latin Square design in which the six treatments were different quantities of nitrogenous and phosphatic fertilizers. Properly randomized and replicated field experiments allow us to partition the variation in yield (and other traits) into differences between cultivars (nature), fertilizers (nurture) and their interaction. In the absence of an interaction, we can say how much of the variation is due to nature and how much is due to nurture; but the question becomes meaningless if we ask about how a high-yielding cultivar, through growth and development, achieves its yield when well-fertilized.

Since the pioneering work at Rothamsted Experimental Station, yield response curves for the application of fertilizers to potatoes have been determined and recommendations made to farmers. In practice, pre-season soil analyses would be done to determine how much of each mineral is already present in the soil (strictly speaking, we want to know how much is available for uptake by the potato plants). A simple quadratic response curve has commonly been used to explore nitrogen fertilizer rates:

$$y = a + bx - cx^2$$

where y is the yield response in tonnes (t) per hectare (ha), x is the nitrogen applied in kilograms (kg) per hectare and the coefficients a, b and c are constants for any one genotype. Hence:

$$y_{ij} = a_i + b_i x_j - c_i x^2{}_j$$

where y_{ij} is the yield of genotype i at nitrogen level j.

In Fig. 7.7, I have created an example of nitrogen response curves for two cultivars. I have assumed that without the added nitrogen fertilizer, there would be no yield so that constant a is zero. In practice there would be nitrogen in the soil from other sources, depending on the quality of the soil, and not all of the applied nitrogen would be available to the plant. Mineral nitrogen is available in the forms of nitrate (NO_3^-) and ammonium (NH_4^+) with the former more common in aerated soils, and ammonium nitrate is an example of a nitrogen fertilizer. I have also assumed that the optimum amount of fertilizer for both cultivars is 200 kg N per hectare, but that cultivar B is higher yielding than A at all nitrogen levels.

As well as the optimum amount of fertilizer for the highest yield, farmers will want to know the economically optimum amount. This is the point at which additional yield does not compensate for the cost of the additional nitrogen fertilizer. It will depend on the cost of fertilizer relative to the sale price of the potato tubers. Results from an experiment in south-eastern Brazil in 2003, which used rates of 0, 50, 100, 200 and 300 kg N ha^{-1}, found that the economically optimum rates were between 86% and 95% of the rates for maximum total tuber fresh-weight yield, depending on cultivar and price conditions [45]. The highest marketable yields for the four cultivars occurred between 168 and 212 kg N ha^{-1}, the marketable yield being tubers of acceptable size and quality. In contrast, the economic optimum N fertilizer rates ranged from 147 to 201 kg N ha^{-1}.

In Fig. 7.8 I have shown the nitrogen use efficiencies (NUE) for my two cultivars. Strictly speaking, I have shown the agronomic NUE, which is the fresh-weight yield of tubers per unit N supplied (kg of tubers harvested per kg of N applied where one tonne is 1000 kg). NUE is highest at low nitrogen levels and steadily decreases as more fertilizer is applied. In the Brazilian experiment, for marketable yield, it decreased from 457–614 to 81–102 kg tuber kg^{-1} N (depending on cultivar) as the N fertilization rate increased from 50 to 300 kg N ha^{-1}. If fertilizer is applied more accurately to where it is needed in a field, less will be required, and agronomic NUE will increase. This is one of the aims of modern precision agriculture, but of more interest to me is the comparison between cultivars. In Fig. 7.7, cultivar B had the higher yields at all nitrogen levels and hence higher nitrogen

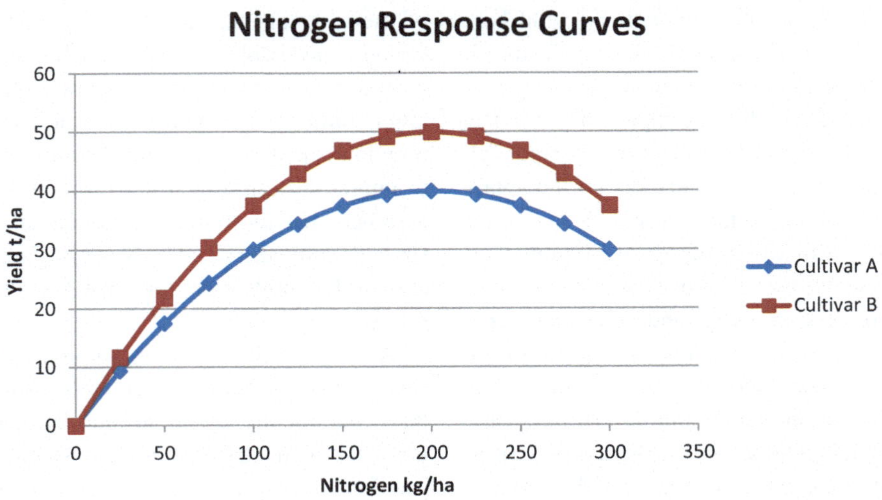

Fig. 7.7 Nitrogen response curves for two cultivars A and B

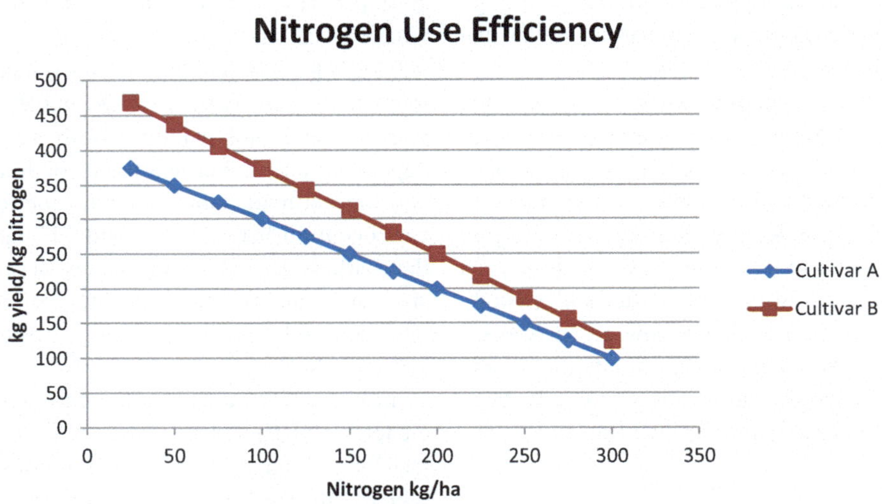

Fig. 7.8 Corresponding nitrogen use efficiency lines for cultivars A and B

use efficiencies (1.25 times that of cultivar A). Again, in the Brazilian experiment, cultivars Agata and Atlantic had the highest marketable yields (33.1 and 33.3 t ha^{-1}) at the optimal fertilizer (168 and 175 kg N ha^{-1}) rate and also the highest NUE (197 and 190 kg tuber kg^{-1} N). However, in selecting cultivars for high agronomic NUE there are some complications. First, early maturing, determinate cultivars usually have a lower NUE than later maturing, indeterminate cultivars, so that maturity class is an issue. Second, NUE has two components, N uptake (acquisition) efficiency and N utilization efficiency. The former is the amount of N accumulated by a crop per unit N supplied (and available), and the latter is the fresh-weight production per unit N accumulated by a crop (strictly speaking, it should be the dry-matter production). The interested reader can look at the 2018 paper by Cohan et al. on the effects of potato cultivar on the components of nitrogen use efficiency [46]. Uptake efficiency is likely to be associated with root traits, whereas high utilization efficiency is more likely to be associated with traits such as optimizing leaf canopy architecture, maintaining N in the shoot and maximizing DM harvest index (the proportion of the plant's dry matter that is in its tubers) [40]. One final point worth making is that the application of high levels of nitrogen (and phosphate) fertilizer can result in leaching of excess nitrate from the potato field into waterways, which can cause an explosive growth of algae and aquatic plants (eutrophication), resulting in a depletion of the dissolved oxygen required by fish and other aquatic animals [41].

Potatoes have an equally high demand for potassium (K) [40]. The two common potassium fertilizers are muriate of potash (potassium chloride) and sulphate of potash (potassium sulphate). However, in recent years a natural mineral fertilizer has been found in England that contains potassium, sulphur, magnesium and calcium. The mineral is called polyhalite and has been found as a layer of rock below North Yorkshire that extends under the North Sea near Whitby for a mineable area of 25,000 ha. The fertilizer, polysulphate, is made by simply crushing and granulating the mineral without chemical processing. My understanding is that it can therefore be used by organic farmers. Two mineshafts are currently under construction by the company Anglo American plc, who, according to their website, expect to start extraction in 2027 (https://www.angloamerican.com/products/polyhalite). In the meantime, *The Scottish Farmer* for 6 April 2024 reported (page 20) that some Scottish farmers have been using polysulphate since 2018.

I now want to move on to phosphorus (P) and the need for new cultivars that make better use of phosphate (P_2O_5) fertilizers [40]. Potatoes recover less than 10% of the P applied to them as fertilizer because of its low solubility [47]. As a consequence, potatoes receive more inorganic P-fertilizer than other crops, and there is a high loss of P from fields and environmental pollution. In field trials done in 2011 in Sao Paulo State, Brazil (23°S, 49°W, 649–744 masl), phosphate fertilizer increased the total fresh-weight yield of tubers up to application rates of 500, 250 and 125 kg P_2O_5 ha^{-1} in the soils with low (14), medium (36) and high (70) P availability (mg P dm^{-3} soil), respectively. Tuber yields were the same (35 t ha^{-1}) in the soils with medium and high P availability but lower (20 t ha^{-1}) in soils with low P availability [48]. Hence, there is a need for new cultivars that use P-fertilizer inputs more efficiently, particularly in soils with low P availability (i.e., soils with high P-fixing capacity). In contrast to NUE, agronomic phosphorus use efficiency (PUE) is defined as crop dry-matter (DM) yield per unit of P available in the soil (g DM g^{-1} P soil). It follows that PUE, like NUE, is the product of uptake efficiency and utilization efficiency, where the former is P acquired by the crop per unit of available P (g P crop g^{-1} P soil) and the latter is DM yield per unit P acquired by a crop (g DM g^{-1} P crop). A literature review of relevant research revealed that genetic variation is more likely to be found for uptake efficiency [47]. A series of trials in Dundee, Scotland (56°28'N, 03°07'W) over the period 2006–2011, with a wide range of cultivated potatoes, demonstrated that increased juvenile root vigour accelerated P-uptake and initial canopy cover and hence increased tuber yields [47]. Increased juvenile root vigour would therefore be a good selection criterion in a breeding program.

Information on fertilizers to supply the rest of the macro- and micro-nutrients required by potatoes can be found in the 2018 review by Philip White, along with a vision of the future in which precision agriculture will take account of variation in nutrient status within fields [40], as well as in a book chapter by Naumann and Pawelzik [49].

7.10 Irrigation and Water Requirements

Potatoes need water, particularly as they have a shallow root system. In some areas of the world potatoes can only be grown under irrigation, whereas in other areas they benefit from some irrigation, and this need could increase in future years with climate change [50]. Back in 2009, farming overall was using 70% of the world's available fresh water, with predicted increased needs in competition with industrial and domestic users as the human population grew in size [51]. Hence, there is interest in making better use of the available water through improved irrigation management. In both China and India, research is being done on alternatives to furrow irrigation, such as drip irrigation, partial root-zone drying furrow irrigation in which half of the root system is irrigated while the other half is kept dry with each irrigation, and drip irrigation with straw mulching. The details need not concern us, but the results are encouraging [52, 53]. In the Indian Punjab (30°56′N, 75°52′E, 247 masl), over two winter seasons, yields were much higher for bed-planted potatoes with drip irrigation and mulch (32.31 t/ha) than for ridge-planted potatoes with furrow irrigation but no mulch (18.28 t/ha) [53]. Furthermore, as with fertilizers, water response curves can be determined and cultivars sought with better water use efficiency.

Soil salinity is a problem associated with irrigation practices, particularly in arid and semi-arid regions. The period from the early 1990s to 2014 saw a global increase in salt-degraded land from 45 to around 62 million hectares, which was 20% of the world's irrigated land and was expected to increase [54, 55]. The problem occurs where rainfall is too low to maintain regular percolation of rainwater and its naturally dissolved minerals through the soil and where irrigation is practised without a natural or artificial drainage system. Water settles on the irrigated land and evaporates, leaving the salt, which can accumulate to dangerous levels in the root zone of crops rather than make its way to the sea. Crop productivity is reduced as a result. The ultimate solution to the problem would be desalination of irrigation water, but this will require sufficient cheap energy. In the meantime, land will either have to be taken out of potato production or potatoes selected with enough salt tolerance to minimize yield losses [55].

Does the same apply to drought tolerance when it is not possible to provide potatoes with adequate supplies of water? Researchers from CIP did some research on the coast of Peru over the 5-year period from 2004 to 2008 at their La Molina experimental station (12.1°S, 76.9°W, 240 masl) which is characterized by an absence of rainfall but high relative humidity and average annual minimum and maximum temperatures of 14.9 °C and 18.0 °C, respectively [56]. They assessed a set of 918 accessions from their potato genebank under irrigation and drought conditions in which the total water applied was 375–414 mm and 128–175 mm, respectively. Yields under drought conditions were 52%, 57% and 63% of those under irrigation conditions for the improved cultivars, genetic stocks and landraces, respectively.

There were 13 accessions with yields greater than 800 g/plant under both irrigated (average 1154 g/plant) and drought (883 g/plant = 77% of irrigated) conditions (Group 1 in Fig. 7.9). In contrast, the 13 accessions with the highest yields under irrigation (average 1278 g/plant) had relatively lower yields under drought conditions (808 g/plant = 63%) (Group 2 in Fig. 7.9) compared with the relatively higher yields of the 13 accessions with the highest yields under drought conditions (886 g/plant = 80% of 1115 g/plant) (Group 3 in Fig. 7.9). These results provide a classic example of genotype × environment interactions that result in a change of rankings of the three groups of genotypes with a change of environment. They also demonstrate that selection in the high-yielding environment (irrigation) results in genotypes that are highly responsive to changes in water availability, whereas selection in the low-yielding environment (drought) results in unresponsive genotypes. Selection based on yield in both environments results in an intermediate response, but in this example, closer to the unresponsive one. The relatively higher yield under drought conditions of the group selected under these conditions could be due to tolerance of dehydration or avoidance of dehydration by mechanisms such as osmotic adjustment.

There are other possibilities for breeders. They could select new cultivars for early maturity to escape later-occurring droughts when lower yields from a shorter growing season are acceptable. They could select cultivars with a root system that can reach deeper moisture zones and thus avoid drought. Finally, they could select cultivars that tolerate drought by reducing transpiration losses, but this would reduce CO_2 fixation and hence yield.

More generally, breeders will have to decide whether they want to select different cultivars for

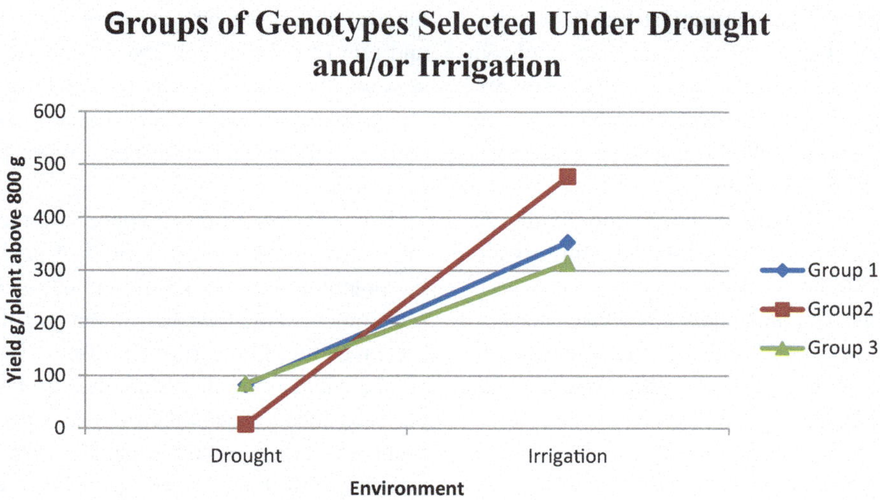

Fig. 7.9 Groups of genotypes selected under drought and irrigation (Group 1), irrigation (Group 2) and drought (Group 3) conditions based on data in a paper by Cabello et al. [56]

different environments defined by the presence or absence of abiotic stresses or try to find a compromise with cultivars that yield relatively well in a wide range of environments. The outcomes will probably be different kinds of plants for these different situations. Whether one calls cultivars resistant to, or tolerant of, abiotic stress can be confusing. If a cultivar is much higher yielding than others in the presence of an abiotic stress, it is tempting to call it resistant. On the other hand, if a cultivar has a high yield in the absence of stress and a relatively small loss of yield in the presence of stress, it is tempting to call it tolerant. However, in practice, it may not be obvious without further experimentation how a cultivar achieves a relatively high yield in various situations and what its yield will be in different situations. I personally like the kind of visual presentation shown in Fig. 7.9 and tend to use the term tolerance, as it describes performance in the presence and absence of stress.

7.11 Heat Stress

Average daily temperatures of above 21 °C have an adverse effect on plant growth and tuber formation. Heat stress is therefore a major limiting factor for potato production in the lowland tropics and subtropics, as well as in the Mediterranean and Sahelian Africa, and may spread to other areas with climate change and rising temperatures [57, 58]. One likely consequence is that the suitable heat-free period for potato production will shorten in these areas unless new cultivars can be bred with some heat tolerance. In my book on potato breeding [31], I reported examples of such breeding in Brazil, Peru and India [59–61], with heat stress occurring in the range 25–32 °C. The breeding involved in vitro (often at 28 °C) and glasshouse screening as well as field tests. The heat-tolerant cultivar Kufri Lima, released in 2018, with early maturity and resistance to PVY and PVX, was developed for early planting in the North Indian plains from a collaboration between ICAR-CPRI in India and CIP in Peru (https://cipotato.org/blog/new-potato-variety-kufri-lima-take-heat-virus-india/) [38].

In the longer term, current research on photorespiration may result in potatoes that can produce higher yields than currently possible under heat stress. Recently (2024) a team of researchers from the University of Illinois (United States) achieved this in field trials of cultivar Desiree with an alternative photorespiratory pathway (AP3) that had been introduced by genetic transformation with *Chlamydomonas reinhardtii* glycolatedehydrogenase (CrGDH), *Cucurbita maxima* malate synthase (CmMS) and RNAi targeting the glycolate–glycerate transporter

PLGG1 [62]. This chloroplast-localized glycolate photorespiratory bypass increased photosynthetic capacity and tuber mass with yields 30% greater than the untransformed control. In contrast, the normal photorespiratory pathway is adversely affected by high temperature, as a key enzyme, RuBisCo, then interacts more with oxygen than carbon dioxide, thus depleting fixation of the latter.

Now that we understand the basic needs of potato plants and some of the stresses they encounter, it is time to consider potential and actual yields.

7.12 Potential and Actual Yields

Potato yields depend on resource capture during the available growing season and resource use in producing tubers. Resource capture is both through the above-ground canopy and the below-ground root system. The available growing season is largely characterized by a temperature window, as potatoes don't like it too hot or too cold. The required resources are solar radiation, carbon dioxide, water and nutrients. Crop physiologists have developed models of crop growth which allow them to predict potential yields based primarily on the effects of intercepted solar radiation (length of growing season and daylengths), temperatures and water availability on resource capture and for carbon fixation and partitioning to tubers, assuming that nutrients are in adequate supply and there are no other abiotic stresses [63–66]. They also recognize that potato crops have a number of growth stages: sprouting of tubers and emergence of shoots, vegetative growth, tuber initiation, tuber bulking and finally maturation and senescence of the foliage. Potato yields will be the highest possible for a given environment if cultivars are available which match their growth cycle from planting until crop maturity with the length of the available growing season [50]. Such cultivars will have a high harvest index of around 0.75, i.e., 75% of their total dry weight will be in their harvestable tubers compared with 40–60% for cereal grains. They will also be adapted to produce tubers in the daylengths of the season, and their tuber initiation and rate of bulking will also match the growing season. The one exception to matching cultivars and growing season is when a quick-growing cultivar is required for sale into a more profitable early market.

It is also important to appreciate that we are interested in the yield of a field of potatoes, not the potential yield of a single plant, and that the seed tubers which farmers plant are the initial resource for plant growth and development. Hence, yield components are the number of eyes per seed tuber, the number of stems per eye, the number of tubers produced per stem and the average weight of these tubers; but as one component can compensate for another, it turns out that the weight of the seed tuber is critical, as more weight means more resources and hence more yield from a better start. According to Dutch physiologist Paul Struik [67], for a good ware crop, farmers plant 50,000 seed tubers, each weighing 50 g, per hectare to produce a stand of 45,000–50,000 plants (150,000 stems), which in turn will produce about 500,000 marketable tubers with an average fresh weight of 120 g, giving a total yield of 60 t ha^{-1} from the 2.5 t ha^{-1} of seed potatoes. The physiological age of the seed tuber is also important, but I am not going into this concept, nor am I going into the yield potential of the seedling tubers produced on plants with single stems grown from true potato seeds.

There is a vast literature on all of the topics just mentioned, as can be seen in a relatively recent review and two books [68–70], but I want to move on to consider how estimated potential yields compare with actual yields. Despite examples in this and previous chapters of some very high yields, globally in 2016, actual yields averaged out at about 50% of potential yields, but ranged from 21% to 72% over various climatic regions [50, 70]. Potential yields depended on the length of the growing season, daylengths and temperatures. Thus, a winter crop of 100 days in the lowland subtropics was not expected to yield as well as a summer crop of 150 days in lowland temperate areas. So, let us look at a few published results. Beukema and van der Waag [70] found that potential yields in tropical, sub-tropical and temperate countries or regions varied from 40 to 140 t/ha depending upon the length of the growing season and temperatures. Average fresh weight yields in these areas in 2016 ranged from 24% to 65% of potential yields, with a mean of around 50%. Haverkort and Struik [50] quoted figures of 72%, 60% and 35% for the most productive areas of the Netherlands (65 compared with 90 t/ha), for Northern Maine, United States (31 compared with 55 t/ha), and for the less productive areas of Chile. Silva et al. [71] reported figures for Dutch farms from 2008 to 2012 of 70.8% (51.4 compared with 72.6 t/ha) for ware potatoes and 60.3% (43.2 compared with 71.6 t/ha) for starch potatoes. Deguchi et al. [25] produced figures of 63% down to 21% in Japan, with an average of 40%, but not weighted by area of potatoes. Summer crops in Hokkaido in the north (45°N) were higher yielding than winter crops in Okinawa in the south (25°N), with actual yields of over 30 t/ha and 15.5 t/ha, respectively. Again, we see in Japan the popularity of old cultivars, with Irish Cobbler (introduced 1908) and May Queen (1912) still the most widely grown table ones in 2011. Finally, in 2020, Yagiz et al. [72] came up with a value of 37% for Turkey (34.93 t/ha compared with 94.42 t/ha).

The more sophisticated crop growth models can make predictions about the effects of climate change on potential yields. The predictions for Japan were that a 2 °C increase in temperature would reduce potential yields in Hokkaido from 56 to 53 t/ha but a 2-week increase in the growing season would increase them from 53 to 60 t/ha. However, as CO_2 concentrations increase to 550 ppm by 2050 (from 380 ppm in 2000), potential yield could increase to 77 t/ha (28% increase) [25]. Likewise, the predictions for Turkey were large, also based on a 2 °C rise in temperature and 550 ppm of CO_2 by 2050 [72]. The average water limited potential yield could increase from 36.31 t/ha to 54.41 t/ha and the potential yield from 94.42 t/ha to 120.63 t/ha (28% increase), based on changes in planting and harvest dates to accommodate new periods suitable for potato growth. The main reason for the predicted higher yields is that rising CO_2 concentrations will increase the rate of photosynthesis in potato plants because they fix carbon dioxide by the C_3 pathway, so called because the 3-carbon molecule 3-phosphoglycerate is the first to be formed in the Calvin cycle. The burning of fossil fuels is also likely to increase the ozone (O_3) concentration in the earth's atmosphere to 60 ppb (from 50 ppb pre-industrialization), which will damage plant leaves and decrease yields. When the desirable and undesirable effects of CO_2 and O_3, respectively, are combined, potential potato yields could still increase by 28.5% [73]. Hence, our considerations of climate change converge on anticipated increases in the potential yields of potatoes by around 28%. However, for these to be realized in practice around the world, there will need to be adequate supplies of water and absences of abiotic stresses such as heat and salinity. More sophisticated models are being used to evaluate larger data sets and to make predictions about climate change, but it is too early to say how accurate these will prove to be in practice, both for individual countries and for the global average. An analysis published in 2020, for example, taking into account CO_2 fertilization and adaptation to climate change through changing growing seasons and new cultivars, put the global average yield increase from the period 1990–2010 to 2040–2060 between 9% and 20%, depending on the climate change model [74]. The model used daily minimum and maximum temperature, precipitation, solar radiation and soil data and considered 102 countries. The clock, however, is ticking towards 2050, which will be just 25 years away when this book is published, so the young farmers of today don't have that long to wait for an answer in terms of actual yields.

7.13 Reasons for the Gap Between Potential and Actual Yields

Potato production in Africa is suboptimal for a number of reasons: shortage of good-quality seed tubers, particularly of improved cultivars; disease and pest problems such as late blight, bacterial wilt, viruses and potato tuber moth; high cost of fertilizers and crop protection chemicals; lack of storage facilities for both seed and ware potatoes; and marketing problems with rapid fluctuations in supply and demand [75]. Furthermore, some of the lowest potato yields in the world are in Sub-Saharan Africa, excluding South Africa, where the average for 2016–2020 was 36.55 t/ha. I am therefore going to look at the reasons for low yields in two countries in the equatorial highlands of East Africa, Burundi and Kenya, for which research results are available.

In 2016 yields in Burundi (10.6 t/ha) were half of the global average (20.4 t/ha), having increased from 3.06 t/ha in 1966 [http://faostat.fao.org]. During the 3 years, 2014, 2015 and 2016, a group of researchers did on-farm experiments at five farms in each of three provinces surrounding the town of Bujumbura for one or two seasons, making 15 or 30 replicates for treatment comparisons [76]. Yield from healthy early-generation seed from a rapid multiplication scheme was 13.89 t/ha compared with 7.69 t/ha from farm-saved seed (80% increase, one season). Diffused light storage of seed tubers was superior to dark storage with yields of 11.76 t/ha compared with 9.15 t/ha (29% increase, one season). Alternating contact and systemic fungicide applications to control late blight increased yields to 9.94 t/ha from 6.68 t/ha with lack of appropriate methods by farmers (49% increase, two seasons). Chemical fertilizers (N, P, K at 60, 90, 60 kg/ha) resulted in yields of 14.4 t/ha compared with 9.05 t/ha from farmers applying manure and compost (59% increase, two seasons). Finally, the new cultivar Victoria was higher yielding than widely grown Ndinamagara (= Cruza 48 from CIP) at 12.28 t/ha compared with 9.57 t/ha (28% increase, three seasons and two harvest dates). The highest average yield achieved in a set of treatments was 15.5 t/ha. The authors concluded that yields could be doubled by planting healthy pre-sprouted seed of a new cultivar and applying chemical fertilizers and fungicides, but reducing the incidence of *Ralstonia solanacearum*, the cause of brown rot of tubers and bacterial wilt of foliage, was more challenging.

In Kenya, Wang'ombe and van Dijk [77] took a different approach. They analysed a small random sample of 402 completed questionnaires from farmers in 2009/2010. They found that healthy planting material (seed tubers), use of fertilizers, fungicides and irrigation in lower rainfall areas all increased yield. The combination of healthy seed tubers and proper use of fertilizers gave the highest yield of 18.35 t/ha compared with the overall average of 12.17 t/ha. However, they also found that healthy seed had the lowest adoption rate (4.5%) despite most farmers (79%) being aware of its desirability (availability was the problem). The adoption rate of irrigation was also low (23%). In contrast, there was widespread use of fertilizers (96%) and fungicides (92%), primarily to control late blight. Interestingly, they found that farmers were discouraged from using animal manure because of the risk of contamination with *R. solanacearum*.

The highest yields reported in Burundi (15.5 t/ha) and Kenya (18.35 t/ha) were below the global average of 21.41 t/ha for 2016–2020 and a long way below the potential yields we encountered in the previous section. More encouragingly, we can argue that for many countries in the world there is scope for increased production from higher potato yields, but this will also depend upon favourable political, economic and societal environments.

Even in countries where yields are high, there is scope for improvement. Interestingly, farmers and agronomists in the Netherlands thought that increased phosphorus (P) and potassium (K) fertilizer application rates would increase yields, but this was not borne out in experiments done on 22 and 24 commercial potato fields across the Netherlands in 2019 and in 2020, respectively [78]. Each year, half of the fields were on sandy soils where cultivar Fontane was grown (average control yields of 64 and 64 t/ha in 2019 and 2020, respectively), and half were on clay soils with cultivar Innovator (65 and 60 t/ha). Although in some fields increased K application had a small positive effect on yield, the overall conclusion was that increasing P and K application rates in the Netherlands would not reduce the gap between potential and actual yields. Increased rates would, however, decrease P and K use efficiencies and hence be undesirable for environmental and economic reasons. The researchers also concluded that they had been unable to identify the causes of the yield gaps between potential and actual yields, and hence more research was required: not an uncommon conclusion!

7.14 Precision Agriculture and Smart Farming

This chapter has been about potato farmers increasing production by achieving the potential yields of their crops. Precision agriculture and smart farming may sound like advertising slogans, but they are serious ideas and emerging technologies that are currently being explored. They may help farmers achieve their objectives of increased but sustainable and economic potato production during a period of challenges from climate change. One can see this vision, for example, in a recent article (13 October 2024) by the editor and publisher of *Potato News Today*, Lukie Pieterse, entitled "An AI-powered potato industry". There is hope that artificial intelligence (AI), coupled with machine learning, can interpret vast amounts of data and provide farmers with sound advice for targeted action on their farms. The key is interpretation, as farmers already have access through mobile phone networks and apps to vast amounts of information that could help them make better decisions on their farms; something that was inconceivable to the first Andean farmers who domesticated the potato.

Here I am just going to give a brief indication of the emerging technologies and what commercial companies are beginning to offer farmers. More detailed accounts can be found in two recent book chapters [79, 80]. Farmers will be able to collect comprehensive real-time data about the soil in their fields, the health and needs of their potato plants, and even the underground development of tubers using soil sensors and unmanned ground (robots) and aerial (drones) vehicles. Drones, for example, with high-resolution cameras and multispectral sensors, will be able to detect changes in leaf colour, uneven growth patterns and the presence of other plants that AI can interpret as nutrient deficiencies, pest infestations and weeds, and recommend action. Likewise, soil sensors will be able to detect changes in moisture status and the need for water. Furthermore, irrigation, fertilizers, pesticides and herbicides will be targeted at specific areas of fields and specific groups of plants, rather than a more expensive and wasteful blanket cover which could lead to environmental concerns. Satellites will be able to monitor much larger areas of crops. Finally, robots equipped with a ground penetrating radar system are being developed to monitor tuber development and provide data on yield and tuber size distribution that can be used to optimize the time of harvest; for example, using the "TuberScan" system currently being developed in Great Britain by B-hive Innovations and partners (https://www.b-hiveinnovations.co.uk/project/tuberscan/).

Lukie Pieterse, in his article, also pointed out that AI can be used in potato marketing and supply chain optimization, and envisaged automating potato planting and harvesting. It remains to be seen how quickly commercial companies make the new technologies available and the extent and impact of their uptake by farmers.

7.15 Conclusions on Ways to Increase Potato Production

Potato production has increased in three ways: by bringing new land into cultivation, by using existing cultivated land for potatoes rather than other crops, and by increasing production per unit area of land. All three ways could be used in the future, and their importance will probably vary from country to country, particularly during a period of climate change. Currently, the emphasis globally appears to be on increasing production per unit area of land.

Where soils are not providing enough nutrients for crops to achieve their potential yields, they are expected to respond to the addition of inorganic (chemical) fertilisers, but farmers may not be able to afford the quantities required for maximum yields. In the longer term, farming systems may change in ways aimed at regenerating, increasing and then maintaining soil structure and fertility without the need for chemical fertilizers. New cultivars may be bred that are better adapted to these new farming systems and which are better at capturing and utilizing nutrient resources.

Where rainfall is inadequate, crops will respond to irrigation, but this needs to be done in ways that will not result in problems from increased soil salinity. There appears to be scope for better irrigation systems that get more water to where it can be utilized by the plants. Again, new cultivars may be bred that are better at capturing and utilizing water resources. Cultivars with drought tolerance may be able to keep the loss of yield and quality to a minimum where it is not possible to provide sufficient irrigation.

Where heat stress is reducing potato growth and tuber production, it may be possible to change planting and harvesting dates to reduce the length of periods of heat stress. Cultivars with heat tolerance may be able to keep loss of yield and quality to a minimum where the periods of heat stress are not too long nor too severe. However, with climate change there may be areas where potatoes can no longer be grown because it is too hot, but also areas where potatoes can be grown because it is no longer too cold.

Where yield and quality are being lost because of pests and diseases, as well as competition from weeds, better control measures will increase production. Use of synthetic chemical pesticides, fungicides and herbicides has proved effective but raises concerns about adverse effects on human health and the environment, as well as affordability. Cultivars with inbuilt resistances to pests and diseases have provided, and will provide, a better solution, and so might biological-control measures.

Lack of healthy seed tubers for planting is a problem in low-income countries with informal seed systems based on farm-saved tubers. The solution adopted in high-income countries has been statutory seed certification schemes in areas where potatoes are grown only for seed and which are geographically and climatically less favourable to the vectors of viruses which cause systemic infections. However, propagation of potato crops from true potato seed (TPS) may be a cheaper and better option in low-income countries, as TPS transmits very few diseases.

The current yield gaps between actual and theoretical potato yields show the potential for increased potato production through higher yields per hectare, but the realization of this potential will require favourable political, economic and societal environments.

References

1. Hawkes JG, Francisco-Ortega J (1993) The early history of the potato in Europe. Euphytica 70:1–7
2. FAO statistics on 6 September 2023 (http://faostat.fao.org)
3. Rhymes J, Stockdale E, Napier B, Williamson J, Morton D, Dearlove E, Staley J, Young H, Thomson A, Evans C (2023) The future of UK vegetable production—Technical Report. https://www.wwf.org.uk/our-reports/future-uk-vegetable-production-technical-report
4. Pevsner N (1954) The buildings of England: Cambridgeshire. Penguin Books Ltd., Harmondsworth
5. Boyce J (2020) Imperial mud: the fight for the fens. Icon Books Ltd., London
6. Rotherham ID (2013) The lost fens. The History Press, Stroud
7. Convention on Wetlands (2021) Global wetland outlook: special edition 2021. Secretariat of the Convention on Wetlands, Gland
8. Riddet WM (1925) The turnip crop. In: Paterson WGR (ed) Farm crops, vol II. Root crops. Gresham Publishing Co. Ltd., London, pp 69–118
9. Fiennes J (2022) Land healer. Witness Books, London
10. Sly R (2010) Soil in their souls. The History Press, Stroud
11. Landmark, Issue 53, Autumn 2023, NIAB, Cambridge
12. Crowhurst R (2024) Peatlands face future cropping challenges. The Vegetable Farmer July 2024, Act Publishing, Maidstone, pp 19–21
13. Whitehead T, McIntosh TP, Findlay WM (1945) The potato in health and disease, 2nd edn. Oliver and Boyd, Edinburgh
14. Tiemens-Hulscher M, van Bueren ETL, Hutten CB (2012) Potato: perspectives to breed for an organic crop ideotype. In: van Bueren ETL, Myers JR (eds) Organic crop breeding. Wiley-Blackwell, Chichester, pp 227–237

15. Reganold JP, Wachter JM (2016) Organic agriculture in the twenty-first century. Nat Plants 2., Article Number: 15221. https://doi.org/10.1038/nplants.2015.221
16. Silva GO, Carvalho ADF, Ragassi CF, Pereira AS, Bortoletto AC, Feldberg NP, Lopes CA, Emygdio BM (2023) Evaluation of potato clones for agronomic and processing traits in conventional and organic crop systems. Hortic Bras 41:elocation e2587. https://doi.org/10.1590/s0102-0536-2023-e2587
17. Landmark, Issue 50, Autumn 2022, NIAB, Cambridge
18. Sarangi SK, Maji B, Sharma PC, Digar S, Mahanta KK, Burman D, Mandal UK, Mandal S, Mainuddin M (2021) Potato (*Solanum tuberosum* L.) cultivation by zero tillage and paddy straw mulching in the saline soils of the Ganges Delta. Potato Res 64(2):277–305
19. Langford S (2022) Rooted: how regenerative farming can change the world. Penguin Books (2023), UK
20. Shehata SA, Abouziena HF, Abdelgawad KF, Elkhawaga FA (2019) Weed control efficacy, growth and yield of potato (*Solanum tuberosum* L.) as affected by alternative weed control methods. Potato Res 62(2):139–155
21. Simmonds NW (1981) Genotype (*G*), Environment (*E*) and *GE* components of crop yields. Expl Agric 17:355–362
22. Simmonds NW (1980) Comparisons of the yields of four potato varieties in trials and in agriculture. Expl Agric 16:393–398
23. Potato Yearbook & Buyers Guide 2024. Act Publishing, Maidstone
24. Potato Yearbook & Buyers Guide 2021. Act Publishing, Maidstone
25. Deguchi T, Iwama K, Haverkort AJ (2016) Actual and potential yield levels of potato in different production systems of Japan. Potato Res 59:207–225
26. Lawson HM (1983) True potato seeds as arable weeds. Potato Res 26:237–246
27. Wellington PS, Silvey V (1997) Crop and seed improvement. The Dorset Press, Dorchester
28. Talbot M (1984) Yield variability of crop varieties in the U.K. J Agric Sci (Camb) 102:315–321
29. Talbot M (1997) Resource allocation for selection systems. In: Kempton RA, Fox PN (eds) Statistical methods for plant variety evaluation. Chapman and Hall, London, pp 162–174
30. Bradshaw JE (2009) A genetic perspective on yield plateau in potato. Potato J 36:79–94
31. Bradshaw JE (2021) Potato breeding: theory and practice. Springer, Cham
32. Bradshaw JE, Dale MFB, Mackay GR (2009) Improving the yield, processing quality and disease and pest resistance of potatoes by genotypic recurrent selection. Euphytica 170:215–227
33. Lucier G, Budge A, Plummer C, Spurgeon C (1990) U.S. potato statistics, 1949–89. USDA, Economic Research Service 829:1–8. U.S. Gov Print Office, Washington, DC
34. Douches DS, Maas D, Jastrzebski K, Chase RW (1996) Assessment of potato breeding progress in the USA over the last century. Crop Sci 36:1544–1552
35. Troyer AF (2006) Adaptedness and heterosis in corn and mule hybrids. Crop Sci 46:528–543
36. Love SL, Pavek JJ, Thompson-Johns A, Bohl W (1998) Breeding progress for potato chip quality in North American cultivars. Am J Potato Res 75:27–36
37. Kunkel R, Campbell GS (1987) Maximum potential potato yield in the Columbia Basin, USA: model and measured values. Am Potato J 64:355–366. https://doi.org/10.1007/BF02853597
38. Luthra SK, Gupta VK, Tiwari JK, Kumar V, Bhardwaj V, Sood S, Dalamu, Kaur RP, Kumar R, Vanishree G, Kumar D, Mhatre P, Chakrabarti SK (2020) Potato breeding in India. CPRI Technical Bulletin No 74 (revised), ICAR-Central Potato Research Institute, Shimla, Himachal Pradesh, India
39. Fuglie KO, Echeverria RG (2024) The economic impact of CGIAR-related crop technologies on agricultural productivity in developing countries, 1961–2020. World Dev 176:106523. https://doi.org/10.1016/j.worlddev.2023.106523
40. White PJ (2018) Improving nutrient management in potato cultivation. In: Wale S (ed) Achieving sustainable cultivation of potatoes. Volume 2: Production, storage and crop protection. Burleigh Dodds Science Publishing, Cambridge, pp 45–67
41. King J (2011) Reaching for the sun: how plants work, 2nd edn. Cambridge University Press, Cambridge
42. Gilbert N (2009) The disappearing nutrient. Nature 461:716–718
43. Fisher RA, Mackenzie WA (1923) Studies in crop variation. II. The manurial response of different potato varieties. J Agric Sci 13:311–320
44. Fisher RA (1935) The design of experiments. Oliver and Boyd, Edinburgh
45. Fontes PCR, Braun H, Busato C (2010) Economic optimum nitrogen fertilization rates and nitrogen fertilization rate effects on tuber characteristics of potato cultivars. Potato Res 53(3):167–179
46. Cohan JP, Hannon C, Houilliez B, Gravoueille JM, Geille A, Lampaert E, Laurent F (2018) Effects of potato cultivar on the components of nitrogen use efficiency. Potato Res 61:231–246
47. White PJ, Bradshaw JE, Brown LK, Dale FMB, Dupuy LX, George TS, Hammond JP, Subramanian NK, Thompson JA, Wishart J, Wright G (2018) Juvenile root vigour improves phosphorus use efficiency of potato. Plant Soil 432:45–63. https://doi.org/10.1007/s11104-018-3776-5
48. Fernandes AM, Soratto RP (2016) Response of potato cultivars to phosphate fertilization in tropical soils with different phosphorus availabilities. Potato Res 59:259–278
49. Naumann M, Pawelzik E (2023) Nutrient management in potato. In: Caliskan ME, Bakhsh A, Jabran

K (eds) Potato production worldwide. Academic, London, pp 101–120
50. Haverkort AJ, Struik PC (2015) Yield levels of potato crops: recent achievements and future prospects. Field Crop Res 182:76–85
51. Bell A (2009) Peak water. Luath Press Limited, Edinburgh
52. Qin J, Antonio DCR, Xie K, Li W, Yactayo W, Jin L, Quiroz R (2018) Is the partial root-zone drying more appropriate than drip irrigation to save water in China?: a preliminary comparative analyses for potato cultivation. Potato Res 61:391–406
53. Brar AS, Buttar GS, Thind HS, Singh KB (2019) Improvement of water productivity, economics and energetics of potato through straw mulching and irrigation scheduling in Indian Punjab. Potato Res 62:465–484
54. Qadir M, Quillérou E, Nangia V, Murtaza G, Singh M, Thomas RJ, Drechsel P, Noble AD (2014) Economics of salt-induced land degradation and restoration. Nat Res Forum. https://doi.org/10.1111/1477-8947.12054
55. Roy SJ, Negrão S, Tester M (2014) Salt resistant crop plants. Curr Opin Biotechnol 26:115–124
56. Cabello R, De Mendiburu F, Bonierbale M, Monneveux P, Roca W, Chujoy E (2012) Large-scale evaluation of potato improved varieties, genetic stocks and landraces for drought tolerance. Am J Potato Res 89:400–410
57. Govindakrishnan PM, Haverkort AJ (2006) Ecophysiology and agronomic management. In: Gopal J, Khurana SMP (eds) Handbook of potato production, improvement and postharvest management. The Haworth Press, Binghampton, pp. 179–229
58. Haverkort AJ, Verhagen A (2008) Climate change and its repercussions for the potato supply chain. Potato Res 51:223–237
59. Benites FRG, Pinto CABP (2011) Genetic gains for heat tolerance in potato in three cycles of recurrent selection. Crop Breed Appl Biotechnol 11:133–140
60. Khan MA, Munive S, Bonierbale M (2015) Early generation in vitro assay to identify potato populations and clones tolerant to heat. Plant Cell Tissue Organ Cult 121:45–52. https://doi.org/10.1007/s11240-014-0677-z
61. Bonierbale MW, Amoros WR, Salas E, de Jong W (2020) Potato breeding. In: Campos H, Ortiz O (eds) The potato crop: its agricultural, nutritional and social contribution to humankind. Springer, ebook, pp 163–218
62. Meacham-Hensold K, Cavanagh AP, Sorensen P, South PF, Fowler J, Boyd R, Jeong J, Burgess S, Stutz S, Dilger RN, Lee M, Ferrari N, Larkin J, Ort DR (2024) Shortcutting photorespiration protects potato photosynthesis and tuber yield against heat-wave stress. Glob Change Biol 30:e17595. https://doi.org/10.1111/gcb.17595
63. Haverkort AJ, Grashoff C (2004) Ideotyping-potato a modelling approach to genotype performance. In: MacKerron DKL, Haverkort AJ (eds) Decision Support Systems In Potato Production. Wagengingen Academic Publishers, Wageningen, pp 199–211
64. Khan MS, Yin X, van der Putten PEL, Struik PC (2014) An ecophysiological model analysis of yield differences within a set of contrasting cultivars and an F1 segregating population of potato (*Solanum tuberosum* L.) grown under diverse environments. Ecol Model 290:146–154
65. Haverkort AJ, Franke AC, Steyn JM, Pronk AA, Kooman PL (2015) A robust potato model: LINTUL-POTATO-DSS. Potato Res 58:313–327
66. Kleinwechter U, Gastelo M, Ritchie J, Nelson G, Asseng S (2016) Simulating cultivar variations in potato yields for contrasting environments. Agric Syst 145:51–63
67. Struik PC (2018) Understanding ageing processes in seed potatoes. In: Wang-Pruski G (ed) Achieving sustainable cultivation of potatoes. Volume 1: Breeding improved varieties. Burleigh Dodds, Cambridge, pp 33–55
68. Frederick CM, Mori M, Bethke PC (2018) Advances in understanding potato plant physiology and growth. In: Wang-Pruski G (ed) Achieving sustainable cultivation of potatoes. Volume 1: Breeding improved varieties. Burleigh Dodds, Cambridge, pp 3–31
69. Vreugdenhil D (ed) (2007) Potato biology and biotechnology advances and perspectives. Elsevier, Oxford
70. Beukema HP, van der Waag DE (1990) Introduction to potato production. Pudoc, Wageningen, pp 23–24
71. Silva JV, Reidsma P, van Ittersum MK (2017) Yield gaps in Dutch arable farming systems: Analysis at crop and crop rotation level. Agric Syst 158:78–92
72. Yagiz AK, Cakici M, Aydogan N, Omezli S, Yerlikaya BA, Ayten S, Maqbool A, Haverkort AJ (2020) Exploration of climate change effects on shifting potato seasons, yields and water use employing NASA and National long-term weather data. Potato Res 63:565–577
73. Jaggard KW, Qi A, Ober ES (2010) Possible changes to arable crop yields by 2050. Philos Trans R Soc B 365:2835–2851
74. Jennings SA, Koehler A-K, Nicklin KJ, Deva C, Sait SM, Challinor AJ (2020) Global potato yields increase under climate change with adaptation and CO_2 fertilisation. Front Sustain Food Syst 4:519324. https://doi.org/10.3389/fsufs.2020.519324
75. Muthoni J, Shimelis H (2023) An overview of potato production in Africa. In: Caliskan ME, Bakhsh A, Jabran K (eds) Potato production worldwide. Academic, London, pp 435–456
76. Bararyenya A, Nahayo P-C, Nduwimana A, Niyonzima P, Nyawakira D, Sindihebura J-P, Vyizigiro E, Ndayishimiye V, Haverkort AJ (2018) Assessment of opportunities for Burundian small-scale potato farmers to increase productivity and income. Potato Res 61:73–88

77. Wang'ombe JG, van Dijk MP (2013) Low potato yields in Kenya: do conventional input innovations account for the yields disparity? Agric Food Secur 2:14. https://doi.org/10.1186/2048-7010-2-14
78. Ravensbergen APP, van Ittersum MK, Kempenaar C, Reidsma P (2023) Current phosphorus and potassium fertiliser application rates do not limit tuber yield and quality in potato production systems in The Netherlands. Potato Res 66:1033–1058. https://doi.org/10.1007/s11540-022-09613-5
79. Unal Z, Kizildeniz T (2023) Smart agriculture practices in potato production. In: Caliskan ME, Bakhsh A, Jabran K (eds) Potato production worldwide. Academic, London, pp 317–329
80. Rawal S, Mankar P, Nare B, Kumar D (2024) Indian sustainable subtropical potato crop management. In: Khurana SMP, Bradshaw JE, Bhardwaj V (eds) Approaches for potato crop improvement and stress management. Springer, Singapore, pp 405–444

8 Improved Nutritional Value

As part of a meal, potatoes are primarily seen as a source of energy in the form of starch, although they also contribute fibre, proteins with a good amino acid balance, minerals, particularly potassium, vitamins, particularly B vitamins and vitamin C, and antioxidants. They can replace wheat and rice in meals and hence have a role as a food security crop. When potatoes form most of the diet, they need to provide nutritionally balanced meals, for which increased amounts of micronutrients are desirable, with a current emphasis on iron and zinc biofortification. The chemical composition of potatoes is affected by storage, cooking and processing. Furthermore, the bioavailability of ingested nutrients for utilization by the body needs to be determined in vivo using human subjects, although there are some in vitro proxies. Only then can one assess the contributions of potatoes to a healthy diet. Other nutritional issues are the need to reduce the glycaemic index of potatoes and the production of acrylamide in fried and roasted products. Finally, their relatively low requirement for land is one reason potatoes could contribute more to feeding the world and why they have often been the staple food of the poor.

8.1 Inconsistent Messages on Nutritional Value

Our perception of the nutritional value of potatoes has changed over the years, as explained by Rebecca Earle in her 2020 book *Feeding the People* [1]. It was favourable in the second half of the eighteenth century, unfavourable in the second half of the nineteenth century, and then favourable again in the twentieth century, as a key component of food security during and after the two world wars [1]. Today, potatoes still appear to receive inconsistent messages about their nutritional value. On 2 August 2008 the front cover of New Scientist had a picture of potatoes under the heading "Super Crop: How the humble potato could feed the world" and a cover story "Let them eat spuds. They're cheaper, more nutritious and easier to grow than grain." [2]. Just 2 years later, in 2010, I picked up a leaflet about children in Bolivia. It showed a picture of five potatoes and a young girl called Aracely: "The only vegetable that Aracely gets to eat is a potato. As a consequence, her diet is lacking all the vital minerals and nutrients that we expect for our own children … we think that every child

J. E. Bradshaw, *Can Potatoes Feed the World?*, Sustainable Development Goals Series,
https://doi.org/10.1007/978-3-031-92890-1_8

deserves the right to a nutritious diet so they can grow healthy." As we in the United Kingdom are encouraged to improve our diet by eating five portions of fruit and vegetables each day, I wondered if potatoes counted. I turned to our National Health Service (NHS) for the answer, which was no! Their relevant website page was updated on 16 September 2022:

> Potatoes are a starchy food and a great source of energy, fibre, B vitamins and potassium.
> In the United Kingdom, we get a lot of our vitamin C from potatoes.
> When eaten as part of a meal, potatoes are generally used in place of other sources of starch, such as bread, pasta or rice. Because of this, they don't count towards your 5 A Day.
> Potatoes play an important role in your diet, even if they don't count towards your 5 A Day. It's best to eat them without any added salt or fat.
> They're also a good source of fibre, so leave the skins on where possible to keep in more of the fibre and vitamins.

These different messages about the nutritional value of potatoes are targeted at different audiences in terms of the importance of potatoes in their diets. As a stable food potatoes are the main source of energy but need to be complemented with other foods for a nutritionally balanced diet, and the same is true for wheat and rice as food staples. If this is not possible, we need to improve their nutritional value. We also want eating potatoes and their products to be an enjoyable experience and would therefore like them to have a nice flavour and texture but after cooking or processing because we are unable to digest their raw ungelatinized starch [3]. This chapter concentrates on nutritional value and deals only briefly with flavour and texture.

We need to be able to work out the nutritional value of potatoes from their chemical composition both after storage and cooking or processing. Then we can consider ways of improving their nutritional value. However, in order to set targets in terms of estimated average requirements (EAR), recommended daily allowances (RDA) and dietary reference intakes (DRI) of nutrients, we need to consider their fate after food digestion, which is not easy to determine. The nutrients must be taken up and utilized by our bodies to be of value. The technical term is bioavailable: the fraction of ingested nutrient that is absorbed and utilized by the body [4]. It needs to be measured in vivo, which is difficult. Hence, an in vitro proxy, bioaccessibility, is often used. This is the fraction (quantity) of ingested nutrient that is released from its food matrix during gastrointestinal digestion (oral, gastric and small intestinal) and is available for assimilation by intestinal epithelium and absorption into the body (bloodstream) [4]. Bioaccessibility is therefore a prerequisite for bioavailability.

Sources of more detailed information and references to original research can be found in the 2007 book chapter by Storey [5], the 2016 book *Advances in Potato Chemistry and Technology, 2nd Edition*, edited by Singh and Kaur [6], the 2019 Special Issue of the American Journal of Potato Research [7], my 2019 book chapter [8], the 2020 book chapter by Burgos et al. [9], the 2020 book *Potato Nutrition and Food Security*, edited by Raigond et al. [10], my 2021 book on potato breeding [11] and the 2022 review by Huey et al. [4] on the bioaccessibility and bioavailability of biofortified food and food products.

8.2 Chemical Composition

Table 8.1 gives values for the chemical composition of a potato weighing 100 g [2, 5–12]. This is typical of a small round potato used for boiling (70 × 60 × 50 mm). A larger oval potato for baking (85 × 70 × 55 mm) would weigh 200 g, and a large baking potato weighs 275–350 g. The range of values given in Table 8.1 covers raw potatoes and those cooked by boiling, baking or microwaving, both with and without their skins. If you check various websites, you may find different values from those given in Table 8.1, but the broad picture will be the same.

Water The water content of a potato tuber is in the range 74–82%. This means that the dry matter (DM) content is in the range 18–26%. The DM content determines specific gravity and affects the texture of cooked potatoes, their suitability

Table 8.1 Typical chemical composition of a raw or cooked potato tuber weighing 100 g [2, 5–12]

Chemical	Content
Water	80 g (74–82 g)
Starch	14 g (13–15 g)
Crude fibre	2 g (about)
Proteins	2 g (about)
Sugars	1 g (about)
Fat/lipids	0.1 g (about)
Potassium (K)	280–500 mg
Phosphorus (P)	27–59 mg
Ca, Mg, Na and Cl combined	71–84 mg
Iron	0.18–0.85 mg
Zinc	0.16–0.99 mg
Vitamin C (ascorbic acid)	6.5–46.0 mg
Vitamin B_1 (thiamine)	0.049–0.232 mg
Vitamin B_3 (niacin)	0.6–1.4 mg
Vitamin B_5 (pantothenic acid)	0.37–0.52 mg
Vitamin B_6 (pyridoxine)	0.36–0.54 mg
Vitamin B_9 (folates)	0.010–0.047 mg
Anthocyanins: White flesh and skin	None
Anthocyanins: Red and purple flesh and skin	5.5–35.0 mg
Carotenoids: White flesh	0.033–0.064 mg
Carotenoids: Yellow flesh	0.058–0.175 mg
Carotenoids: Deep yellow (orange)-fleshed diploid potatoes	Up to 2.135 mg

for processing and their susceptibility to mechanical damage and bruising [13]. A high DM content is desirable in a processing potato because it reduces the energy required to remove water, increases the product yield per unit of fresh weight and lowers oil absorption [14]. An often-quoted average figure for the water content of potato tubers is 80%.

Starch Starch is the major component of the dry matter at 65–75%. Hence, the potato is seen as a major source of carbohydrate and energy (300–400 kJ/100 g or 72–96 kcal) in the human diet. Starch occurs in tuber cells in organelles called amyloplasts, which form single or complex granules. It is a polymer of glucose molecules in a linear chain in its amylose form (17–25% of starch) and as highly branched chains in its amylopectin form (75–83%). The amylose to amylopectin ratio is one of the factors that affects the functional properties of starch for both human consumption and industrial uses (some countries, particularly in North America and Europe, have a well-established potato starch industry). When we come to consider nutritional value and human health, it is useful to classify starch as rapidly digestible (20 min), slowly digestible (20–120 min, 7–37% of total) and resistant (remainder at 120 min, 12–27% of total), based on the amount hydrolysed during in vitro incubation with α-amylase.

Crude Fibre The cell wall and intercellular polysaccharides, mainly cellulose, hemicelluloses, pentosans and pectic substances, are termed crude fibre and comprise about 2% of the fresh weight. Dietary fibre comprises the insoluble and soluble polysaccharides of the cell wall together with lignin and resistant starch. It is not digested in our small intestine and hence is considered unavailable but desirable for a healthy digestive system. The tuber skin contains more dietary fibre than the flesh.

Proteins Proteins make up about 2% of the fresh weight of tubers. This is about 10% on a dry-weight basis and comparable to most cereals, including rice and wheat. However, many of the food chemists and nutritionists of the second half of the nineteenth century considered the potato a poor substitute for meat because of its low protein content [1]. The main potato storage protein is patatin, a group of high molecular weight (40–45 kDa) glycoproteins that represents 40% of the soluble protein found in tubers. Protease inhibitors are the second main group of proteins, accounting for a further 30–40%. The biological value of the tuber protein is between 90 and 100, greater than most non-animal proteins and close to the whole chicken egg protein standard with a value of 100. The biological value is the proportion of absorbed protein from a food that is incorporated into the proteins of the human body relative to a standard. Potato proteins are relatively low in the sulphur amino acids, methionine and cysteine, and tryptophan, but relatively high in lysine, which is low in rice and wheat (methionine, tryptophan and lysine are essential amino acids, which humans are unable to make). Free amino acids account for around 50% of the total nitrogen. Overall, it is reasonable to conclude that as a staple food, the potato provides the human diet with a significant amount of protein of good amino acid balance.

Sugars Although starch is the major carbohydrate in potatoes, the small but varying amounts of glucose and fructose (reducing sugars) and sucrose are important for processing quality (total sugars about 1% of fresh weight). High levels of reducing sugars are associated with the production of unacceptably dark and bitter-tasting fry products by non-enzymic browning, as well as with the production of acrylamide, a Group 2A carcinogen, as discussed later.

Fat/Lipids Lipids make up only a tiny fraction of the fresh weight of a potato (as low as 0.1%) and comprise phospholipids, glycolipids and neutral lipids in forms containing esterified fatty acids. In other words, potato tubers don't contain lipid reserves, and hence their lipids are unlikely to have much impact on human health.

Minerals A review published in 2009 [12] confirmed previous reports that potato tubers are high in potassium (280–500 mg/100 g fresh weight), which can help to control high blood pressure, and identified considerable genetic variation between modern potato cultivars for the 10 minerals examined: nitrogen (N), phosphorus (P), potassium (K), sulphur (S), calcium (Ca), magnesium (Mg), copper (Cu), iron (Fe), manganese (Mn) and zinc (Zn). Again, the food chemists and nutritionists of the second half of the nineteenth century disliked the low phosphorus content of potatoes compared with meat [1]. Today, however, when the potato is viewed as a staple food, the focus is on its contents of the micronutrients iron (0.18–0.85 mg/100 g fresh weight) and zinc (0.16–0.99 mg/100 g fresh weight) because of extensive global deficiencies in our human diet and the consequent health problems.

Vitamins As a staple food, the potato is an important source of vitamin C (ascorbic acid) (6.5–46.0 mg/100 g fresh weight), as well as the B vitamins, albeit at lower concentrations (individually in the range 0.01–1.4 mg/100 g fresh weight). The most abundant in potato tubers are B_1 (thiamine), B_3 (niacin), B_5 (pantothenic acid), B_6 (pyridoxine) and B_9 (folates). Currently there is interest in selecting potatoes for higher concentrations of folate and vitamin C.

Anthocyanins The pink, red, blue and purple pigments in the tuber skin, and sometimes the flesh, are anthocyanins, which are antioxidants (as is vitamin C and chlorogenic acid) and hence classed as phytonutrients. Antioxidants prevent oxidative reactions, often by scavenging reactive oxygen species, before they can damage cells. There is interest in red-fleshed and purple-fleshed tubers as speciality potatoes; for example, AmaRosa and Purple Pelisse, which are promoted as a healthy food because of their antioxidant content [15, 16]. Total anthocyanin contents in red-fleshed and purple-fleshed potatoes are in the range of 5.5–35.0 mg/100 g fresh weight.

Carotenoids The yellow flesh colour of tubers is due to the presence of carotenoids, which are also antioxidants. There is particular interest in lutein and zeaxanthin as they can help prevent macular eye degeneration [17]. The yellow-fleshed, cultivated diploid species of the Andes have high concentrations of these carotenoids (0.396 and 1.106 mg/100 g fresh weight, respectively) [18]. Furthermore, total carotenoid levels can reach 2.135 mg/100 g fresh weight in these diploids.

So, what can we tentatively conclude about the nutritional value of potatoes? The New Scientist article of 2 August 2008 produced similar figures to those in Table 8.1 for 100 g of baked potato including skin and compared the values with 100 g of boiled fresh pasta and 100 g of boiled long grain rice. The article concluded that overall potatoes compared favourably with wheat and rice as a staple food, although there were differences [2]. This statement is correct in implying that potatoes could replace cereals as a staple food (main energy source). However, if these staples are not complemented with other foods to provide a balanced diet, iron, zinc and vitamin A deficiencies can be a problem. The statement that Aracely's potato diet lacked all vital minerals and nutrients is correct (if all is replaced by some) for rural communities in the high Andes of Bolivia and Peru where potatoes make up a very large part of their diet. Finally, the UK National Health Service (NHS) website is correct about the nutritional value of potatoes in the United Kingdom, where we eat them primarily as a source of energy in the form of starch as part of an overall meal.

At this point it is worth considering two research papers from the Netherlands, which were published in 2002 [19, 20]. They examined the effects of food consumption patterns on land requirements for their production. The authors pointed out that a diet of only wheat (bread) could prevent starvation for a few weeks but would cause malnutrition in the long run because it lacks many essential nutrients. In contrast, they quoted a diet from the Netherlands Nutritional Council in 1998 that could meet a person's basic daily physiological needs for energy (10 MJ) and nutrients: bread (200 g), potatoes (200 g), vegetables (175 g), fruits (200 g), milk and milk products (375 g), cheese (30 g), raw meat (100 g), meat products (23 g), low-fat spread (30 g) and margarine (15 g). The authors called this a subsistence-level diet because it provides bodily health for a total life span. Here the total chemical composition of the potatoes would be used to work out their contribution to the whole diet. In practice, the food packages that we eat are also influenced by social and cultural demands and vary between countries and between people, often depending on their affluence. More affluent diets containing greater amounts of meat, fats and beverages require more land for their production. In contrast, land requirements for potatoes are low (0.2 m^2 year kg^{-1}; i.e., area per kg crop year) compared with butter (13.8), coffee (15.8), vegetable oil (20.7), beef (20.9), margarine (21.5) and tea (35.2). Although in 1990 a Dutch household was consuming only 1.8 kg of tea per year compared with 153.6 kg of potatoes, the land required for potato production (30.72 m^2) was less than that for tea (63.36 m^2). The land requirement for beef was even higher (175.56 m^2) with household consumption at 8.4 kg per year. The low requirement for land is one reason potatoes could contribute more to feeding the world and why they have often been the staple food of the poor, but the two research papers also reveal the complexity of food production systems. More information on food systems can be found in the second edition of *The Atlas of Food*, published in 2008 [21], but it is now time for us to look at the effect of storage on the chemical composition of potatoes.

8.3 Effect of Storage on Chemical Composition

Although it is nice to eat freshly dug potatoes, a sufficient supply between crops means storage, and this can be for as much as 6 months when only one crop is grown each year. Although it is sometimes possible to store potatoes at ambient temperatures for a few weeks [22], refrigeration is the best option for long-term storage, as recognized in India. A study published in 2019 identified a need for an additional 16–24 million tonnes of cold storage capacity in India to meet a projected increase in potato demand of 20–30 million tonnes by 2030 [23]. Potatoes are either stored in boxes (Fig. 8.1) or in bulk, the latter being common for processing.

Low temperature (ideally 2–4 °C) and high relative humidity (95%) reduce respiration and transpiration, respectively, and hence physiological loss of tuber mass and quality, but some starch breakdown and sugar conversions can occur [24]. Seed potatoes for the next crop can be stored at the ideal temperature of 2–4 °C because they are planted rather than eaten or processed. In contrast, potatoes for table use are stored at 4–6 °C and those for processing at warmer temperatures of 6–8 °C for French fries and 7–12 °C for crisps [24]. The warmer temperatures above 4 °C reduce or prevent (above 8 °C) the production of glucose and fructose (cold-induced sweetening), which would be a particular problem in potatoes for processing. However, warmer temperatures allow sprout growth to occur once tuber dormancy ends. Hence, potatoes for processing are treated with sprout inhibitors because their dormancy period is usually too short to prevent sprout growth. For over 50 years the major globally used sprout suppressant has been chlorpropham (CIPC), but not in the European Union since 1 January 2020 because of concerns over residues in tubers. Alternatives now being used are ethylene (Fig. 8.1), maleic hydrazide, 1,4-dimethylnaphthalene and spearmint oil [25]. In some crisping cultivars stored at 7–8 °C, ethylene increased respiration rates and the breakdown of starch to reducing sugars, resulting in unacceptably dark fry colours. However, at this year's Potatoes in Practice in Dundee (8 August 2024), I heard that a company, "Restrain", specializing in potato storage solutions, had solved the problem with their new Precision Ethylene Treatment (PET) technology. It utilizes ultra-low doses of ethylene to optimize potato storage.

Much has been published on cold-induced sweetening, but I am going to give just one example from India [26]. The researchers studied three cultivars, which are popular with Indian farmers: Kufri Chipsona-1 (for processing), Kufri Jyoti and Kufri Pukhraj. They stored them for 180 days at both 4 and 12 °C. The potatoes stored at 12 °C were treated twice with CIPC. The reducing sugar (glucose + fructose) content of Kufri Chipsona-1 increased from 180 mg/100 g fresh weight (0.18%) to 712 mg/100 g fresh weight (0.71%) after 180 days of storage at 4 °C but remained below 0.33% at 12 °C. In contrast, cultivars Kufri Jyoti and Kufri Pukhraj exceeded this level at 12 °C as well as at 4 °C. Hence, these results confirmed the suitability of Kufri Chipsona-1 for processing, provided it was stored at 12 °C with sprout suppressants. A reducing

Fig. 8.1 Box storage of potatoes (left) in the Netherlands with ethylene as a sprout suppressant (right) (22 January 2020). (From Bradshaw JE (2021) Potato Breeding: Theory and Practice. Springer, Cham with permission)

sugar content below 0.33% is considered acceptable for processing, with as low as 0.1% ideal for the manufacture of crisps [14]. Interestingly, the advice in the United Kingdom to consumers is to store the potatoes they buy from the supermarket in a fridge rather than at room temperature. That way they will last more than three times as long before they lose their quality through sprouting and weight loss [27]. In the days before refrigerators, potatoes were stored on racks in a dry, dark room (larder). If potatoes were stored in the light, greening would occur as a result of the formation of chlorophylls a and b. These are harmless and tasteless to humans but may indicate the production of toxic glycoalkaloids.

Vitamin C (ascorbic acid) is an essential micronutrient in the human diet, with scurvy a well-known symptom of vitamin C deficiency. Potatoes are noted for their high content of the vitamin, but this is the one micronutrient where storage can definitely have an adverse effect on the concentration. One recent review concluded that potatoes can lose up to 60% of their vitamin C content during cold storage, depending on temperature, length of storage and cultivar [28]. I am going to present some results from researchers at the International Potato Centre (CIP) in Lima, Peru, on the ascorbic acid concentration of native Andean potatoes grown in the central Peruvian Andes between 3280 and 3800 masl [29]. The 25 potatoes belonged to Groups Andigena (13), Stenotomum/Goniocalyx (9), Chaucha (2) and Tuberosum (1). The researchers determined the ascorbic acid content of peeled potatoes at 2, 9, 18 and 26 weeks after storage in a dark, well-ventilated room with a mean temperature of 13 °C and RH of 69%, typical of the storage used by Andean farmers. The content decreased with storage for all 23 varieties actually investigated. The retention after 26 weeks ranged from 22.0% to 76.8%, which was consistent with the wide range of values that can be found in an albeit limited literature. The Goniocalyx cultivar Maria Cruz had the highest content at 2 (100% retention), 9 (77% retention) and 18 (42% retention) weeks and the second highest content at 26 weeks (34% retention). Its content 2 weeks after harvest did not vary much between the three Andean sites: 34.0, 36.9 and 33.6 mg/100 g fresh weight. In contrast, the Stenotomum cultivar Morada Taruna had the best retention (100%, 85%, 73% and 77% at 26 weeks) but one of the lowest contents across the three sites: 11.0, 9.1 and 10.6 mg/100 g fresh weight. Cultivar Maria Cruz also retained the highest content after boiling (20.6 mg/100 g fresh weight) despite the lowest retention of 54% (range 54–97%). The boiling was done 2 weeks after harvest in a stainless-steel pot until the peeled tubers were cooked (easy penetration of a stainless-steel probe), which took between 25 and 35 min depending on variety. The tubers had been stored at 5 °C and then placed in the pot and covered with cool water, which was brought to the boil. The researchers concluded that 100 g of boiled potatoes of cultivar Maria Cruz could provide adults with 17–20% of their RDA of ascorbic acid and that this percentage could be more given the high level of potato consumption in the highlands of Peru (200 g per day for children and 800 g per day for women). The percentages are based on a suggested RDA for ascorbic acid of 100–120 mg per day for adults to remain healthy. Finally, 4 weeks after harvest, three cooking methods were compared on six of the cultivars, including Maria Cruz. Retention of ascorbic acid in unpeeled tubers was lower with microwaving for 4–5 min (24%) and with baking in an oven at 160–180 °C for 39–57 min (32%) than with boiling (75%), the figures in brackets being for Maria Cruz. The message is clear: vitamin C content is highest in freshly dug potatoes, but some can be lost in cooking.

8.4 Effect of Cooking and Processing on Chemical Composition

Andean farmers must have baked their fresh potatoes in the embers of a fire or cooked them in an earth oven on hot stones because ceramics were not developed until at least four millennia after the start of domestication [30]. Much later in Europe, we learn from John Gerard's *Herball* of 1597 that boiling in water can be added to baking in embers as a cooking method. Boiling is quicker, taking as little as 15–20 min at 100 °C, compared to today with 60–90 min of baking at

180–220 °C in an oven. Nevertheless, hot baked (in jacket, i.e., in their skin) potatoes were popular on the streets of London in the 1850s [31] and are still popular today when served with a variety of fillings and toppings, or at least with the addition of butter. Tables providing information on the nutritional value of potatoes commonly present data for boiled or baked potatoes, either with or without their skins.

Table 8.2 shows the effect of cooking on the chemical composition of 100 g of potatoes. The figures are derived from American values (USDA) in a 2019 review paper by Katherine Beals in the American Journal of Potato Research [32]. Although there were differences between boiled, baked and microwaved potatoes with and without their skins, I think that the average values given are a fair reflection of the effect of cooking. There is a slight reduction in the water content, increases in the energy, carbohydrate (mainly starch) and fibre contents, and a reduction in protein content. There is a small increase in potassium content, increases in magnesium and zinc contents and reductions in calcium and iron contents. There is a reduction in vitamin C content but varying increases in the B vitamins (although another review states a reduction in B_6 [9]).

It can be argued that boiling and baking, along with steaming and microwave cooking, provide the healthiest potatoes for eating because they do not use fats or oils. They can be considered part of the solution to the United Kingdom's current "Junk Food Cycle", in which 80% of processed food is cheap and popular but unhealthy because it is high in salt, refined carbohydrates, sugar and fats, and low in fibre [33]. The United Kingdom is not alone in facing this problem.

The two most popular types of processed potato products are French fries (chips in the United Kingdom) and chips (crisps in the United Kingdom). They started out as types of fried potatoes. Pommes frites (French fries) became a culinary item in France and Belgium around 1800 [3], although the Friet Museum in Bruge suggests that they could have originated in the Mosane region of Belgium during the harsh winters of the 1750s when the rivers were frozen. In the absence of fish, the local inhabitants cut potatoes into slices resembling small fish and fried them. In 1860, Joseph Malin opened the first shop in London to fry fish and chips (French fries) together, thus establishing "fish and chips" as a good cheap meal for poorer families in Britain [34]. Today cookbooks recommend home production of French fries in a two-stage cooking

Table 8.2 Effect of cooking (average of boiled, baked and microwaved potatoes with and without their skins) on chemical composition of 100 g of mainly russet potatoes [32]

Chemical composition	Raw	Cooked	Cooked-raw
Energy	312 kJ	393 kJ	+26%
Water	78.3 g	74.5 g	−5%
Carbohydrate	17.6 g	21.5 g	+22%
Fibre	1.36 g	1.86 g	+37%
Protein	2.71 g	2.13 g	−21%
Fat	0	0	
Potassium (K)	421 mg	445 mg	+6%
Calcium (Ca)	13.6 mg	8.79 mg	−35%
Magnesium (Mg)	22.4 mg	26.7 mg	+19%
Iron (Fe)	0.746 mg	0.633 mg	−15%
Zinc (Zn)	0.271 mg	0.351 mg	+30%
Vitamin C (ascorbic acid)	18 mg	12 mg	−33%
Vitamin B_1 (thiamine)	0.081 mg	0.094 mg	+16%
Vitamin B_2 (riboflavin)	0.020 mg	0.032 mg	+60%
Vitamin B_3 (niacin)	1.1 mg	1.5 mg	+36%
Vitamin B_6 (pyridoxine)	0.14 mg	0.29 mg	+107%
Vitamin B_9 (folates)	0.0163 mg	0.0178 mg	+9%

process involving deep-frying in vegetable oil (rapeseed or sunflower) in a wire basket in a chip pan. The oil is commonly heated to 190 °C in the second stage, but a lower temperature of 170–175 °C is better to reduce production of acrylamide, a Group 2A carcinogen (see later). The lower temperature results in French fries that are an attractive light-yellow colour rather than golden brown.

The British passion for crisps (chips to the rest of the world) can now be dated as early as 1817. A crisp recipe has been found in an 1817 book sold at auction in London in 2023 for £420, namely *The Cook's Oracle* by William Kitchiner: "Peel large potatoes, and slice them the thickness of a two-penny piece. Dry them well in a clean cloth, flour them, and fry them in lard." The auction was reported in an article in *The Times* of 11 February 2023, which informed its readers that the name crisps was first associated with potatoes in 1897, that they were first produced commercially by Mikesell's Potato Chip Company in America in 1910, and that flavoured crisps first appeared in the 1950s [35]. Today the big, but in a sense trivial, debate is whether cheese and onion are more popular than salt and vinegar crisps.

Since the 1950s, potato processing into crisps, French fries and other frozen products, has grown into a global industry, which continues to expand to meet the demand for convenience food in the home, fast food in restaurants and snacking. By the early 2000s, between 50% and 60% of the potato crop was being processed in North America and some European countries, and we are now seeing increases in countries where the potato is primarily a staple food [36, 37].

The big differences between boiled and processed potatoes can be seen in Table 8.3, which shows the information on the packets of potatoes and potato products from my local supermarket. As we move from boiled to French fries to crisps, the water content goes down while the energy, starch, fibre and protein contents go up. Interestingly, the sugar content comes down. Not surprisingly, the fried products have a much higher fat content than boiled potatoes, as well as higher salt contents, although the boiling was done in unsalted water. There is no doubt that crisps are a very popular energy- and nutrient-dense snack, but their high fat content has raised health concerns with the global increase in obesity. In response, crisp manufacturers have almost eliminated the use of saturated and trans-unsaturated fats, which are associated with heart disease, as well as reducing salt content. The roast potatoes that complete a traditional Sunday lunch in Britain have a chemical composition much closer to the French fries than the boiled potatoes in Table 8.3, with a slightly higher water content but a slightly lower fat content than the fries.

8.5 After-Cooking Blackening

An unsightly non-enzymic discolouration of the tuber flesh, known as after-cooking blackening (ACB), can occur during the cooling of boiled and steamed potatoes and, to a lesser extent, in

Table 8.3 Chemical composition of 100 g of boiled and processed potatoes given on packets

	Anya potatoes boiled in unsalted water	Oven baked McCain French fries	Walkers lightly salted crisps
Energy	280 kJ	908 kJ	2174 kJ
Water	82.3 g	53.5 g	2.43 g
Starch	12.5 g	29.2 g	52.6 g
Fibre	1.8 g	4.5 g	6.5 g
Protein	1.8 g	3.3 g	6.4 g
Sugars	1.1 g	0.8 g	0.4 g
Fat (of which saturates)	<0.5 g (<0.1 g)	8.2 g (1.0 g)	31.0 g (2.5 g)
Salt	<0.01 g	0.55 g	0.67 g

baked potatoes and French fries [14]. During cooking, chlorogenic acid combines with iron to form a complex, which oxidizes on cooling to the ferri-dichlorogenic acid, which is responsible for the blue-grey discolouration at normal pH. However, chelating agents in a potato tuber, such as citric acid, compete with the chlorogenic acid for the iron to form a colourless complex. Hence, cultivar differences in ACB have been correlated with their contents of chlorogenic acid and, to a lesser extent, citric acid [38], and you sometimes see a squeeze of lemon juice in the water as a suggested solution to the problem.

8.6 Flavour and Texture

We want to enjoy eating cooked potatoes and potato products and hence would like them to have a nice flavour and texture, as well as being nutritious. Although some cultivars, such as Anya (Table 8.3), are marketed as having a distinctive nutty flavour, most are fairly bland. We are more likely to notice differences in texture both when cooking and eating potatoes. What is not in doubt is the large number of potato recipes now available in books, of which I will mention two examples. My first is the 2009 edition of the book *Potato, the Definitive Guide to Potatoes and Potato Cooking* by Alex Barker with over 150 recipes by Sally Mansfield [39]. The book provides a directory of 168 potato cultivars from around the world, dating back into the nineteenth century, and their suitability for different cooking methods. The recipes cover soups, starters and snacks, salads, side dishes, meat and poultry dishes, fish dishes, vegetarian dishes, and breads and scones. If you want even more recipes, you can turn to my second example and find another 70 in the 2019 book *Posh Potatoes* by Rebecca Woods [40]. Here I am going to provide a few brief comments on flavour and texture.

Flavour is a combined sense of taste and smell, which can be assessed in organoleptic tests by a professionally trained panel of sensory assessors, commonly referred to simply as a taste panel. The sense of taste (sweet, sour, salt, bitter and umami) is perceived on our tongues, whereas the sense of smell is perceived when volatiles come into contact with receptors in our nasal passages, either by sniffing or chewing, as explained in a review by McKenzie and Corrigan [41]. Cooking releases a wide range of volatile compounds, with the method determining the type and extent of flavour-producing reactions. Sensitive volatile analysis of boiled potatoes can be done by entrainment of headspace volatile organic compounds followed by various types of gas chromatography (Fig. 8.2). Sugars, lipids and amino acids are common precursors of over 100 compounds involved in the production of flavour [41].

Texture depends on many structural and chemical properties of the tuber, including starch

Fig. 8.2 Entrainment of headspace volatile organic compounds (left) and texture profile analysis using probe and load cell (right). (From Bradshaw JE (2021) Potato Breeding: Theory and Practice. Springer, Cham with permission; source: SCRI)

and dry matter (DM) content [42]. Sensory panellists can assess the components of texture, including mealiness (waxy/not mealy to very mealy/floury). Floury potatoes tend to have a high DM content and are prone to disintegration on boiling and thus require a shorter cooking time. Potatoes are often grouped into four cooking types: A, B, C and D, where A is a quite firm salad type and D is a very mealy, mushy type [43]. Textural qualities can be seen using scanning electron microscopy and confocal scanning laser microscopy. They can also be analysed using probes and suitable load cells that measure compression, puncture and shear forces (Fig. 8.2).

I now want to turn to four recent topics concerning health problems and opportunities associated with potatoes.

8.7 Acrylamide and Chips

Concern about acrylamide in starchy foods was first raised in Sweden in 2002 [44]. It is found in a wide range of foods such as French fries, crisps, roasted potatoes and root vegetables, toast, cakes, biscuits, cereals and coffee. It is formed during frying, baking, roasting, grilling and toasting above 120 °C from chemical reactions starting with reducing sugars and free asparagine, a nitrogen-rich amino acid [45]. The reactions are known as non-enzymic Maillard browning and affect the formation of flavour as well as colour. Early studies indicated that acrylamide-forming potential in heated wheat and rye flour was determined by the free asparagine concentration unless the flour was derived from sulphur-deprived wheat [46]. In contrast, reducing sugars were the main determinant in most data sets for potatoes [47], although free asparagine could sometimes contribute to the variation [48, 49]. One piece of research reported that free asparagine had an effect when its ratio with the reducing sugars (free asparagine/reducing sugars) fell below 2.26 [46]. We shall shortly see that reducing asparagine levels does, in fact, have a big effect on potatoes.

Acrylamide is classified as a Group 2A carcinogen (probably carcinogenic to humans) as a result of laboratory tests, in which animals (rodents) developed cancer. Reducing dietary intake of acrylamide has become the concern of governments, and the following is a brief indication of what has been happening in Europe. A "Scientific Opinion on acrylamide in food" was published by the European Food Safety Authority (EFSA) on 4 June 2015 [50]. Then on 11 April 2018, European Union mandatory mitigation measures came into force, which set benchmark levels of 750 μg kg^{-1} for potato crisps (chips) and 500 μg kg^{-1} for French fries (British chips). My understanding is that these are desirable rather than enforceable maximum limits. Here in the United Kingdom, as a precaution, the Food Standards Agency (FSA) recommends that we reduce the amount of acrylamide we consume [51]. Their advice when frying, baking, toasting or roasting starchy foods is to aim for a golden yellow or lighter colour. Furthermore, in the United Kingdom: "All food businesses operators are required to put in place simple practical steps to manage acrylamide within their food safety management systems. This ensures that acrylamide levels are as low as reasonably achievable in their food. Best practice, mitigation measures and benchmark levels for the reduction of the presence of acrylamide in food have been established as set out in the retained (after Brexit) EU Regulation 2017/2158."

A detailed account of acrylamide formation and mitigation methods in fried potato products can be found in a 2018 review by De Meulenaer et al. [45]. Mitigation requires removal of the acrylamide precursors or interference with the Maillard reaction. Many chemicals that have been added during processing in a laboratory setting have reduced acrylamide levels in the finished potato products, but their effects have yet to be confirmed on an industrial scale [45]. Hence, mitigation is focused on reducing acrylamide-forming potential in raw potatoes and during storage, where cold-induced sweetening can occur below 8 °C and is marked by 4 °C. Acrylamide-forming potential is affected by N, P, K and S fertilizer levels, but in a complicated way [45]. In one study, nitrogen application (200 kg/ha) increased acrylamide-forming potential by varying amounts in 9 of 13 cultivars but decreased it in two, whereas sulphur application (15 kg/ha) reduced glucose levels and mitigated the effect of high nitrogen in some potatoes classified as suitable for French fries [52]. In contrast, the desired outcome can be achieved in new cultivars by both conventional breeding and genetic modification.

Genetic variation exists in potato tubers for both reducing sugars and asparagine concentrations. At the Scottish Crop Research Institute, we examined the genetic variation in the cross between processing clone 12601ab1, known to be resistant to cold-temperature sweetening [53], and table cultivar Stirling. A subset of 43 clones and 5 controls from a replicated trial in 2007 were stored at 4 °C for 4 months. We then determined the concentrations of the reducing sugars and asparagine in their tubers, followed by the fry colour and acrylamide content of their crisps, which had been fried at 180 °C. Fry colour was mainly explained by reducing sugar content (71.4% of variation), whereas about half (53.4%) of the variation in acrylamide content was explained by a combination of reducing sugars and asparagine, with no correlation between them. The clone with the lowest acrylamide content was third lowest for reducing sugars, fourth lowest for asparagine and second best for fry colour (light colour). We therefore concluded that selecting for lower levels of both reducing sugars and asparagine were worthwhile breeding objectives [48]. In Scotland, Cygnet Potato Breeders have produced a crisping cultivar, Olympus, from the cross between Atlantic and 12601ab1, which is suitable for long-term storage and produces extremely low levels of acrylamide during processing [54]. Likewise, in 2015, breeders in the United States released Payette Russet, a new processing cultivar that gave an 81% reduction in acrylamide content in its French fries, compared with Ranger Russet and Russet Burbank, after 8 months' storage at 9 °C. It had lower concentrations of both asparagine and reducing sugars [49].

In the United States, the genetic transformation approach is being pursued by the J R Simplot Company, but it is not yet clear if their new genetically modified (GM) Innate™ potatoes will be acceptable to companies like McDonald's. Monsanto had been the first company to commercialize GM processing cultivars in North America from 1995. However, the leading processing and fast-food outlet companies in North America were reluctant to purchase them because of consumer concerns about GM potato products. As a consequence, Monsanto stopped marketing them in 2001 [55]. There was also considerable resistance to GM potatoes in Europe, and none are currently being marketed. Returning to the United States, on 20 March 2015, the authorities declared the first Innate™ potatoes as safe and nutritious as their conventional counterparts. They had lower levels of asparagine in their tubers, which resulted in greatly reduced acrylamide production, as well as desirable resistance to black spot bruise. The second group of Innate™ potatoes was approved for planting in March 2017. They have lower levels of reducing sugars, which reduces acrylamide production as well as allowing colder storage. The area of both groups of Innate™ potatoes is slowly increasing [55].

The lower levels of tuber asparagine and reducing sugars were the results of a big research effort over more than 15 years [11]. The biosynthesis of asparagine is complicated (Fig. 8.3), but the final key step is the amidation of aspartate, as can be seen by consulting a modern textbook on plant biochemistry [56]. A group of scientists at Simplot Plant Sciences found that silencing the *asparagine synthetase 1* (*StAS1*) gene in potato tubers but not elsewhere in the plant reduced asparagine levels by 60–80% and resulted in a 70% reduction in acrylamide-forming potential but without affecting the tuber shape and yield of field-grown potatoes [57]. The enzyme asparagine synthetase 1 is not produced, and as a consequence, aspartate is not converted to asparagine. It was not necessary to silence a second *asparagine synthetase* (*StAS2*) gene. The gene silencing was done by a technique known as RNA interference because the genetic transformation results in double-stranded RNA being produced that triggers the degradation of targeted gene transcripts (i.e., post-transcriptional gene silencing).

Photosynthesis

↓

3-Phosphoglycerate

↓

Phosphoenolpyruvate

↓

Oxaloacetate

↓

Aspartate → Asparagine

(amidation: C-NH_2 replaces C=O)

↓

Threonine

Isoleucine

Lysine

Methionine

Fig. 8.3 Aspartate as a precursor of five other amino acids, including asparagine [56]

Starch breakdown is also a complicated process, but one of the factors that determines the levels of the reducing sugars glucose and fructose is their reversible conversion to sucrose, catalysed by the enzyme invertase (sucrose ←→ glucose + fructose) [56]. A group of researchers from China and the United States silenced the vacuolar invertase gene *VInv* and the two asparagine synthetase genes, *StAS1* and *StAS2*, in cultivar Russet Burbank [58]. Again, they used an RNA interference construct for the gene silencing. They found that the fried potato products from the best two lines contained 84 and 89 μg/kg acrylamide, respectively, compared with 1233 μg/kg for the Russet Burbank control, well below the benchmark value of 500 μg/kg. There was, however, one concern needing further research. The field plants grown from transplants were normal in a trial in 2013, but those grown in 2014 from seed tubers were stunted and chlorotic, although their tubers developed normally. Interestingly, the acrylamide level was almost as low (88 μg/kg) in the best line previously developed from Russet Burbank with just *VInv* silenced. The authors concluded that the production of cultivars with low reducing sugars may be an adequate solution to the acrylamide problem in processed potatoes. The scientists at Simplot Plant Sciences successfully targeted other genes to lower the level of reducing sugars in their second group of Innate™ potatoes [59]. Finally, a group of Australian researchers has recently shown in cultivars Atlantic (a processing cultivar) and Desiree (a much-researched cultivar) that by using a number of guide RNAs, the CRISPR-Cas9 gene editing system can also simultaneously "knock out" all alleles of the genes encoding vacuolar invertase (*VInv*) and asparagine synthetase 1 (*StAS1*) [60]. Again, the researchers achieved up to 80% reduction in acrylamide concentration when both genes were edited but up to 86% in one example of just *VInv* being edited. Clearly there is more than one solution to the acrylamide problem, and time will tell which proves to be the most commercially successful one.

In this section, we have seen that genetic transformation can be used to silence genes and stop the production of undesired acrylamide. In the next section, we shall see the reverse, namely the introduction of genes to produce a desirable chemical.

8.8 Golden Potatoes and Rice

A major cause of poor eyesight and blindness in children in Asia, Africa and Latin America is vitamin A deficiency, which results from a dietary lack of its precursors known as provitamin A, including β-carotene [61]. Provitamin A is lacking in milled rice grains and is either absent or present in low concentrations in potatoes and their wild relatives.

Golden Rice was conceived in the late 1980s by Professors Ingo Potrykus and Peter Beyer from Zurich and Freiburg, respectively. An understanding of carotenoid biosynthesis allowed them to genetically engineer rice to produce some provitamin A and hence acquire a new golden-yellow colour. Not surprisingly, it became known as Golden Rice [62]. Subsequent genetic modification of a temperate japonica rice cultivar, Kaybonnet, added a combination of two transgenes, one from maize (phytoene synthase,

Zmpsy1) and the other from a common soil bacterium, *Erwinia uredovora* (carotene desaturase, *crtI*). This combination of genes substantially increased carotenoid accumulation in what became known as Golden Rice 2 (GR2) [63]. About 30 μg of β-carotene per gram of endosperm was achieved and considered adequate for a provitamin ingested with food to be nutritionally effective. Then a non-scientific problem was encountered. GR2 was developed for humanitarian purposes, but the underpinning research required procedures and technology protected by 70 Intellectual Property and Technology Protection Rights, which were owned by 32 different companies and universities [64]. Eventually, they all agreed to forgo profits provided the annual income of recipient small farmers remained below US$10,000. There were, however, still obstacles to overcome. The GR2 provitamin A trait needed to be transferred into locally adapted cultivars for use in developing countries. This was done for three popular Asian rice cultivars, with transformant GR2E, by a breeding technique known as marker-assisted backcrossing [65]. There was also the need to overcome the traditional habit of eating white rice [61]. Eventually, on 23 July 2021, the Philippines became the first country in the world to approve GR2E for planting. The necessary developmental work had been done by the Philippine Rice Research Institute (PhilRice) and the International Rice Research Institute (IRRI). According to IRRI, GR2E is expected to provide up to 50% of the estimated average requirement for vitamin A in young children (https://www.irri.org). A case for extending approval of GR2E to other countries as quickly as possible was made by scientists in an "Opinion: Allow Golden Rice to save lives" in the Proceedings of the National Academy of Sciences of the United States on 21 December 2021 [66]. However, on 17 April 2024 the Court of Appeals in the Philippines revoked the permit for golden rice as a result of a case brought by Greenpeace Southeast Asia and other groups. The judges found that the nation's constitution required the government to follow the precautionary principle of waiting to approve new crops until there was a scientific consensus on their safety for humans and the environment. My understanding is that the Philippine's government will appeal the decision to their Supreme Court [67]. The arguments go on.

Golden Potatoes (GP) have also been produced, but I would not like to predict when they will be grown by farmers for human consumption. The starting point for the synthesis of carotenoids is geranylgeranyl diphosphate (also known as geranylgeranyl pyrophosphate), but you cannot tell from looking at the pathways which ones will predominate in potato tubers (Fig. 8.4). That will depend on the control of gene expression and enzyme levels. Genetic transformations have therefore involved an element of trial and error, or if you prefer, hypothesis testing.

Diretto et al. reported success in 2007 [68] from the transformation of cultivar Desirée (light yellow flesh) with a mini-pathway of three genes from the bacterium *Erwinia herbicola* that encoded the enzymes phytoene synthase (*CrtB*), phytoene desaturase (*CrtI*) and lycopene beta-cyclase (*CrtY*). Use of a tuber-specific (patatin) promoter resulted in tubers with deep yellow flesh, but without any adverse leaf phenotypes. Carotenoid levels in one transformant (GP17) had increased from 5.8 to 114.4 μg g^{-1} DW and β-carotene from 0.013 to 47.4 μg g^{-1} DW compared with "Golden Rice 2" at 31 μg g^{-1} DW. The effects on other carotenoids were increases of phytoene from 0 to 13.8, α-carotene from 0 to 6.2 (also a provitamin A carotenoid), lutein from 1.0 to 23.1, zeaxanthin plus antheraxanthin from 1.9 to 11.0, violaxanthin from 0.7 to 5.6 and neoxanthin from 0.6 to 5.7 μg g^{-1} DW (Fig. 8.4). Another similarly produced transformant (GP30) had the second highest β-carotene content at 26.4 μg g^{-1} DW.

Ten years later, Chitchumroonchokchai et al. assessed the potential of golden potatoes to improve diets in developing countries and published their results in 2017 [69]. They used in vitro digestion to investigate the bioaccessibility of carotenoids and α-tocopherol (αTC, an isomer of vitamin E) in boiled samples of Desirée, GP17 and GP30 potatoes. They confirmed the

Geranylgeranyl diphosphate

↓ **Phytoene synthase from *Erwinia herbicola***

Phytoene (colourless) **[13.8]**

↓ **Phytoene desaturase from *Erwinia herbicola***

Lycopene

Lycopene epsilon-cyclase ↓ ↓ **Lycopene beta-cyclase from *Erwinia herbicola***

δ-Carotene (γ-Carotene)

↓ ↓

α-Carotene **[6.2]** **β-Carotene (deep yellow) [47.4]**

↓ ↓ β-Carotene hydroxylase

α-Cryptoxanthin β-Cryptoxanthin

↓ ↓ β-Carotene hydroxylase

Lutein **[23.1]** Zeaxanthin (orange)

↓ ↓ Zeaxanthin epoxidase

Lutein-5,6-epoxide Antheraxanthin (yellow) [+ Zeaxanthin **11.0]**

↓ Zeaxanthin epoxidase

Violaxanthin (yellow) **[5.6]**

↓

Neoxanthin (yellow) **[5.7]**

Fig. 8.4 Carotenoids in golden potato (GP17) with a total of 114.4 $\mu g\ g^{-1}$ DW compared with 5.8 $\mu g\ g^{-1}$ DW in untransformed control (δ-carotene pathway also produces yellow compounds). (Data from Ref. [68])

high levels of β-carotene (36.9 and 55.8 $\mu g\ g^{-1}$ DW) and α-carotene (31.4 and 34.8 $\mu g\ g^{-1}$ DW) in raw samples of GP17 and GP30, respectively, making provitamin A totals of 68.3 and 90.6 $\mu g\ g^{-1}$ DW; although the actual values differed from those previously published. They also found that α-tocopherol had unexpectedly increased from 4.96 in Desirée to 78.3 and 90.4 in GP17 and GP30, respectively. Retention in the boiled samples for provitamin A was 90% and 96% for GP17 and GP30, respectively, and 100% in both for α-tocopherol. Then further retention on in vitro digestion for GP17 and GP30 was 12% and 15% for provitamin A (α-carotene plus β-carotene) and 24% and 25% for α-tocopherol. Uptake by human intestinal cells during incubation for 4 h was around 15% for provitamin A and 53% for α-tocopherol for both GP17 and GP30. On the basis of these figures, the researchers concluded that for children and women of reproductive age, 150 g of boiled GP30 (better than GP17) potatoes could contribute 42% and 23%, respectively, of their daily requirement of provitamin A in terms of retinol activity equivalents (RAE), as well as 34% and 17%, respectively, of their daily vitamin E requirement.

8.9 Carotenoid-Rich Potatoes

Although Golden Potatoes remain just a possibility, the diploid yellow-fleshed potatoes of the Andes can be eaten for other health benefits associated with high concentrations of carotenoids

(Table 8.1). There is particular interest in lutein and zeaxanthin, as they can help prevent macular eye degeneration [17], as well as more general interest in higher levels of carotenoids as antioxidants. Selection in a breeding program for intense yellow flesh does result in higher concentrations of carotenoids but does not guarantee higher levels of particular carotenoids such as lutein and zeaxanthin, as these depend on the action of a number of key enzymes [70]. The main carotenoids present in cultivated potatoes are the ones below the dashed line in Fig. 8.4: lutein, lutein-5,6-epoxide, zeaxanthin, antheraxanthin, violaxanthin and neoxanthin. Underlined are three key enzymes, which are encoded by three genes: *β-carotene hydroxylase 2* (*Chy2* on chromosome 3), *zeaxanthin epoxidase* (*Zep* on chromosome 2) and *lycopene epsilon cyclase* (*Lcye* on chromosome 12) [71–73]. Combinations of alleles (variants) at these genetic loci affect flesh colour and carotenoid concentrations. Our current understanding is that one copy of *Chy2* allele 3 is sufficient for yellow flesh, whereas both alleles in a diploid need to be c*hy2* for white flesh (i.e., c*hy2* is a recessive allele). One copy of another *Chy2* allele known as *Or* results in orange flesh and large amounts of zeaxanthin (1426 μg compared with 304 μg 100 g^{-1} FW in white potatoes), but not increased amounts of lutein (146 μg 100 g^{-1} FW in both orange and white potatoes) [72]. One copy of *Chy2* allele 3 combined with two copies of *zep* allele 1 (a recessive) also results in orange flesh and high levels of zeaxanthin [73, 74]. The *zep* allele 1 effect can also be achieved by inhibiting the conversion of zeaxanthin to violaxanthin through tuber-specific downregulation (gene silencing) of *Zep* by RNA interference [75]. In contrast to these results, none of the *Lcye* alleles found to date have much effect on flesh colour [73]. I think that we can conclude that there is still more to learn about the control of carotenoid biosynthesis.

The above results suggest that it might be difficult to achieve orange flesh and high levels of zeaxanthin in tetraploid potatoes if four copies of the *zep* allele 1 are required. However, in Japan, the new tetraploid cultivar Nagasaki Kogane, with high carotenoid content, yellowish/orange flesh and other desirable characteristics, has been bred from the diploid cultivar Inca-no-mezame, which in turn was derived from crosses including a Phureja clone [76]. Reported zeaxanthin concentrations in Nagasaki Kogane were 801.5 μg 100 g^{-1} FW compared with 1131.7 μg 100 g^{-1} FW in Inca-no-mezame.

The effect of cooking and digestion on the lutein and zeaxanthin concentrations of native Andean potatoes have also been studied [18, 70]. Lutein and zeaxanthin concentrations were not affected by boiling, whereas those of violaxanthin and antheraxanthin were significantly reduced in all (CIP) accessions examined. The bioaccessible concentrations of lutein and zeaxanthin after in vitro digestion were estimated for seven yellow-fleshed diploid accessions. Amarilla Tumbay (Goniocalyx accession 701862) had the highest bioaccessible lutein concentration of 281 out of 396 μg 100 g^{-1} FW, but the lowest zeaxanthin concentration. Yema de Huevo (Phureja accession 704218) and another unnamed Phureja accession (703566), both deep-yellow-fleshed, had the highest bioaccessible zeaxanthin concentrations of 608 out of 1106 and 665 out of 1196 μg 100 g^{-1} FW, respectively, but Yema de Huevo had the lowest lutein concentration. The researchers worked out that 701862 would provide 14% of the lutein intake suggested for health benefits, and 703566 and 704218 would provide 50% more than the suggested zeaxanthin intake, assuming a mean potato intake in the Andes of 500 g per day. In other words, boiled, deep-yellow-fleshed potatoes provide a significant amount of bioaccessible zeaxanthin to the human diet.

In this and the previous sections, we have seen that genetic transformations resulting in genetically modified potatoes have increased our understanding of plant biochemistry. They show us what can be done by conventional potato breeding because the necessary genetic variation is available through sexual hybridizations of cultivated potatoes and their wild relatives. They also show us where genetic transformation is required for novel biochemistry not available from such hybridizations. Where both methods are possible, we can choose the one that is fastest, most efficient and most effective.

8.10 Resistant Starch and Glycaemic Index

Glycaemic index (GI) measures the effect of food ingestion on blood glucose levels over a period of 2 h. Foods with a high GI result in a rapid increase and corresponding insulin responses [77]. They can therefore contribute to obesity, glucose intolerance, type-2 diabetes and cardiovascular disease but could help athletes who need carbohydrate sources to enhance muscle glycogen storage and to deliver carbohydrate to muscle and other organs during strenuous exercise [78]. GI is measured on a scale of 0–100, where glucose is the reference food with a score of 100. A value of 70 or more is high and 55 or less is low. As the form of energy storage in potatoes is almost entirely starch, they are assumed to have a high GI, although values from 56 to 94 have been reported in Great Britain [79]. If you eat a 100 g portion of boiled potatoes with a GI of 75 and containing 15 g of available starch, the glycaemic load (GL) is said to be 11.25 [75 (GI) × 15 (g starch) ÷ 100 (not the portion size)], but there are reservations about giving nutritional advice based solely on the GI/GL of individual food items. Potatoes are consumed in mixed meals with other foods, and GI/GL can be affected by these as well as the cooking method [80–84].

Despite the wide range of values found for potatoes, they are usually considered a high GI food, and this has raised concerns about their nutritional value and led to the search for low-GI ones [82, 85]. Clinical evaluation of GI is expensive, but values correlate well with in vitro glucose release from starch digestion using the enzyme α-amylase [86]. Starch can be classified into rapidly digestible (first 20 min), slowly digestible (20–120 min) and resistant (remainder at 120 min) [85]. Research found that the amount of rapidly digestible starch (RDS) was positively correlated with GI whereas the amount of resistant starch (RS) was negatively correlated, irrespective of whether the potatoes were uncooked, cooked or boiled and cooled for 48 h at 4 °C [85]. Interestingly, cool-storing after cooking has been shown to increase the amount of slowly digestible and resistant starch and decrease GI [87, 88].

Genetic variation has been found for starch digestibility and GI among breeding material and cultivars [81, 83, 84]. Nicola (GI = 69), for example, was identified as medium-GI and Carisma™ (a waxy type with GI = 53) as low-GI [87]. It was launched as a low-GI potato in Australia in 2010 and then marketed and grown in North America as a speciality potato.

Resistant starch is not digested in our small intestine and hence reaches the large intestine essentially intact. It is therefore considered part of our dietary fibre with similar physiological effects that contribute to a healthy digestive system [89, 90]. In this section we have therefore been considering how dietary fibre can be increased through more resistant starch. An additional way in the future could be to add a different kind of fibre through genetic transformation.

In 2000, Hellwege et al. [91] reported they had developed transgenic potato tubers of cultivar Désirée which, in place of some starch, could synthesize the full range of inulin molecules that occur naturally in globe artichoke (*Cynara scolymus*). The two genes they transferred from globe artichoke were *1-SST* and *1-FFT*. The former encodes sucrose:sucrose 1-fructosyltransferase and the latter encodes fructan:fructan 1-fructosyltransferase. Inulin is a storage carbohydrate and member of the fructan group of polysaccharides. It comprises a mixture of linear fructose polymers with different chain lengths and a glucose molecule on the end. The inulin (fructan) made up about 5% of the combined sugar, fructan and starch content (in hexose equivalents) of the transgenic tubers. No adverse effects were observed in the greenhouse-grown transgenic plants.

8.11 Mineral Biofortification

There are different ways of making sure that we get the micronutrients we need, both minerals and vitamins. The one I want to look at in this section is biofortification of iron and zinc in potatoes as a staple food.

Iron (Fe) deficiency anaemia is a global problem that is particularly common among women

(20%) and children (49%) in the rural highlands of Peru, together with zinc (Zn) deficiency and also vitamin A deficiency, which we considered earlier [92, 93]. Iron (Fe) and Zinc (Zn) were therefore chosen by the International Potato Centre (CIP) in 2004 as potato targets in the HarvestPlus Biofortification Challenge [94]. HarvestPlus and its partners work across the CGIAR (Consultative Group for International Agriculture Research) as part of the International Food Policy Research Institute (IFPRI) to "fight hidden hunger worldwide by scaling staple food crops that are bred to be rich in essential vitamins and minerals". These crops "should sustainably improve the health and livelihoods of smallholder farming households and low-income consumers", particularly in Africa, Asia and Latin America. According to their 20th anniversary report in 2023, iron varieties have been released for bean, lentil and pearl millet; zinc varieties for maize, rice and wheat; iron and zinc varieties for cowpea and sorghum; and vitamin A varieties for banana/plantain, cassava, maize and orange sweet potato. The report says that "over 100 million people in farming households are eating biofortified crops". Work on potatoes is a little behind these other staple food crops. Iron-zinc potato varieties are being tested in Peru, Bolivia, Nepal, Bhutan, Rwanda, Kenya, Ethiopia, Uganda, Malawi, China and India, but none have yet been released (https://www.harvestplus.org). So, let's look at the progress that has been made (progress with other crops can be found in the 2017 review by Bouis and Saltzman [95]).

Research has shown that genetic biofortification of Zn can be complemented with agronomic biofortification using foliar or soil fertilizers, whereas results are mixed for Fe [96–99]. Here, however, I am going to concentrate on the genetic biofortification of iron and zinc. The cultivated diploid potatoes of the Andes proved good starting material, as high concentrations were found with moderate heritabilities and hence good prospects for further increases [94, 100]. Concentrations in raw, peeled tubers ranged from 9 to 37 mg Fe kg^{-1} DW and from 8 to 20 mg Zn kg^{-1} DW. After two generations of selection in a breeding program, clones were found with values as high as 42.5 mg Fe kg^{-1} DW and 38.9 mg Zn kg^{-1} DW. The researchers estimated that their best material could provide significant contributions in Peru to the estimated average requirements (EAR) of children aged 4–8 years and female adults aged 19–30: 56% and 28%, respectively, for Fe and 37% and 22% for Zn. The calculations were based on potato consumption in Peru of 253 g per adult male equivalent per day. By 2017 CIP had more than 50 clones with the capacity to absorb and contain concentrations greater than 35 mg Fe and 35 mg Zn kg^{-1} tuber flesh DW (approximately 0.7 mg per 100 g FW) [100]. Clones with high contents, good general combining ability, and the ability to produce some diploid pollen were used to pollinate tetraploid potatoes [101]. New tetraploid hybrids were selected with higher contents than modern tetraploid clones; for example, iron contents of 0.40–0.49 mg per 100 g FW compared with 0.3 mg in current modern cultivars. Three biofortified clones were identified as potential cultivars after field and laboratory assessments conducted from 2018 to 2022 [101]. Thirty-seven tetraploid hybrids have been tested in Rwanda, and ten were selected for further evaluation [102]. However, as I have mentioned throughout this chapter, proof of nutritional value requires an assessment of bioavailability.

The first research published in 2015 used in vitro gastrointestinal digestion to assess the bioaccessibility of Fe from boiled samples of 12 Andean potatoes [103]. The bioaccessibility was high at between 64% and 79%. Furthermore, the complexity of bioaccessibility was revealed. There was a negative correlation between bioaccessible Fe and phenolic concentrations in the samples (chlorogenic acid is the principal phenolic acid), but a positive one with ascorbic acid concentrations, high levels of ascorbic acid being a known promoter of Fe absorption [104]. Hence, a breeder might want to increase ascorbic acid concentration as well as iron content, but not chlorogenic acid concentration, which is also associated with ACB (after cooking blackening). The high bioaccessibility of Fe is encouraging, but in vivo confirmation of bioavailability was needed.

Two related studies were published in 2023 on the bioavailability of Fe and Zn [92, 93]. The subjects were women living in the city of Huancavelica in the Peruvian Highlands (3600 masl). The iron study used 28 participants with moderate iron stores (mean plasma ferritin of 21.3 μg/L) in what is known as a single-blinded, randomized crossover design. Each subject consumed a sequence of biofortified and non-biofortified potato meals, which were labelled with two different stable isotopes of iron (0.3 mg iron sulphate with ^{58}Fe or ^{57}Fe given in 50 mL water with the meal). The isotopic labels were absorbed and incorporated into the blood erythrocytes and reached a stable plateau after 14 days. The fractional iron absorption (FIA) and total iron absorption (TIA) of the two types of potatoes were determined in a single blood sample taken 14 days after the last administration of the test meals, which was on day 40 of the study. Each of the 20 test meals (10 of each type) comprised 460 g of boiled, peeled and mashed potato divided into two equal portions. The results were as follows, with the biofortified values given first (for simplicity, I have given mean values without their standard deviations, i.e., the participant variation).

Iron	0.63 v 0.31 mg per 100 g of meal [biofortified 103% higher]
Phytic acid	39.34 v 3.10 mg per 100 g of meal
Ascorbic acid	7.65 v 3.74 mg per 100 g of meal [16.22 v 9.92 before storage][a]
Chlorogenic acid	15.14 v 22.52 mg per 100 g of meal
FIA	12.1% v 16.6% [biofortified 27% lower]
	[0.35 ÷ (0.63 × 4.6) v 0.24 ÷ (0.31 × 4.6)]
TIA	0.35 v 0.24 mg per 460 g meal [biofortified 46% higher]

[a]The mash was stored at −20 °C for 5 months until COVID-19 restrictions were lifted

The iron- and zinc-biofortified potato was diploid clone CIP311623.75, and the non-biofortified potato was the commercial variety Peruanita (diploid Group Goniocalyx), both grown at the same site in Peru and both yellow-fleshed. Although the iron content of the biofortified potato was 103% higher than the non-biofortified one, the total iron absorption (TIA) was only 46% higher, partly because the fractional iron absorption (FIA) was 27% lower. The researchers attributed the lower FIA to the higher phytic acid concentration, as phytic acid is a known inhibitor of iron absorption. Furthermore, the ascorbic acid concentration (an enhancer) was not particularly high in the biofortified potato because COVID-19 restrictions meant that mash samples had to be stored for 5 months at −20 °C before use, with consequent loss of over 50% of ascorbic acid; nor was the chlorogenic acid concentration particularly low (an inhibitor). Nevertheless, the researchers concluded that "iron biofortification of potatoes through conventional breeding is a promising approach to improve iron intake in iron-deficient women". As women in the rural areas of the highlands of Peru, and elsewhere in Latin America, consume meals containing only potatoes, their potato consumption can be 500 g per day. In these circumstances the researchers estimated that iron-biofortified potato would provide 27% of their iron requirement of 1.41 mg per day [(0.35 × 500 ÷ 460) ÷ 1.41], if they had moderate iron stores in their bodies, and this figure could rise to 50% if their iron stores were low (plasma ferritin levels around 10 μg/L). However, we do not know if these results apply in other circumstances. We have seen that iron absorption depends on the concentrations of iron promoters and inhibitors in the diet. Hence, meals that combine potatoes with other foods, such as beans in rural Africa, could give different results.

The zinc study used 37 participants in a single-blinded, randomized crossover design that was similar to the iron one but with an important difference. The iron- and zinc-biofortified yellow-fleshed diploid clone CIP311623.75 was compared with yellow-fleshed cultivar Amarilla (diploid Group Goniocalyx). The study with one type of potato started on day 1 and was followed by the other type on day 29. With each type of potato, the participants consumed 1000 g of boiled, peeled and mashed potato in two meals on two separate occasions within the same day. After the first 250 g portion of the meal, the first isotype, ^{67}Zn, was given orally as zinc sulphate in 50 mL water, followed 2 h later with the second isotope, ^{70}Zn, as an intravenous infusion of zinc chloride in 50 mL of saline. Two hours later they were given the second 250 g portion of the meal following the same procedure. Zinc absorption was measured using 100 mL urine samples taken before eating on day 1 and then after eating on days three to seven. I am now going to the results without explaining the details.

Zinc	0.48 v 0.32 mg per 100 g of meal [biofortified 50% higher]
Phytic acid	22.6 v 42.1 mg per 100 g of meal
Ascorbic acid	5.55 v 9.59 mg per 100 g of meal
Chlorogenic acid	15.15 v 12.16 mg per 100 g of meal
FZA	20.8% v 25.5% [biofortified 18% lower]
TZA	0.49 v 0.40 mg per 500 g meal [biofortified 22.5% higher]

Although the zinc content of the biofortified potato was 50% higher than the non-biofortified one, the total zinc absorption (TZA) was only 22.5% higher, partly because the fractional zinc absorption (FZA) was 18% lower. Although phytic acid is a known inhibitor of zinc absorption, concentrations were lower in the biofortified potato, so this is not the explanation. The researchers calculated the contribution of biofortified potato to zinc requirements in the poor rural areas of the Peruvian Andes. These could be between 42% and 56% of the EAR (estimated average requirement) for women of reproductive age eating 600–800 g of potatoes per day and between 31% and 38% of the EAR for children aged 6–36 months eating 165–200 g of potatoes per day.

8.12 Making Potatoes More Nutritious

My four examples have dealt with potential and actual health issues: the presence of a Group 2A carcinogen (acrylamide) in roast potatoes and fried potato products; poor eyesight and even blindness from vitamin A deficiency, iron deficiency anaemia and also zinc deficiency; the role of lutein and zeaxanthin in the prevention of macular eye degeneration; and the role of resistant starch in lowering glycaemic index and preventing obesity, glucose intolerance, type-2 diabetes and cardiovascular disease. Production of potatoes with altered chemical composition could help to solve these health issues, either by conventional breeding or use of genetic transformation. Additional nutritional benefits could come from increased protein quantity and quality (currently low in methionine and cysteine) and increased concentrations of vitamin B_9 (folate) to over 0.04 mg/100 g tuber and of vitamin C (ascorbic acid) to over 20 mg/100 g tuber [8, 11]. However, additional nutritional benefits in new cultivars will need to be accompanied by resistance to the major diseases and pests that occur in a production area, together with tolerance of abiotic stresses that also occur in the area; otherwise, farmers will not want to grow the more nutritious potatoes.

References

1. Earle R (2020) Feeding the people. Cambridge University Press, Cambridge
2. MacKenzie D (2008) Let them eat spuds. NewScientist 2 August 2008:30–33
3. Burton WG (1989) The potato, 3rd edn. Longman Scientific and Technical, Harlow
4. Huey SL, Mehta NH, Konieczynski EM, Bhargava A, Friesen VM, Krisher JT, Mbuya MNN, Monterrosa E, Nyangaresi AM, Boy E, Mehta S (2022) Bioaccessibility and bioavailability of biofortified food and food products: current evidence. Crit Rev Food Sci Nutr. https://doi.org/10.1080/10408398.2022.2142762
5. Storey M (2007) The harvested crop. In: Vreugdenhil D (ed) Potato biology and biotechnology advances and perspectives. Elsevier, Oxford, pp 441–447
6. Singh J, Kaur L (2016) Advances in potato chemistry and technology, 2nd edn. Elsevier, Amsterdam
7. Jansky S, Navarre R, Bamberg J (2019) Introduction to the special issue on the nutritional value of potato. Am J Potato Res 96(2):95–97
8. Bradshaw JE (2019) Improving the nutritional value of potatoes by conventional breeding and genetic modification. In: Qureshi AMI, Dar ZA, Wani SH (eds) Quality breeding in field crops. Springer Nature, Cham, pp 41–84
9. Burgos G, Felde TZ, Andre C, Kubow S (2020) The potato and its contribution to the human diet and health. In: Campos H, Ortiz O (eds) The potato crop: its agricultural, nutritional and social contribution to humankind. Springer, ebook, pp 37–74
10. Raigond P, Singh B, Dutt S, Chakrabarti SK (2020) Potato nutrition and food security. Springer, Cham
11. Bradshaw JE (2021) Potato breeding: theory and practice. Springer, Cham

12. White PJ, Bradshaw JE, Dale MFB, Ramsay G, Hammond JP, Broadley MR (2009) Relationships between yield and mineral concentrations in potato tubers. HortScience 44:6–11
13. Dale MFB, Mackay GR (1994) Inheritance of table and processing quality. In: Bradshaw JE, Mackay GR (eds) Potato Genetics. CAB International, Wallingford, pp 285–315
14. Storey RMJ, Davies HV (1992) Tuber quality. In: Harris P (ed) The potato crop. Chapman & Hall, London, pp 507–569
15. Brown CR, Vales I, Yilma S, James S, Charlton B, Culp D, Hane D, Shock C, Feibert E, Pavek M, Knowles R, Novy R, Whitworth J, Stark J, Miller JC, Holm D, Quick R, Navarre R (2012) "AmaRosa," a red skinned, red fleshed fingerling with high phytonutrient value. Am J Potato Res 89:249–254. https://doi.org/10.1007/s12230-012-9248-1
16. Vales MI, Brown CR, Yilma S, Hane DC, James SR, Shock CC, Charlton BA, Karaagac E, Mosley AR, Culp D, Feibert E, Stark JC, Pavek MJ, Knowles NR, Novy RG, Whitworth JL (2012) Purple Pelisse: a specialty 'fingerling' potato with purple skin and flesh and medium specific gravity. Am J Potato Res 89:306–314
17. Snodderly DM (1995) Evidence for protection against age-related macular degeneration by carotenoids and antioxidant vitamins. Am J Clin Nutr 62(Suppl):1448–1461
18. Burgos G, Muñoa L, Sosa P, Bonierbale M, zum Felde T, Díaz C (2013) In vitro bioaccessibility of lutein and zeaxanthin of yellow fleshed boiled potatoes. Plant Foods Hum Nutr 68:385–390
19. Gerbens-Leenes PW, Nonhebel S (2002) Consumption patterns and their effects on land required for food. Ecol Econ 42:185–199
20. Gerbens-Leenes PW, Nonhebel S, Ivens WPMF (2002) A method to determine land requirements relating to food consumption patterns. Agric Ecosyst Environ 90:47–58
21. Millstone E, Lang T (2008) The atlas of food, 2nd edn. Earthscan, London
22. Nyankanga RO, Murigi WW, Shibairo SI (2018) Effect of packaging material on shelf life and quality of ware potato tubers stored at ambient tropical temperatures. Potato Res 61:283–296
23. Scott GJ, Petsakos A, Suarez V (2019) Not by bread alone: estimating potato demand in India in 2030. Potato Res 62(3):281–304
24. Gottschalk K, Ezekiel R (2006) Storage. In: Gopal J, Khurana SMP (eds) Handbook of potato production, improvement and postharvest management. The Haworth Press, Binghamton, pp 489–522
25. Briddon A, Cunnington A, Harper G (2018) Post-harvest storage of potatoes. In: Wale S (ed) Achieving sustainable cultivation of potatoes. Volume 2: Production, storage and crop protection. Burleigh Dodds Science Publishing, Cambridge, pp 119–135
26. Raigond P, Mehta A, Singh B (2018) Sweetening during low-temperature and long-term storage of Indian potatoes. Potato Res 61:207–217
27. The Vegetable Farmer September 2022 p 6. ACT Publishing, Maidstone
28. Navarre DA, Moehninsi M, Lin S, Hellman H (2018) Nutritional properties and enhancement/biofortification of potatoes. In: Wang-Pruski G (ed) Achieving sustainable cultivation of potatoes. Volume 1: Breeding improved varieties. Burleigh Dodds Science Publishing, Cambridge, pp 191–221
29. Burgos G, Auqui S, Amoros W, Salas E, Bonierbale M (2009) Ascorbic acid concentration of native Andean potato varieties as affected by environment, cooking and storage. J Food Compos Anal 22:533–538
30. Hawkes JG (1990) The potato: evolution, biodiversity and genetic resources. Belhaven Press, London
31. Reader J (2008) Propitious esculent. William Heinemann, London
32. Beals KA (2019) Potatoes, nutrition and health. Am J Potato Res 96:102–110
33. England: The National Food Strategy: The Plan-July 2021. nationalfoodstrategy.org
34. Rayner J (2005) Enduring Love (not the film!). The Guardian, 3 November 2005
35. Malvern J (2023) Buyer bags slice of history as first crisp recipe is sold. The Times 11 February 2023, p 25
36. Li X-Q, Scanlon MG, Liu Q, Coleman WK (2006) Processing and value addition. In: Gopal J, Khurana SMP (eds) Handbook of potato production, improvement, and postharvest management. Food Products Press, New York, pp 523–555
37. Kirkman MA (2007) Global markets for processed potato products. In: Vreugdenhil D (ed) Potato biology and biotechnology advances and perspectives. Elsevier, Oxford, pp 27–44
38. Hughes JC, Evans JL (1967) Studies on after-cooking blackening in potatoes. IV Field experiments. Eur Potato J 10:16–36
39. Barker A (2009) Potato. Hermes House, London
40. Woods R (2019) Posh potatoes. Quadrille, London
41. McKenzie M, Corrigan V (2016) Potato flavour. In: Singh J, Kaur L (eds) Advances in potato chemistry and technology, 2nd edn. Elsevier, Amsterdam, pp 339–368
42. Singh J, Kaur L, Rao MA (2016) Textural characteristics of raw and cooked potatoes. In: Singh J, Kaur L (eds) Advances in potato chemistry and technology, 2nd edn. Elsevier, Amsterdam, pp 475–501
43. Tiemens-Hulscher M, Delleman J, Eising J, van Bueren ETL (2013) Potato breeding. Aardappelwereld BV, The Hague
44. Swedish National Food Administration (2002) Swedish National Food Administration: analysis of acrylamide in food. https://www.slv.se/acrylamide
45. De Meulenaer B, Vinci RM, Mestdagh F (2018) Acrylamide formation in fried potato products and

its mitigation. In: Wale S (ed) Achieving sustainable cultivation of potatoes. Volume 2: Production, storage and crop protection. Burleigh Dodds Science Publishing, Cambridge, pp 137–176

46. Muttucumaru N, Powers SJ, Elmore JS, Dodson A, Briddon A, Mottram DS, Halford NG (2017) Acrylamide-forming potential of potatoes grown at different locations, and the ratio of free asparagine to reducing sugars at which free asparagine becomes a limiting factor for acrylamide formation. Food Chem 220:76–86
47. De Wilde T, De Meulenaer B, Mestdagh F, Govaert Y, Ooghe W, Fraselle S, Demeulemeester K, Van Peteghem C, Calus A, Degroodt J-M, Verhé R (2006) Selection criteria for potato tubers to minimize acrylamide formation during frying. J Agric Food Chem 54(6):2199–2205
48. Shepherd LVT, Bradshaw JE, Dale MFB, McNicol JW, Pont SDA, Mottram DS, Davies HV (2010) Variation in acrylamide producing potential in potato: segregation of the trait in a breeding population. Food Chem 123:568–573
49. Novy RG, Whitworth JL, Stark JC, Schneider BL, Knowles NR, Pavek MJ, Knowles LO, Charlton BA, Sathuvalli V, Yilma S, Brown CR, Thornton M, Brandt TL, Olsen N (2017) Payette russet: a dual-purpose potato cultivar with cold-sweetening resistance, low acrylamide formation, and resistance to late blight and potato virus Y. Am J Potato Res 94:38–53
50. EFSA (2015) Scientific opinion on acrylamide in food. EFSA J 13(6):4104
51. FSA (2022) Acrylamide. https://www.food.gov.uk/safety-hygiene/acrylamide (last updated 26 April 2022)
52. Muttucumaru N, Powers SJ, Elmore JS, Mottram DS, Halford NG (2013) Effects of nitrogen and sulfur fertilization on free amino acids, sugars, and acrylamide-forming potential in potato. J Agric Food Chem 61:6734–6742. https://doi.org/10.1021/jf401570x
53. Bradshaw JE, Hackett CA, Pande B, Waugh R, Bryan GJ (2008) QTL mapping of yield, agronomic and quality traits in tetraploid potato (*Solanum tuberosum* subsp. *tuberosum*). Theor Appl Genet 116:193–211
54. Cygnet Potato Breeders, Variety Open Day 28 & 29 August 2024
55. ISAAA (2019) Global status of commercialized biotech/GM crops in 2019. ISAAA Brief 55
56. Heldt HW, Piechulla B (2021) Plant biochemistry, 5th edn. Academic, London
57. Chawla R, Shakya R, Rommens CM (2012) Tuber-specific silencing of *asparagine synthetase-1* reduces the acrylamide-forming potential of potatoes grown in the field without affecting tuber shape and yield. Plant Biotechnol J 10:913–924
58. Zhu X, Gong H, He Q, Zeng Z, Busse JS, Jin W, Bethke PC, Jiang J (2016) Silencing of vacuolar invertase and asparagine synthetase genes and its impact on acrylamide formation of fried potato products. Plant Biotechnol J 14(2):709–718
59. Rommens CM, Ye J, Richael C, Swords K (2006) Improving potato storage and processing characteristics through all-native DNA transformation. J Agric Food Chem 54:9882–9887
60. Ly DNP, Iqbal S, Fosu-Nyarko J, Milroy S, Jones MGK (2023) Multiplex CRISPR-Cas9 gene-editing can deliver potato cultivars with reduced browning and acrylamide. Plants 12:379. https://doi.org/10.3390/plants12020379
61. Hahlbrock K (2009) Feeding the planet: environmental protection through sustainable agriculture. Haus Publishing, London
62. Ye X, Al-Babili S, Klöti A, Zhang J, Lucca P, Beyer P, Potrykus I (2000) Engineering the provitamin A (β-Carotene) biosynthetic pathway into (carotenoid-free) rice endosperm. Science 287:303–305. https://doi.org/10.1126/science.287.5451.303
63. Paine JA, Shipton CA, Chaggar S, Howells RM, Kennedy MJ, Vernon G, Wright SY, Hinchliffe E, Adams JL, Silverstone AL, Drake R (2005) Improving the nutritional value of Golden Rice through increased pro-vitamin A content. Nat Biotechnol 23:482–487. https://doi.org/10.1038/nbt1082
64. Slater A, Scott NW, Fowler MR (2008) Plant biotechnology: the genetic manipulation of plants, 2nd edn. Oxford University Press, Oxford
65. Mallikarjuna Swamy BP, Marundan S, Samia M et al (2021) Development and characterization of GR2E Golden rice introgression lines. Sci Rep 11:2496. https://doi.org/10.1038/s41598-021-82001-0
66. Wu F, Wesseler J, Zilberman D, Russell RM, Chen C, Dubock AC (2021) Opinion: Allow Golden Rice to save lives. Proc Natl Acad Sci USA 118(51):e2120901118. https://doi.org/10.1073/pnas.2120901118
67. Normile D (2024) What a Philippine court ruling means for transgenic Golden Rice, once hailed as a dietary breakthrough. Plant engineered to counter a debilitating vitamin A deficiency in developing countries has faced fierce opposition. Science 3 May 2024. https://doi.org/10.1126/science.zl7058i
68. Diretto G, Al-Babili S, Tavazza R, Papacchioli V, Beyer P, Giuliano G (2007) Metabolic engineering of potato carotenoid content through tuber-specific overexpression of a bacterial mini-pathway. PLoS One 2(4):p.e350
69. Chitchumroonchokchai C, Diretto G, Parisi B, Giuliano G, Failla ML (2017) Potential of golden potatoes to improve vitamin A and vitamin E status in developing countries. PLoS One 12(11):e0187102. https://doi.org/10.1371/journal.pone.0187102
70. Burgos G, Amoros W, Salas E, Muñoa L, Sosa P, Díaz C, Bonierbale M (2012) Carotenoid concentrations of native Andean potatoes as affected by cooking. Food Chem 133:1131–1137

71. Bonierbale MW, Plaisted RL, Tanksley SD (1988) RFLP maps based on a common set of clones reveal modes of chromosomal evolution in potato and tomato. Genetics 120(4):1095–1103
72. Brown CR, Edwards CG, Yang CP, Dean BB (1993) Orange flesh trait in potato: inheritance and carotenoid content. J Am Soc Hortic Sci 118:145–150
73. Wolters AMA, Uitdewilligen JGAML, Kloosterman BA, Hutten RCB, Visser RGF, van Eck HJ (2010) Identification of alleles of carotenoid pathway genes important for zeaxanthin accumulation in potato tubers. Plant Mol Biol 73:659–671. https://doi.org/10.1007/s11103-010-9647-y
74. Sulli M, Mandolino G, Sturaro M, Onofri C, Diretto G, Parisi B, Giuliano G (2017) Molecular and biochemical characterization of a potato collection with contrasting tuber carotenoid content. PLoS One 12(9):e0184143. https://doi.org/10.1371/journal.pone.0184143
75. Römer S, Lübeck J, Kauder F, Steiger S, Adomat C, Sandmann G (2002) Genetic engineering of a zeaxanthin-rich potato by antisense inactivation and co-suppression of carotenoid epoxidation. Metab Eng 4:263–272. https://doi.org/10.1006/mben.2002.0234
76. Sakamoto Y, Mori K, Matsuo Y, Mukojima N, Watanabe W, Sobaru N, Tamiya S, Nakao T, Hayashi K, Watanuki H, Nara K, Yamazaki K, Chaya M (2017) Breeding of a new potato variety 'Nagasaki Kogane' with high eating quality, high carotenoid content, and resistance to diseases and pests. Breed Sci. https://doi.org/10.1270/jsbbs.16168
77. Monro J, Mishra S (2009) Nutritional value of potatoes: digestibility, glycemic index, and glycemic impact. In: Singh J, Kaur L (eds) Advances in potato chemistry and technology. Academic, Burlington, pp 371–394
78. Kanter M, Elkin C (2019) Potato as a source of nutrition for physical performance. Am J Potato Res 96(2):201–205
79. Henry CJK, Lightowler H, Strik C, Storey M (2005) Glycaemic index values for commercially available potatoes in Great Britain. Br J Nutr 94:917–921
80. Nayak B, Berrios JDJ, Tang J (2014) Impact of food processing on the glycemic index (GI) of potato products. Food Res Int 56:35–46
81. Akilen R, Deljoomanesh N, Hunschede S, Smith CE, Arshad MU, Kubant R, Anderson GH (2016) The effects of potatoes and other carbohydrate side dishes consumed with meat on food intake, glycemia and satiety response in children. Nutr Diabetes 6:e195. https://doi.org/10.1038/nutd.2016.1
82. Borch D, Juul-Hindsgaul N, Veller M, Astrup A, Jaskolowski J, Raben A (2016) Potatoes and risk of obesity, type 2 diabetes, and cardiovascular disease in apparently healthy adults: a systematic review of clinical intervention and observational studies. Am J Clin Nutr 104:489–498. https://doi.org/10.3945/ajcn.116.132332
83. Tian J, Chen J, Ye X, Chen S (2016) Health benefits of the potato affected by domestic cooking: a review. Food Chem 202:165–175
84. Sagili VS, Chakrabarti P, Jayanty S, Kardile H, Sathuvalli V (2022) The glycemic index and human health with an emphasis on potatoes. Foods 11(15):2302. https://doi.org/10.3390/foods11152302
85. Pinhero RG, Waduge RN, Liu Q, Sullivan JA, Tsao R, Bizimungu B, Yada RY (2016) Evaluation of nutritional profiles of starch and dry matter from early potato varieties and its estimated glycemic impact. Food Chem 203:356–366
86. Ek KL, Wang S, Brand-Miller JC, Copeland L (2014) Properties of starch from potatoes differing in glycemic index. Food Funct 10:2509–2515
87. Ek KL, Wang S, Copeland L, Brand-Miller JC (2014) Discovery of a low-glycaemic index potato and relationship with starch digestion *in vitro*. Br J Nutr 111:699–705
88. Monro J, Mishra S, Blandford E, Anderson J, Genet R (2009) Potato genotype differences in nutritionally distinct starch fractions after cooking, and cooking plus storing cool. J Food Compos Anal 22:539–545
89. Birt DF, Boylston T, Hendrich S, Jane JL, Hollis J, Li L, McClelland J, Moore S, Phillips GJ, Rowling M, Schalinske K, Whitley EM (2013) Resistant starch: promise for improving human health. Adv Nutr 4:587–601
90. Lockyer S, Nugent AP (2017) Health effects of resistant starch. Nutr Bull 42:10–41
91. Hellwege EM, Czapla S, Jahnke A, Willmitzer L, Heyer AG (2000) Transgenic potato (*Solanum tuberosum*) tubers synthesize the full spectrum of inulin molecules naturally occurring in globe artichoke (*Cynara scolymus*) roots. PNAS 97(15):8699–8704. https://doi.org/10.1073/pnas.150043797
92. Burgos G, Liria R, Zeder C, Kroon PA, Hareau G, Penny M, Dainty J, Al-Jaibaji O, Boy E, Mithen R, Hurrell RF, Salas E, zum Felde T, Zimmermann MB, Fairweather-Tait S (2023) Total iron absorbed from iron-biofortified potatoes is higher than that from nonbiofortified potatoes: a randomized trial using stable iron isotopes in women from the Peruvian Highlands. J Nutr 153(6):1710–1717
93. Liria-Domínguez R, Penny M, Kroon PA, Burgos G, Dainty J, Zeder C, Zimmermann MB, King J, Mithen R, Boy E, Al-Jaiballi O, Fairweather-Tait S (2023) Biofortified yellow-fleshed potatoes provide more absorbable zinc than a commonly consumed variety: a randomized trial using stable isotopes in women in the Peruvian Highlands. J Nutr 153(10):2893–2900
94. Paget M, Amoros W, Salas E, Eyzaguirre R, Alspach P, Apiolaza L, Noble A, Bonierbale M (2014) Genetic evaluation of micronutrient traits in diploid potato from a base population of Andean Landrace Cultivars. Crop Sci 54:1949–1959
95. Bouis H, Saltzman A (2017) Improving nutrition through biofortification: a review of evidence from

HarvestPlus, 2003 through 2016. Glob Food Sec 12:49–58
96. Kromann P, Valverde F, Alvarado S, Vélez R, Pisuña J, Potosí B, Taipe A, Caballero D, Cabezas A, Devaux A (2017) Can Andean potatoes be agronomically biofortified with iron and zinc fertilizers? Plant Soil 411:121–138. https://doi.org/10.1007/s11104-016-3065-0
97. Ierna A, Pellegrino A, Mauro RP, Leonardi C (2020) Micronutrient foliar fertilization for the biofortification of raw and minimally processed early potatoes. Agronomy 10:1744. https://doi.org/10.3390/agronomy10111744
98. White PJ, Broadley MR, Hammond JP, Ramsay G, Subramanian NK, Thompson J, Wright G (2012) Bio-fortification of potato tubers using foliar zinc-fertiliser. J Hortic Sci Biotechnol 87:123–129
99. White PJ, Thompson JA, Wright G, Rasmussen SK (2017) Biofortifying Scottish potatoes with zinc. Plant Soil 411:151–165. https://doi.org/10.1007/s11104-016-2903-4
100. Burgos G, Amoros W, Morote M, Stangoulis J, Bonierbale M (2007) Iron and zinc concentration of native Andean potato cultivars from a human nutrition perspective. J Sci Food Agric 87:668–675. https://doi.org/10.1002/jsfa.2765
101. Lindqvist-Kreuze H, Bonierbale M, Grüneberg WJ, Mendes T, De Boeck B, Campos H (2024) Potato and sweetpotato breeding at the international potato center: approaches, outcomes and the way forward. Theor Appl Genet 137:12. https://doi.org/10.1007/s00122-023-04515-7
102. Rukundo P, Ndacyayisenga T, Vandamme E, Nshimiyimana JC, Thiago M (2023) Performance of tetraploid bio-fortified potato clones in Rwanda. Crop Sci. https://doi.org/10.1002/csc2.20984
103. Andre CM, Evers D, Ziebel J, Guignard C, Hausman J-F, Bonierbale M, zum Felde T, Burgos G (2015) In vitro bioaccessibility and bioavailability of iron from potatoes with varying vitamin C, carotenoid, and phenolic concentrations. J Agric Food Chem 63:9012–9021. https://doi.org/10.1021/acs.jafc.5b02904
104. Fairweather-Tait S (1983) Studies on the availability of iron in potatoes. Br J Nutr 50:15–23

9 Conventional Breeding

The cultivars being grown today are the products of deliberate hybridizations and selections by breeders, starting in the nineteenth century with potatoes of unknown pedigrees, but tracing back to the cultivated landraces of South America. This conventional breeding was taken forward in the twentieth century by national and international research institutes, universities and commercial companies. Since 1961 these organizations have been able to secure Plant Breeders' Rights and royalty income on their new cultivars. Furthermore, since its foundation in Peru in 1971, the International Potato Centre (CIP) has done pre-breeding with potato wild relatives and cultivated landraces and made improved germplasm available to developing countries, mainly in the global south, for release as cultivars or for further breeding through their own National Agricultural Research Systems (NARS). The focus is on cultivars for the tropical highlands and subtropical lowlands across South and Southeast Asia, East Africa, Latin America and the Caribbean. These global efforts should result in new cultivars that will help potato production both to increase globally and to make a greater contribution to food security and human health. Genetic research is making conventional breeding more efficient and effective through the use of molecular markers and diploid F_1 hybrids.

9.1 Emergence of Potato Breeding in the Nineteenth Century

The potatoes being grown around the world at the end of the eighteenth century were the ones that had been selected, kept and planted the next season by generations of farmers, dating back to domestication in the Andes a few thousand years ago. In contrast, the ones being grown today are the products of deliberate hybridizations and selections by breeders, starting in the nineteenth century. We now talk of conventional breeding, but back then it was new and modern, and in essence it hasn't changed that much over the years. It is therefore worth starting this chapter with a brief look at the emergence of potato breeding in the nineteenth century through some examples.

Thomas Andrew Knight (1759–1838), from Herefordshire in England, was one of the world's first plant breeders, working on fruit (e.g., apples) and vegetables (e.g., the garden pea), and possibly the world's first potato breeder [1–6]. He was the second president (1811–1838) of the Royal Horticultural Society (RHS), which was founded in 1804 as the Horticultural Society of London. Today the RHS is best known as the organizer of

J. E. Bradshaw, *Can Potatoes Feed the World?*, Sustainable Development Goals Series,
https://doi.org/10.1007/978-3-031-92890-1_9

the annual Chelsea Flower Show in London. Knight read his first paper to the Horticultural Society on 6 January 1807, titled "On raising new and early varieties of the potato" (in those days spelt Potatoe). It was clear from his subsequent communications that he was a strong advocate of breeding new cultivars from true botanical seed for two reasons. First, he realized that growers in his native Herefordshire needed cultivars better adapted to the growing season than the ones coming over from Ireland. Second, he could see that continual propagation from tubers resulted in loss of vigour and quality, and the appearance of a disease known as "the curl" (in fact, a viral disease). He found that he could get early varieties to flower by letting their fibrous roots enter the soil, but not their stolons so that developing tubers could be removed. In later papers he explained that when he made crosses between late-maturing varieties, with tall luxuriant growth and large tubers, and early-maturing ones with dwarfish growth and small tubers, he could produce new high-yielding varieties that matured in time for harvest from the end of August to the beginning of September, thus allowing farmers enough time to sow wheat as their next crop. In his garden in 1832, he grew one of his new early varieties, from a cross of two early varieties made in 1830, and found that the internal part of the plot yielded a remarkable 86.5 t/ha. However, more realistic yields came from another source. Back in 1811, he had argued the need for a garden to compare the merits of different varieties of potato and other vegetables, which the Horticultural Society duly did with the establishment of its first official garden in Chiswick in 1822. On 15 November 1831, John Lindley, the assistant secretary of the RHS, presented the results of a trial of 20 new unnamed varieties from Knight along with 25 well-known varieties from the collection of the Society [7]. The potatoes were grown as widely spaced plants (120 cm apart in both directions) and harvested in October. The average yield of Knight's potatoes was 25.31 t/ha compared with 12.12 t/ha for the named varieties. However, the spacing may have been too wide for maximum yields, as 20 of the 25 varieties were also grown in rows of eight plants with 30 cm between plants and 60 cm between rows. Their average yield was 21.70 t/ha, which Lindley said was typical of growers in the London area. It is also very close to the global average potato yield given in Chap. 1 of 21.41 t/ha for the period 2016–2020!

We assume that Knight made his crosses by artificial hybridizations, as we know he had a greenhouse for his breeding work on fruit and vegetables and had published a paper on the pollination of vegetables [1]. However, I don't know for certain who was the first breeder in Britain to produce successful cultivars from deliberate cross-pollinations. William Paterson in Dundee, Scotland, was typical in propagating new cultivars from botanical seed in naturally occurring berries produced from open-pollination, a common practice that continued into the twentieth century. He started this propagation work in 1853, and his famous cultivar Victoria was made available to growers from 1860 [8]. Victoria was a natural seedling of the Fluke, a cultivar that bore natural berries and was itself a natural seedling from a berry taken from a field of potatoes, in 1841, on Langley Hall Farm, in Lancashire, by J. Turner, who was a handloom weaver and occasional farm worker [9].

James Clark in England did do the painstaking work involved in cross-pollination, as well as selecting some cultivars from naturally occurring berries, such as Magnum Bonum, released in 1876 [10]. Hence, we do not know if the commercially successful Magnum Bonum was a natural seedling of Early Rose by self-pollination or cross-pollination with Victoria [9]. In contrast, his successful cultivar Epicure came from the cross of Magnum Bonum with Early Regent and was introduced in 1897 after his death. The Sutton family multiplied, marketed and distributed Magnum Bonum and all of his later and successful cultivars [10, 11]. Suttons, as it is now called, was established as a seed company in England in 1806 by John Sutton, and from 1837 expanded as John Sutton & Son, concentrating on bulbs, flowers, vegetables, grass seed and seed potatoes [11]. Salaman, in his 1926 book *Potato Varieties*, lists all of the potato cultivars introduced by

Rough Purple Chili (1851)

↓ Open-pollination

Garnet Chili (1857) selected by Reverend Chauncey E. Goodrich in 1853

↓ Open-pollination

Early Rose (1867) selected by Albert Breese in 1861; crossed with European Tuberosum

↓ Open-pollination

Late Rose (1870), and Burbank (1876) selected by Luther Burbank in 1873

↓ Mutation/sport

May's Netted Gem (1902), but became widely known as Russet Burbank

Fig. 9.1 The importance of Rough Purple Chili: introduction or date of release in brackets [14–16]

Sutton and Sons from 1876 to 1924, a total of 69 cultivars from a number of sources [12]. Suttons was one of a steady stream of seed companies that were established in Europe and North America between the middle of the eighteenth century and the end of the nineteenth century. The first Russian potato-breeding companies, for example, appeared in the period 1850–1860, one of which was founded by E.A. Grachev (1826–1877) on his farm near St. Petersburg and created more than 20 new commercial cultivars [13].

Rough Purple Chili proved to be an important introduction of Chilean Tuberosum into the United States from South America via Panama, made by the Reverend Chauncey E. Goodrich of New York State in 1851 [14]. Like Knight, he wanted to restore vigour to potatoes through sexual reproduction. North America's most popular potato cultivar during the twentieth century, namely Russet Burbank released in 1914, was descended from Rough Purple Chili through three generations of open-pollination, which could have been either self- or cross-pollination (Fig. 9.1) [15].

The descendants of intermediate Garnet Chili were widely used as female parents in crosses with European Tuberosum at the end of the nineteenth century, as shown in Fig. 9.1 [15]. We now know that Early Rose and Russet Burbank were derived from Garnet Chili and Early Rose, respectively, by cross-pollination as they contain novel allozyme markers not present in their female parent [16].

By 1900, plant breeding in general, and potato breeding in particular, was established both as an academic discipline and a profession. It was taken forward in the twentieth century, initially by research institutes and universities, and later also by international institutes and private companies. I will start with the research institute where I spent my whole career, as I'm familiar with its work, which is typical of twentieth-century potato breeding.

9.2 Scottish Plant Breeding Station and Scottish Crop Research Institute

I started work at the Scottish Plant Breeding Station (SPBS) in 1975. When I retired 34 years later in 2009, I wrote a history of potato breeding at SPBS and SCRI (Scottish Crop Research Institute) from 1920 to 2008 [17]. The period started with a British food crisis that resulted from the First World War and ended with a global one in which there was a realization that potatoes could help poor countries achieve food security. There was also an awareness back in 1920 of the opportunity to develop scientific breeding methods from the new science of genetics, and the same is true today of gene editing.

Potato breeding started with the appointment of William (Bill) Black in 1926. He went on to be the longest-servicing potato breeder at SPBS, retiring after 42 years in 1968. He made his first cross (Abundance × Majestic) in 1927 and

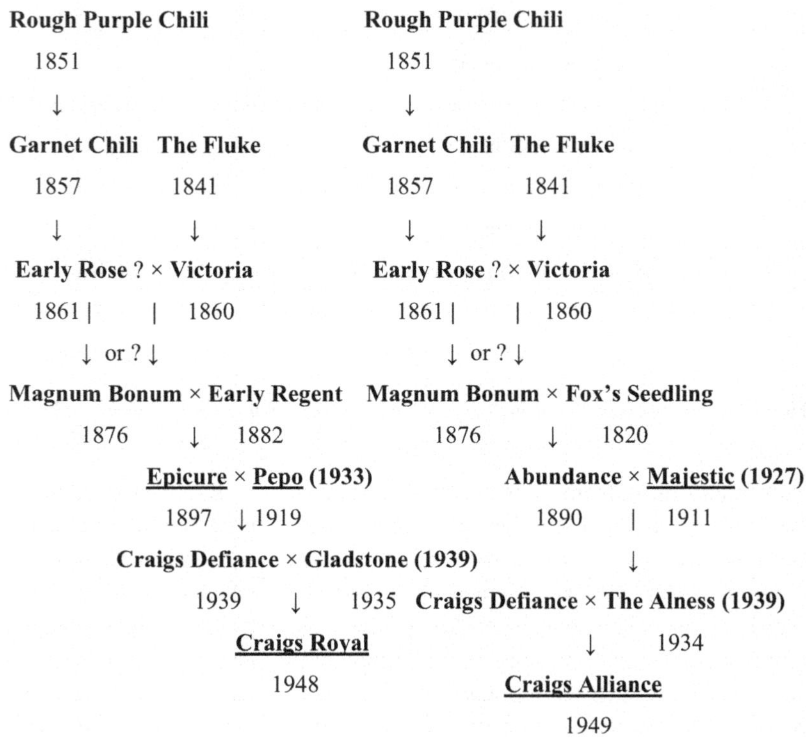

Fig. 9.2 Pedigrees of SPBS cultivars: The Alness (1934), Craigs Defiance (1939), Craigs Royal (1948) and Craigs Alliance (1949), with the female parent first and the year of the cross shown in bold in brackets. Cultivars known to have Tuberosum cytoplasm are underlined [10, 12, 17]

released his first cultivar (The Alness) 7 years later in 1934. Three others released before 1950 (Fig. 9.2) were Craigs Defiance (1939), Craigs Royal (1948) and Craigs Alliance (1949) [17]. As they all came from hybridizations between two cultivars, they inherited half of their genes from each parent.

We can trace the ancestry of the four cultivars back into the nineteenth century but no further [9–12, 15–18]. More generally, the parentages of the cultivars used in the early breeding work at SPBS trace back, either to old European/North American cultivars or to Rough Purple Chili, but with one exception. Dr. McIntosh was a cross between Herald (Abundance × Majestic) and a hybrid of Herald and a diploid Phureja potato [17, 18]. Rough Purple Chili was one of the sources of the Chilean Tuberosum cytoplasm, which is present in most modern cultivars [19].

The steps involved in artificial hybridizations are shown in Fig. 9.3: the tubers are planted on bricks, the lower leaves are removed from the plants, and the stems are tied to canes, for ease of removing unwanted daughter tubers as they form to encourage flowering. Each new cultivar started life as a botanical seed from which a seedling was grown in a glasshouse. A tuber was then harvested from the seedling for the start of multiplication (vegetative reproduction) over a number of clonal generations. During this period, other clones were discontinued as inferior to the one (or few) eventually selected as a new cultivar, from hybridization to release of the new cultivar taking from 6 to 10 years.

Although it is not possible to trace the ancestry of modern cultivars beyond the nineteenth century, there is a continuity from the past. We saw in Chap. 4 that the DNA sequences of whole chloroplast genomes and parts of nuclear genomes for 88 potato plants gave clear-cut results, which are worth repeating [20]. The Andean landraces of *S. tuberosum* from Colombia, Ecuador, Peru, Bolivia and Argentina had Andean cytoplasm, as did the six European herbarium specimens from between 1660 and 1781. The three herbarium specimens from Chile had Chilean cytoplasm. Of the 19 European herbarium specimens from between

Fig. 9.3 Artificial hybridization of potatoes at SCRI. (**a**) Three tubers of the same parent on bricks to be covered with compost (20 May 2008). (**b**) Lower leaves removed from plants and stems tied to canes (26 June 2008). (**c**) Pollen being taken from the male parent (28 July 2008). (**d**) Emasculated female parent being pollinated. (**e**) Pollinated flowers have been labelled. (**f**) Developing berries in a cellophane bag (28 July 2008). (From Bradshaw JE (2021) Potato Breeding: Theory and Practice. Springer, Cham with permission)

1810 and 1891, six (1847–1879) had Andean cytoplasm and 13 (1810–1891) had Chilean cytoplasm. Analysis of the nuclear genomes showed that European potatoes became progressively more similar to the historical Chilean potatoes. Likewise, 23 post-1850 North American cultivars have been shown to cluster with two Chilean landraces based on sequence variation in a diversity panel of wild species, landraces and cultivars [21].

S. rybinii (Ru.159) × *S. demissum* 1937

↓

735 × **Gladstone** 1939

↓

885(2) × **Pepo** 1940

↓

Craigs Defiance × 997a44 1942

↓

Pentland Ace 1951

Fig. 9.4 Pedigree of Pentland Ace [17]

The first of the Pentland series of cultivars was Pentland Ace (Fig. 9.4), which became the R3 blight differential with gene *R3a* (and $R3b^{G1696/G3111}$) from *S. demissum* [22]. In 1937, Black secured a few tetraploid seedlings, including clone 735, by hybridizing diploid *S. rybinii* (Ru.159 from Russia, = Group Phureja) with hexaploid *S. demissum*, despite their difference in endosperm balance number (2 and 4, respectively). Three backcrosses to cultivars of Group Tuberosum, which he was using at the time (Gladstone, Pepo and Craigs Defiance in Fig. 9.2), were achieved in the period 1939–1942; three different cultivars were used to avoid inbreeding depression. In each backcross generation, there was simply selection of seedlings for resistance to late blight, one of which would be successfully used as the parent for the next backcross. The aim was to retain a gene for resistance to late blight from the wild species *S. demissum* whilst more and more genes came from the cultivars: 50% from Gladstone in 885(2), 75% from Gladstone (25%) and Pepo (50%) in 997a44, and 87.5% from Gladstone (12.5%), Pepo (25%) and Craigs Defiance (50%) in Pentland Ace, assuming no selection for or against genes from the wild species, apart from selection for blight resistance. If the process was continued for more backcross generations, one would eventually end up with all of the genes from cultivars, apart from the blight resistance gene and any tightly linked to it, a phenomenon known as linkage drag. The resistance gene would have been introgressed into cultivated potatoes from the wild species. Today, this could be done in three backcross generations through the use of molecular markers, as explained in my book on potato breeding [23].

As the breeding programme developed over the years, it increased in size in order to assess more traits in more environments, and by the 1960s, could be viewed as shown in Fig. 9.5. Hybridizations (crosses) were made to generate genetic variation from which new cultivars were sought over a number of generations of clonal selection. These hybridizations were between two cultivars, or between a cultivar and a breeder's clone that did not become a cultivar, or between two such clones. The outcome was a few new cultivars, at best, and some improved clones, which were also used as parents in a new cycle of hybridizations. Ideally these parents would be identified in earlier clonal generations than the new cultivars, thus reducing the number of years between one cycle of hybridizations and the next. Parental clones also came from the introgression of major dominant genes for disease and pest resistance from landraces and wild tuber-bearing *Solanum* species. Typically, 100,000 seedlings were raised each year in a glasshouse from 200 to 300 hybridizations, and their number reduced to a few potential cultivars over as many as eight clonal generations, as described in my book on potato breeding [23]. The number of traits assessed and used as selection criteria increased over the clonal generations, as the number of potential cultivars decreased, thus allowing the production of more tubers of those remaining for evaluation in more trials with larger plots. Pedigrees became complicated because parents used in one year came from hybridizations made in different previous years.

By 2005, a total of 66 Group Tuberosum cultivars, from 91 different parents, had been bred for use in Great Britain, of which the last was Vales Everest [17]. It was separated from the first cross in 1927 (Abundance × Majestic) by 78 years and 11 generations. *S. demissum* (late blight resistance) featured in the pedigrees of 55 cultivars, *S. vernei* (potato cyst nematode resistance) in 19, *S. microdontum* (*Potato virus Y* resistance) in 14 and *S. multidissectum* (white potato cyst nematode resistance) in 9. Furthermore, the *H1* gene

Cultivar × Cultivar or Cultivar × Clone or Clone × Clone 200 crosses

↓ ↓ ↓ 8 clonal generations

Cultivars and Clones Cultivars and Clones Cultivars and Clones

Introgression Breeding → Clones with *R*-genes for disease and pest resistance

Cultivar × *R*-Clone or *R*-Clone × *R*-Clone 200 crosses

↓ ↓ 8 clonal generations

R-Cultivars and *R*-Clones *R*-Cultivars and *R*-Clones

R-Cultivar × *R*-Cultivar or *R*-Cultivar × *R*-Clone or *R*-Clone × *R*-Clone 200 crosses

↓ ↓ ↓ 8 clonal generations

R-Cultivars and *R*-Clones *R*-Cultivars and *R*-Clones *R*-Cultivars and *R*-Clones

Fig. 9.5 Potato breeding since the 1960s

(golden potato cyst nematode resistance) from an Andigena accession in the CPC (1673) was present in 15 out of 50 cultivars starting with Pentland Javelin in 1968. Hence, relatively few wild species and landraces contributed to the cultivars which were released, and some cultivars did commercially better than others. Breeding for resistance to viruses and potato cyst nematodes was successful, but not for late blight, as new races of *P. infestans* overcame the resistance genes. The SPBS/SCRI breeding programme was typical of those established about the same time in Europe and North America and later in India and China. In the next section, I justify this statement with detailed examples for the interested reader.

9.3 Origins of Breeding Programs Around the World

Russia The All-Russian Potato Research Institute was founded in 1930 in the Moscow region. An account of its breeding work from 1930 to 2005 can be found in a review by Simakov et al. [13]. The Institute was a reorganization of the Korenevskaya Experimental Station, which was established in 1920 after the October Revolution of 1917. The Station had inherited an expanded collection of potato cultivars and clones dating back to those of Grachev (about 250 samples) from the period 1850–1860, as mentioned back in Chap. 4. It was during the period 1930–1940 that Russian potato breeders started to use some of the wild species collected on the scientific expeditions of Bukasov and Yuzepchuk from 1925 to 1927. The cultivar Fitoftoroustoichiviy 8670, released in 1936, was the first with blight resistance introgressed from *S. demissum*. Introgressions of resistance to viruses PVY (from *S. stoloniferum* and *S. chacoense*) and PVX (from *S. acaule*) followed from 1965, with some of the resistant clones coming from breeders in Scotland and Hungary. In total, 38 cultivars with one or more resistances were bred. Finally, from 1977, resistance to the golden potato cyst nematode (*G. rostochiensis*) was sought. Initially three German cultivars were used as sources of the *H1* gene; then resistance was found in the *S. chacoense* with resistance to PVY. The review finished with progress in breeding new cultivars for processing use over the period 1980–2005. More than 90 cultivars were bred over the 75 years from 1930 to 2005.

The Netherlands The history of Dutch potato breeding from 1888 to 2018 was reviewed in 2023 by van Loon et al. [24]. Geert Veenhuizen

(1857–1930) is considered the founder of potato breeding in the Netherlands. The owner of a flower and tree nursery in Sappemeer, he made his first hybridizations in 1888, aimed at cultivars for the starch industry, as well as ones for the export of seed tubers and for ware potatoes. By 1929, his cultivars occupied 66% of the potato-growing area in the Netherlands, with Eigenheimer, his most famous cultivar. The early breeding work was done by enthusiastic hobby breeders. They were helped by the establishment in 1938 of a germplasm collection of over 200 cultivars, mainly from Europe but also from the United States and India. The first breeding companies appeared in the 1930s, initially cooperatives, then private companies. Today the two largest companies are Agrico, established in 1973 from a merger of several regional cooperatives, and HZPC, formed in 1999 from the merger of the private company Hettema and the cooperative ZPC. Teaching and research in plant breeding, based in Wageningen, began at the end of the nineteenth century and developed over the years through various organizations such as IVP and SVP to the present-day Wageningen University and Research (WUR), which was formed in 2016. Scientific potato breeding started in 1943, including the use of potato wild relatives, particularly after a Dutch expedition to South America in 1955 resulted in the Dutch-German Potato Collection (CGN, Wageningen, the Netherlands), already mentioned in Chap. 2. The initial emphasis was on resistance to late blight from *S. demissum* and resistance to the potato cyst nematodes. Today, the Netherlands is the world's largest exporter of seed potatoes.

United States The US Department of Agriculture (USDA) potato breeding programme started in 1910, followed by the establishment of a National Potato Breeding Programme in 1929. The release of cultivars Katahdin and Chippewa in 1933 signalled the start of modern potato breeding in the United States [25, 26]. Cultivars had been introduced from Europe before 1850, and from 1850 to 1910, cultivars had been bred by private breeders, mainly with seeds from open pollination, but with some hybridizations (e.g., cultivar Snowflake from the cross between Ruby (Early Rose) and Excelsior made by Cyrus Guernsey Pringle in 1876) [25]. Katahdin and Chippewa both came from the cross between USDA 40568 and USDA 24642; the former traces back through Busola to Rough Purple Chili and Daber (imported into Germany around 1830 from South America, possibly Chile). Busola was introduced from Russian Poland in 1914 [26, 27]. Katahdin is a progenitor of 88% of the cultivars released up until 1985. Love analysed the pedigrees of 44 cultivars that were prominent in the United States in 1996 and traced their pedigrees back over 12 generations, albeit with limited information [27]. The cultivars had been released between 1914 and 1995, with seven pre-1954. Twenty founding clones were found: ten of North American, three of English, four of German, one of Russian and two of South American origin. The two of South American origin were Daber and Rough Purple Chili. Another introduction from Chile to the United States was Villareola. It appears in the pedigree of clone 11–79 from Australia [USDA 41956 × (Katahdin × Snowflake)], which was the male parent of Pentland Crown, Pentland Envoy and Pentland Javelin in Scotland. After that global tour, let us return to the United States. Plaisted and Hoopes looked at the pedigrees of 126 cultivars released from 1956 to 1985 and found that only three wild species occurred to any extent: *S. demissum* in 74 cultivars, *S. chacoense* in 18 and *S. acaule* in 17 [26]. One of the first and major sources of *S. demissum* germplasm was a hybrid with *S. tuberosum*, which was produced by USDA and sent to Germany in 1908, from which blight resistant "W races" were produced by backcrossing. These were imported back into the United States and used to produce clone 3895-13, which was crossed to Earlaine to produce USDA X96-56, one of the parents of Kennebec, released in 1948. Kennebec was the first cultivar with blight-resistance to be widely grown in the United States. *S. chacoense* and *S. acaule* were sources of resistance to viruses PVY and PVX, respectively.

India In 1935, the Potato Breeding Station in the cool hills at Kufri (Shimla, about 30°N, 2276 masl) initiated a programme to breed cultivars suitable for subtropical conditions in India. This was the forerunner of the programme started in 1949 with the establishment of the Central Potato Research Institute (ICAR-CPRI), whose headquarters were moved to Shimla in 1956. By 2020, ICAR-CPRI had released 64 improved cultivars from hybridizations. Crosses were made in the hills and progenies were evaluated and maintained in the plains during October to December, when aphid and hence virus populations were low. An account of the breeding work can be found in a revised Technical Bulletin published in 2020 [28]. There is little recorded about the introduction of potatoes to India, probably early in the seventeenth century by Portuguese traders or British missionaries, and their early history, although they were grown commercially from 1830 onwards. The earliest introductions resembled short-day-adapted Group Andigena potatoes, but results from genetic analyses published in 2005 found that 21 "Kufri" cultivars released after 1958, together with two early indigenous cultivars, Darjeeling Red Round and Phulwa, clustered with the Chilean and not the Andean landraces included in the study. However, Phulwa lacked the 241-bp deletion in its plastid (chloroplast) DNA characteristic of Chilean cytoplasm. Hence, it could be an Andean-Chilean hybrid [29]. Long-day-adapted European cultivars were introduced between 1924 and 1945 but proved unsuitable for Indian conditions [28]. Nevertheless, crosses involving cultivars from Europe and the United States did result in cultivars suitable for growing as a winter crop in the subtropical plains, where most of India's potato production now takes place. Three examples will suffice: (1) Kufri Sheetman (released 1968) and Kufri Dewa (1973) came from Craigs Defiance × Phulwa (Indian); (2) Kufri Neela (1963) came from Katahdin × Shamrock and Kufri Kumar (1958) from Lumbri (Indian) × Katahdin; and (3) Kufri Jyoti came from Scottish clones and is related to Lady Balfour (3069d4 × 2814a1 gave Kufri Jyoti and the reciprocal cross gave a grandparent of Lady Balfour). The Kufri Jyoti story is interesting. In the 1960s, SPBS took part in an international collaborative programme on late blight resistance with CPRI in India and the Rockefeller Foundation programme in Mexico. Progenies were exchanged, initial screening for resistance was done in India, and selected genotypes were sent to Edinburgh and Mexico for confirmation of resistance. Genotypes showing a high degree of resistance in Mexico and Edinburgh were evaluated for adaptation under Indian conditions. As a result, Kufri Jyoti was released in 1968 and in 2020 was still the most popular cultivar in the northwestern and northeastern hills of India and parts of the North Indian plains. It remained almost immune to late blight until the year 1974 but subsequently started to get infected and by 1986 had completely succumbed to the disease [28]. So, germplasm from a late blight resistance breeding programme provided one of the most successful organic cultivars in the United Kingdom and one of the most successful table cultivars in India, two very different environments. In summary, of the 64 cultivars released by CPRI, 25 have both parents of Indian origin, 28 have one parent of Indian origin and 11 have both parents of non-Indian origin. In total, 98 parents have been used in the breeding program. In 2015, the most widely cultivated cultivar in India was Kufri Pukhraj (33% of total area), particularly in the North Indian plains and the Plateau regions, followed by Kufri Jyoti (21%) and Kufri Bahar (17%). Kufri Pukhraj was released in 1998, from a cross between Craig's Defiance and Indian clone JEX/B-687, and is noted for its early maturity, drought tolerance and high yield, whereas Kufri Bahar was released in 1980 and despite its susceptibility to late blight, is the most popular cultivar in Uttar Pradesh. Economic analysis has produced a "good" value of 48% (45% to 50%) for the internal rate of return (IRR) from the research investment that resulted in Kufri Pukhraj [30]. The authors of the analysis also quoted a study in Bihar that showed a 53% increase in yield when Kufri Pukhraj replaced the local cultivars that were being grown. They also emphasized the importance of micro-irrigation technologies and good-quality seed-tubers for planting, lack of which would limit yield. Most importantly, Kufri Pukhraj shows that cultivars can be bred to deal with the shortening growing window for potatoes in the Indo-Gangetic Plains due to increasing temperatures and more frequent droughts.

China Potatoes were introduced from Europe and America by various routes during the seventeenth century. Although there were large areas of potato cultivation by the beginning of the nineteenth century, it was the expansion of production over the last 48 years that made China the largest producer in the world from 1993. Since the 1970s, most cultivars have come from the National Breeding Programme, now involving over 80 agencies [31]. In his 2008 book, John Reader relates how potato production in China improved after Deng Xiaoping's social and economic reforms in 1978 [32]. One outcome was that breeders at the Crop Research Institute in Kunming, Yunnan Province, were able to use seed provided by the International Potato Centre (CIP) to produce the aptly named cultivar Cooperation 88. It was high-yielding (up to 60 t/ha) and soon accounted for one-quarter of production in Yunnan. A recent account of potato breeding in China is the 2018 book chapter by Professor Jin, who is head of the National Potato Breeding and Genetics Programme in China, based in the Institute of Vegetables and Flowers (IVF) in Beijing, which is part of the Chinese Academy of Agricultural Sciences (CAAS) [31]. I learnt additional information from visiting her in 2010. My visit was confined to the National Programme run by IVF-CAAS from Beijing and the research station at Zhangbei to the north (there are also provincial programs with research in universities and local institutes). The breeding targets were table, processing and special purpose (e.g., purple flesh) cultivars. I was struck more by the similarities than the differences between the breeding programme at Zhangbei and ours at SCRI. The IVF-CAAS germplasm for evaluation and use as parents in breeding is held in tissue culture in their labs in Beijing. It comprised 1500 accessions (clones) from the United States, Europe and CIP in Peru (600–700, including drought-resistant ones). Also included were 23 clones from SCRI. These were in their second year of trials at Zhangbei (44°N) and looked as good as they did in the trial in Dundee, Scotland (57°N). Nevertheless, about 60% of parents used in breeding were Chinese cultivars and clones tracing back to the introductions from Europe and America from the seventeenth century onwards. In total it looks as though over 500 cultivars have been released from the National Programme since the 1960s.

Global Outcome The cycles of crossing, backcrossing and clonal selection in breeding programs around the world resulted in a large number of cultivars, each adapted to a particular region within an agro-ecological zone and for a particular end use, which included consumer preferences. Processors like round to round-oval potatoes for crisps (chips) but long-oval to long-oblong ones for French fries. Supermarkets prefer round-oval potatoes for pre-packing but oval to long-oval ones for baking. Large tubers are required for baking and French fries, whereas baby new potatoes are small in size. Consumers around the world also have preferences for skin colour (e.g., white or red), skin appearance (e.g., russet or non-russet) and flesh colour (e.g., white or yellow), with processors wanting yellow flesh for crisps (chips) but yellow or white (e.g., Russet Burbank) for French fries depending on the market. It is difficult to accurately assess the number of cultivars available. The 2014 World Catalogue of Potato Varieties [33] listed more than 4800 cultivars from 100 countries. Checking all of the databases of modern cultivars would probably give a larger number. At "Gardening Scotland" on 2 June 2013, there were 667 cultivars on display (Fig. 9.6), and my local garden centre offers 100 different cultivars each year: multiply that by 100 countries, and you get 10,000 cultivars, possibly not all different. Then there are over 4000 Andean landraces, but just remember that a large breeding program will start with 100,000 seedlings in a glasshouse, each one a potential new cultivar. One final thought. I think that modern breeding programs need to produce locally adapted cultivars in the way that farmers selected locally adapted landraces, even though breeders would like their cultivars to be grown as widely as possible. That would certainly maintain biodiversity.

Fig. 9.6 "Gardening Scotland" on 2 June 2013: 667 cultivars on display

9.4 Plant Breeders' Rights

The International Union for the Protection of New Varieties of Plants (UPOV, from the French name) was established in Paris in 1961 by the adoption of the International Convention for the Protection of New Varieties of Plants. It led to plant breeders' rights and royalty income for breeders, making breeding attractive to commercial companies. In the United Kingdom, for example, the 1964 Plant Varieties and Seeds Act established Plant Breeders' Rights lasting 25 years with a possible extension to 30 years, and the United Kingdom joined UPOV in 1968. Plant Breeders' Rights are administered by the Plant Variety Rights Office (PVRO). Today in Europe most potato breeding is done by private companies, whereas in the United States, with one exception (Fritolay), it is done either by the land-grant universities or the USDA. Interestingly, India is not a member of UPOV, but new cultivars have received legal protection under "The Protection of Plant Varieties and Farmers Right Act (2001)" since the Indian government became a signatory of the "Trade-Related Aspects of Intellectual Property Rights, 1994" [28]. However, I don't think that royalties are actually collected, and the same is true in China, which officially joined UPOV in 1999 so that the owners of new cultivars could be given legal protection.

UPOV headquarters are in Geneva (Switzerland), and by February 2024, 79 countries were members [34]. Out of the top 15 potato-producing countries listed in Table 1.1, only Bangladesh and India are not members. Under UPOV rules, the unauthorized commercial propagation of cultivars is forbidden, with two important exceptions. Farmers retain their traditional right to sow seed saved from their own crops, and breeders can use protected cultivars as parents for further breeding to ensure future progress. Some significant changes to the rules were made in the 1991 UPOV Act. One change restricted the right of farmers to freely use (botanical) seeds or other propagation material (tubers) for further cultivation by making it an optional exception that is limited to countries, which make it a special provision. A second change allowed plants and cultivars in some countries to be patented as well as have Plant Breeders' Rights. The use of patents to protect biotech crops prevents their use as parents in further breeding, which I think is a cause for concern. A third change stipulated that anyone using a protected variety in breeding has to make major changes, or else the "new" variety will be considered an "essentially derived" variety, falling to the ownership of the first breeder. The aim was to discourage insignificant changes in the variety's characteristics from being passed off as true innovation, but the definition of changes was ambiguous.

In order for a new cultivar to be granted Plant Breeders' Rights, it must be shown to be Distinct, Uniform and Stable (DUS) when compared with existing cultivars, and its proposed name must be accepted. The sample of tubers that the breeder submits for DUS testing becomes the definitive stock of the new cultivar and is used for registration of the new cultivar and certification of seed tubers when it is multiplied for sale; in other words, to make sure that the grower is sold the correct cultivar. The procedures depend on the regulatory system of the country or area in which the breeder wishes to register the new cultivar but need not concern us here.

9.5 The International Potato Centre (CIP)

The International Potato Centre (CIP) was founded in Lima, Peru, in 1971; Peru being a centre of biodiversity for tuber-bearing *Solanum* species. Since its foundation, it has been home to the world's collection of potato germplasm, comprising wild species and the cultivated landraces of South America. CIP does research on this germplasm and uses it to produce populations of clones which have been improved for desirable traits in a process known as pre-breeding. It makes this material available to developing countries, mainly in the global south, for release as cultivars or for further breeding through their own National Agricultural Research Systems (NARS). The focus is on cultivars for the tropical highlands and subtropical lowlands across South and Southeast Asia, East Africa, Latin America and the Caribbean during a period of climate change and the need for increased resource-use efficiency (water and nutrients). CIP also recognizes the need for breeding objectives to be increasingly more demand-driven to achieve a reduction in poverty, better nutrition and a reduction in gender inequality in poor rural communities [35].

We encountered CIP germplasm with high and stable resistance to late blight in Chap. 5, with virus resistance in Chap. 6, with heat tolerance in Chap. 7, and with increased iron and zinc contents in Chap. 8. We can add germplasm with a shorter crop duration in the highland tropics (from 180–150 days down to 100–90 days). CIP's germplasm is freely available to researchers and other breeders, although material transfer agreements and intellectual property (IP) regulations are in place. The approaches taken in the breeding work, the outcomes and the way forward have recently been reviewed by CIP's scientists [36]. Up until July 2023, over 300 cultivars originating from CIP crosses had been registered by NARS partners.

Since 2015, CIP has partnered with HZPC, with the support of the Syngenta Foundation for Sustainable Agriculture (SFSA), to breed short-cycle, disease-resistant cultivars for the fresh and processing markets of Southeast Asia and beyond. The project combines HZPC germplasm with globally relevant table and processing traits and CIP germplasm with relevant traits for adaptation, climate resilience and disease resistance. Part of the profits from joint commercialization of cultivars will go to the International Treaty on Plant Genetic Resources and CIP's breeding program. This project is in marked contrast to the early days of CIP (1972–1992), when they did contract research on the breeding and exchange of germplasm with Cornell University in the United States.

It is reasonable to ask about the impact of CIP's work in developing countries. At the beginning of 2024, Keith Fuglie and Ruben Echeverria published an assessment of the economic impact of CGIAR-related crop technologies on agricultural productivity in developing countries from 1961 to 2020 [37]. For potatoes this means the work of CIP from 1971. I found their statistics on the introduction of modern cultivars interesting. They defined a modern cultivar as "one that was released through a formal breeding program in a country after 1960". They distinguished modern cultivars that could be attributed to CIP and those from national programs and private companies. They estimated the percentage of the potato area which was occupied by modern cultivars in 2016–2020 (first figure) and the percentage of these cultivars which could be attributed to CIP (second figure) as follows: developing countries overall (74.1% and 20.2%), Asia (94.0% and 17.8%), Latin American countries (68.6% and 22.0%) and Sub-Saharan Africa (25.4% and 72.3%). Furthermore, in Asia these were the second generation of new cultivars, whereas in Sub-Saharan Africa, 30% were second-generation cultivars, and in Latin American countries, all were first-generation cultivars. One of the reasons why only 20.2% of modern cultivars could be attributed to CIP was the percentages in China (96.4% and 25.1%), India (93.0% and 5.2%) and Brazil (95.8% and 0.0%). In other words, these countries were doing their own breeding work. Nevertheless, it is clear that CIP has facilitated a greater use of the wild tuber-bearing *Solanum* species and the cultivated potatoes of South America in breeding programs worldwide. The review also produced figures suggesting that yields may have increased by about 50% as a consequence of adopting the first and then second generations of new cultivars. In Asia, for example, potato yields in 1961 were low at 8.43 t/ha. The first generation of new cultivars increased this by 30.5% and then the second generation by a further 19.5%, where the increases were in net yield (gross yield minus change in input costs, such as fertilizer, so that they approximate to gains from cultivar and not from other factors).

9.6 Diploid F_1 Hybrid Breeding

In Chaps. 4 and 6, we encountered true potato seed (TPS) as a way to produce largely disease-free planting material. CIP, from its foundation in 1971, saw the potential of TPS [36]. As potatoes can produce large quantities of TPS, the availability of planting material would be increased and its cost reduced. Furthermore, it could easily be stored and transported without loss from diseases and would allow more flexible planting times than with seed tubers of a certain physiological age. CIP therefore explored ways of breeding tetraploid TPS cultivars from both tetraploid ($4x$) and diploid ($2x$) parental clones ($4x \times 4x$, $4x \times 2x$ and $2x \times 2x$ crosses), while recognizing that the outcome would be genetically heterogeneous hybrids. This might restrict their use in markets where uniformity was paramount, such as the processing industry, but not as a staple crop where food security was important. Collaboration with NARS allowed the testing of TPS cultivars in a number of countries, including Bangladesh, China, Egypt, India, Indonesia, Nepal, Nicaragua, Peru, the Philippines, southern Italy, Uzbekistan and Vietnam [36, 38–40]. However, the biological advantages of using TPS only translated into economic benefits for farmers when healthy seed tubers were costly or not available, or seed tuber quality (uniformity) was not as serious as might have been expected [41]. As a consequence, there was a lack of investment from the private sector, and CIP suspended their TPS breeding in 2007 [36].

The possibility of producing genetically identical seeds through apomixis is an area of active research in some laboratories around the world but is beyond the scope of this book, and as far as I know, is yet to deliver results that can be put into commercial practice for potatoes. In simplistic terms, apomixis can be viewed as a type of asexual reproduction that gives the appearance of sexual reproduction.

Today, there is renewed interest in propagation from TPS because genetically uniform diploid F_1 hybrid cultivars can be produced in the same way that proved possible with maize and other outbreeding crops, where F_1 hybrids became one of the big success stories of twentieth-century plant breeding. I discussed TPS back in Chap. 6, so here I am just going to outline the breeding method developed by the Solynta team in the Netherlands for potatoes, which is simple in principle if not in practice [42]. With sexual reproduction, the selection and multiplication of a desirable potato is done indirectly through botanical seed from inbreeding and crossbreeding, rather than directly by vegetative reproduction from tubers. This became possible from 2008 in normally self-incompatible diploid potatoes with the introduction (by backcrossing) of a self-incompatibility inhibitor gene (*Sli*) discovered in *S. chacoense* back in 1998 [43–45].

Diploid potato germplasm suitable for the objectives of a breeding program could now be chosen and made self-compatible for the start of the inbreeding process. Genetically uniform inbred lines could be produced by five to seven generations of self-pollination (inbreeding) with selection for acceptable vigour and fertility, to counteract inbreeding depression, and for other desirable characteristics. Pairs of inbred lines could then be hybridized to produce genetically uniform F_1 hybrids, displaying hybrid vigour, for assessment as potential cultivars (crossbreeding). Finally, large quantities of F_1 seed of the chosen hybrids could be produced for growing them commercially. Since 2016, diploid F_1 hybrids have been trialled in the tropical highlands of East Africa, as well as in Europe, and there is interest in China and India [42, 46]. Further progress requires continuing cycles of inbreeding and crossbreeding. It is currently difficult to predict how much of future potato production will be from TPS, rather than from seed tubers, and using diploid F_1 hybrid cultivars. One determining factor will be the relative performance of such hybrids compared with the current tetraploid cultivars grown from seed tubers.

Interestingly, the Japanese group of researchers who first discovered the self-incompatibility inhibitor gene (*Sli*) in *S. chacoense* back in 1998 continued their research on its use to produce completely homozygous diploid potatoes. They eventually succeeded and published their results in 2020 [47]. After 10 generations of self-pollination (S_{10}), they produced a plant (17H145-5) that was homozygous for all 18,579 genome-wide single nucleotide polymorphism (SNP) markers under study. It was derived from a hybrid plant (97H32-6) that was 75% Group Phureja and 25% *S. chacoense* and considered the S_0 (non-inbred) generation of the selfing series. Its genome-wide percentage of heterozygous SNP markers was 12.64%, falling to 1.37% at S_6 and finally to 0.00% at S_{10} in 17H145-5. Whereas 97H32-6 was a vigorous plant that naturally produced tens of berries and many tubers under long days, 17H145-5 suffered severe inbreeding depression in terms of vigour and fertility but was self-fertile and could produce some berries as well as tubers in long days. In contrast, the completely homozygous doubled monohaploid (DM) of Group Phureja used to sequence the potato genome was male sterile. The researchers also produced an S_9 plant (C) of *S. chacoense* that was 97.8% homozygous [48]. The researchers went on to use 17H145-5, DM and C as three of the parents in a study of the genome-wide percentage heterozygosity in a range of diploid parents and their hybrids, which were also evaluated for agronomic traits including yield. Hybrid vigour was found for fecundity, growth and tuber yield, and there was a moderate correlation between heterozygosity and tuber yield [48]. The genetic basis of heterosis (hybrid vigour) is still a topic of research, along with the relationship between yield and heterozygosity.

Before leaving F_1 hybrids, it is worth recalling that it was Darwin who reported the effects of cross and self-fertilization in 57 species of flowering plants from 52 genera and 30 families in his 1876 book on the topic [49]. His most important conclusion was that cross-fertilization is generally beneficial and self-fertilization (inbreeding) injurious. The exceptions to inbreeding depression were species such as the garden pea (*Pisum sativum*), which naturally reproduce by self-pollination but can be artificially cross-pollinated. The garden pea therefore proved a good choice by Mendel for his experiments in plant hybridization, which he published in 1865 [50]. However, his work was not appreciated until it was rediscovered in 1900 and an English translation published in 1901, by the Royal Horticultural Society of London in volume 26 of their journal, and promoted by William Bateson in his book *Mendel's Principles of Heredity*, published in 1902 [51]. It was Mendel's 1865 paper on the laws and mechanisms of inheritance and Darwin's 1876 book on mating systems that laid the foundations of modern scientific plant breeding.

9.7 Genetics

I mentioned at the start of this chapter that when the Scottish Plant Breeding Station was created in 1920, there was an awareness of the opportunity afforded by the new science of genetics. Salaman had published his seminal paper on potato genetics in the first issue of the Journal of Genetics in 1910 [52] and would summarize the results of his breeding work and genetics experiments in his 1926 book *Potato Varieties* [12]. However, some 50 years later, in 1970, when I was studying genetics as an undergraduate, potatoes did not feature. Nor did they feature in H.L.K. Whitehouse's 1973 book *Towards an Understanding of the Mechanism of Heredity*, which dealt with the development of ideas about heredity through the testing of hypotheses with experimental evidence [53]. The most frequently mentioned organisms were the fruit fly *Drosophila melanogaster*, the fungus *Neurospora crassa*, the colon bacterium *Escherichia coli* and its T even phages (DNA viruses), the maize plant *Zea mays* and the mouse *Mus musculus*, as well as humans (*Homo sapiens*). The potato had not been used as a model organism for studying genetics because the principal cultivated type displayed complex inheritance patterns because it is an autotetraploid with four sets of chromosomes. This was not appreciated until the period 1937–1960, when the consequences of this tetrasomic inheritance for potato breeding were explored [54–59]. I was eventually introduced to the genetics of autotetraploids as a postgraduate student in applied genetics in 1972 through R.W. Allard's well-known 1960 book *Principles of Plant Breeding* [60].

Knowledge of potato genetics did not, however, have a big impact on practical potato breeding over the next 30 years (1960 to 1990) because there was a sense in which it was not required. After the initial hybridizations, no more sexual reproduction was required to produce a new cultivar, just selection between clones over the generations of reproduction from seed tubers (Fig. 9.5). As a consequence, any impact of genetics needed to be through the production and use of parental material of known genetic constitution and predictable offspring, but in practice, parental material was usually potential cultivars of unknown breeding value for quantitative traits [61]. However, it was during this period that genetic studies and breeding work became possible at the diploid level through the production of diploid potatoes from tetraploid ones by using special clones of Group Phureja as pollinators [62, 63]. These diploid potatoes (strictly speaking, haploids but often referred to as dihaploids) could be crossed with cultivated diploid landraces (Groups Phureja and Stenotomum) and many diploid wild species (2EBN species), as found in the United States by Professor Peloquin and his coworkers [64–66]. They developed a novel breeding strategy both to introgress specific characteristics from diploid germplasm and to broaden the genetic base of potato breeding. Return to the tetraploid level was possible through the production of either diploid pollen or eggs (unreduced $2n$ gametes) in crosses with tetraploid tuberosum cultivars, as I explained in Chap. 3 for the origin of Group Andigena potatoes. However, genetic studies remained difficult because diploid potatoes are usually self-incompatible and many dihaploids are male sterile (this will now change with the advent of self-compatible inbred lines).

It was the availability of molecular markers and maps of their chromosomal locations, from 1988 onwards, that started to have an impact on potato breeding [67, 68]. Most of the mapping was done with diploid potatoes, but the results could be used in tetraploids, as the gene order is the same. Genes affecting qualitative traits, such as tuber skin and flesh colour, were mapped directly, as individuals fall into distinct categories for both markers and traits. In a recent book chapter, I listed most of the mapped major genes of interest to potato breeders, a total of around 45 genes for approximately 30 traits [69]. In contrast, locations affecting quantitative traits, such as yield, were mapped indirectly through associations between molecular markers and trait scores, both in traditional biparental populations and more recently in germplasm collections of interest to breeders through a technique known as Genome Wide Association Studies (GWAS). A summary of progress to date with GWAS can be found in a recent (2024) review by Yuan et al. [70]. Furthermore, advances in statistical techniques and computer software have allowed this mapping to be done in tetraploid potatoes, although it is still easier in diploids. The chromosomal locations are known as quantitative trait loci (QTLs). Today the segregation of single base differences (the A, T, G, C letters of DNA), known as single nucleotide polymorphisms (SNPs), are the markers of choice for mapping because of their abundance. Further progress in potato genetics is taking place with the sequencing of potato genomes, the first potato reference genome having been published by the International Potato Genome Sequencing Consortium (PGSC) in 2011 [71]. One target of research is finding the candidate genes (DNA sequences in the genome) for QTLs and determining their functions. A review of progress since 2011 can be found in a recent (2023) book chapter by Marhadour and Prodhomme [72]. I will now finish this chapter with the use of molecular markers in potato breeding.

Researchers at the James Hutton Institute have developed "diagnostic Resistance gene enrichment Sequencing (dRenSeq)" so that it can identify and confirm the sequence integrity of multiple known functional plant nucleotide-binding, leucine-rich repeat resistance genes (NLRs) in potato cultivars and breeding lines [22]. The technique is able to identify and validate NLRs effective against *Potato virus X*, the white potato cyst nematode *Globodera pallida* and the late blight pathogen *Phytophthora infestans*. It will therefore allow breeders to choose parents and make hybridizations and selections that will allow them to stack desirable combinations of resistance genes in their new cultivars.

Molecular markers can be used to make potato breeding faster, more efficient and more effective. This is because DNA can be extracted from small seedlings in a growth room or glasshouse and assessed for molecular markers associated with the desirable trait, such as disease resistance, good fry colour and even high yield. Molecular-marker-assisted selection can then be practised without having to wait for the clonal generation where it is possible to assess disease resistance, fry colour and yield. Details can be found in my book on potato breeding but need not concern us here [23]. The point is that the number of clonal generations in Fig. 9.5 can be cut from eight to one, thus considerably reducing the cycle time in years (T) between the generations of sexual reproduction. The rate of progress per year (R, response to selection) for a quantitative trait such as yield also depends on the percentage of seedlings/clones selected (i, the intensity of selection), the correlation (r) between the true breeding values of the seedling/clones and those estimated from the molecular markers, and the square root of the variance of the breeding values (σ_A, the usable genetic variation among the seedlings/clones). These are all parameters in what is known as the breeders' equation.

$$R = ir\sigma_A / T$$

The usable genetic variation is the key to success. If breeders are to deliver new generations of cultivars that will allow potato production to increase globally and to make a greater contribution to food security, they must choose the right breeding objectives and then the right germplasm, which were the topics of previous chapters. In the immediate future they will achieve this through conventional breeding aided by the use of molecular markers. In the longer term, gene editing and genetic transformation may make major contributions.

References

1. Knight TA (1799) An account of some experiments on the fecundation of vegetables. Phil Trans R Soc A 89:195–204
2. Knight TA (1807) On raising of new and early varieties of the potato. Trans Hort Soc Lond 1:57–59
3. Knight TA (1810) On potatoes (*Solanum tuberosum*). Trans Hort Soc Lond 1:187–193
4. Knight TA (1811) On two early varieties of the potato and the best method of forcing them. Trans Hort Soc Lond 1:244–247
5. Knight TA (1831) On the potato. Trans Hort Soc Lond Series 2(1):93–100
6. Knight TA (1833) On the culture of the potato. Trans Hort Soc Lond Ser 2(1):415–418
7. Lindley J (1831) The result of some experiments on the growth of potatoes, tried in the garden of the society in the year 1831. Trans Hort Soc Lond Ser 2(1):153–161
8. Paterson W (1871) On propagating new varieties of potatoes. Trans High Ag Soc 1871:135–139
9. McIntosh TP (1927) The potato. Oliver and Boyd, Edinburgh
10. Wilson A (1993) The story of the potato through illustrated varieties. Balding and Mansell, England
11. Suttons seeds a history 1806–2006 (2006) Earley Local History Group, Reading
12. Salaman RN (1926) Potato varieties. Cambridge University Press, Cambridge
13. Simakov EA, Yashina IM, Sklyarova NP (2007) Potato breeding in Russia: history, general trends and achievements. In: Haverkort AJ, Anisimov BV (eds) Potato production and innovative technologies. Wageningen Academic Publishers, Wageningen, pp 311–327
14. Goodrich CE (1863) The origination and test culture of seedling potatoes. Trans NY State Agric Soc 23:89–134
15. Bethke PC, Nassar AMK, Kubow S, Leclerc YN, Li X-Q, Haroon M, Molen T, Bamberg J, Martin M, Donnelly DJ (2014) History and origin of Russet Burbank (Netted Gem) a sport of Burbank. Am J Potato Res 91:594–609. https://doi.org/10.1007/s12230-014-9397-5
16. Douches DS, Ludlam K, Freyre R (1991) Isozyme and plastid DNA assessment of pedigrees of nineteenth century potato cultivars. Theor Appl Genet 82:195–200
17. Bradshaw JE (2009) Potato breeding at the Scottish Plant Breeding Station and the Scottish Crop Research Institute: 1920–2008. Potato Res 52:141–172
18. Glendinning DR (1983) Potato introductions and breeding up to the early 20th century. New Phytol 94:479–505
19. Provan J, Powell W, Dewar H, Bryan G, Machray GC, Waugh R (1999) An extreme cytoplasmic bottleneck in the modern European cultivated potato (*Solanum tuberosum*) is not reflected in decreased levels of nuclear diversity. Proc R Soc Lond B Biol Sci 266:633–639
20. Gutaker RM, Weiß CL, Ellis D, Anglin NL, Knapp S, Fernández-Alonso JL, Prat S, Burbano HA (2019) The origins and adaptation of European potatoes reconstructed from historical genomes. Nat Ecol Evol 3:1093–1101
21. Hardigan MA, Laimbeer FPE, Newton L, Crisovan E, Hamilton JP, Vaillancourt B, Wiegert-Rininger K, Wood JC, Douches DS, Farré EM, Veilleux RE, Buell CR (2017) Genome diversity of tuber-bearing *Solanum* uncovers complex evolutionary history and targets of domestication in the cultivated potato. PNAS 114(46):E9999–E10008. https://doi.org/10.1073/pnas.1714380114
22. Armstrong MR, Vossen J, Lim TY, Hutten RC, Xu J, Strachan SM, Harrower B, Champouret N, Gilroy EM, Hein I (2019) Tracking disease resistance deployment in potato breeding by enrichment sequencing. Plant Biotechnol J 17(2):540–549. https://doi.org/10.1111/pbi.12997
23. Bradshaw JE (2021) Potato breeding: theory and practice. Springer, Cham
24. van Loon JP, Lammerts van Bueren ET, van Cruyningen PJ, Wiskerke JSC (2023) The history of Dutch potato breeding 1888–2018: from hobby to industry. Potato Res. https://doi.org/10.1007/s11540-023-09661-5
25. Jansky SH, Spooner DM (2018) The evolution of potato breeding. Plant Breed Rev 41:169–214
26. Plaisted RL, Hoopes RW (1989) The past record and future prospects for the use of exotic potato germplasm. Am Potato J 66:603–627
27. Love SL (1999) Founding clones, major contributing ancestors, and exotic progenitors of prominent North American cultivars. Am J Potato Res 76:263–272
28. Luthra SK, Gupta VK, Tiwari JK, Kumar V, Bhardwaj V, Sood S, Dalamu, Kaur RP, Kumar R, Vanishree G, Kumar D, Mhatre P, Chakrabarti SK (2020) Potato breeding in India. CPRI Technical Bulletin No 74 (revised), ICAR-Central Potato Research Institute, Shimla, Himachal Pradesh, India
29. Spooner DM, Nunez J, Rodriguez F, Naik PS, Ghislain M (2005) Nuclear and chloroplast DNA reassessment of the origin of Indian potato varieties and its implications for the origin of the early European potato. Theor Appl Genet 110:1020–1026
30. Pandey NK, Kharumnuid P, Kumar S, Chakrabarti SK, Bhardwaj V (2024) Economic impact of early bulking and drought-tolerant potato cultivar Kufri Pukhraj in India. Potato Res 67:1469–1482
31. Jin L (2018) Potato production and breeding in China. In: Wang-Pruski G (ed) Achieving sustainable cultivation of potatoes. Volume 1: Breeding improved varieties. Burleigh Dodds, Cambridge, pp 259–270
32. Reader J (2008) Propitious esculent. William Heinemann, London
33. Pieterse L, Judd J (2014) World catalogue of potato varieties, 6th edn. Agrimedia GmbH, Clenze

34. UPOV (2024) Website 27 March 2024 (www.upov.int)
35. Ojwang SO, Okello JJ, Otieno DJ, Mutiso JM, Lindqvist-Kreuze H, Coaldrake P, Mendes T, Andrade M, Sharma N, Gruneberg W, Makunde G (2023) Targeting market segment needs with public good crop breeding investments: a case study with potato and sweetpotato focused on poverty alleviation, nutrition and gender. Front Plant Sci 14:1105079
36. Lindqvist-Kreuze H, Bonierbale M, Grüneberg WJ, Mendes T, De Boeck B, Campos H (2024) Potato and sweetpotato breeding at the international potato center: approaches, outcomes and the way forward. Theor Appl Genet 137:12. https://doi.org/10.1007/s00122-023-04515-7
37. Fuglie KO, Echeverria RG (2024) The economic impact of CGIAR-related crop technologies on agricultural productivity in developing countries, 1961–2020. World Dev 176:106523. https://doi.org/10.1016/j.worlddev.2023.106523
38. Almekinders CJM, Chilver AS, Renia HM (1996) Current status of the TPS technology in the world. Potato Res 39:289–303
39. Ortiz R (1997) Breeding for potato production from true seed. Plant Breed Abstr 67:1355–1360
40. Simmonds NW (1997) A review of potato propagation by means of seed, as distinct from clonal propagation by tubers. Potato Res 40:191–214
41. Chilver A, Walker TS, Khatana V, Fano R, Suherman R, Rizk A (2005) On-farm profitability and prospects for true potato seed (TPS). In: Razdan MK, Mattoo AK (eds) Genetic improvement of solanaceous crops, Volume I: Potato. Science Publishers Inc, Enfield, pp 39–63
42. Lindhout P, de Vries M, ter Maat M, Ying S, Viquez-Zamora M, van Heusden S (2018) Hybrid potato breeding for improved varieties. In: Wang-Pruski G (ed) Achieving sustainable cultivation of potatoes, Volume 1: Breeding improved varieties. Burleigh Dodds, Cambridge, pp 99–122
43. Hosaka K, Hanneman RE (1998) Genetics of self-compatibility in a self-incompatible wild diploid potato species *Solanum chacoense*. 1. Detection of an S-locus inhibitor (Sli) gene. Euphytica 99:191–197. https://doi.org/10.1023/A:1018353613431
44. Hosaka K, Hanneman RE (1998) Genetics of self-compatibility in a self-incompatible wild diploid potato species *Solanum chacoense*. 2. Localization of an S-locus inhibitor (*Sli*) gene on the potato genome using DNA markers. Euphytica 103:265–271. https://doi.org/10.1023/A:1018380725160
45. Lindhout P, Meijer D, Schotte T, Hutten RCB, Visser RGF, van Eck HJ (2011) Towards F1 hybrid seed potato breeding. Potato Res 54:301–312
46. Vries M, ter Maat M, Lindhout P (2016) The potential of hybrid potato for East-Africa. Open Agric 1(1):151–156. https://doi.org/10.1515/opag-2016-0020
47. Hosaka K, Sanetomo R (2020) Creation of a highly homozygous diploid potato using the *S* locus inhibitor (*Sli*) gene. Euphytica 216:169. https://doi.org/10.1007/s10681-020-02699-3
48. Hosaka K, Sanetomo R (2024) Agronomic performance of diploid inbred potatoes and their F1 hybrid potatoes. Euphytica 220:49. https://doi.org/10.1007/s10681-024-03305-6
49. Darwin C (1876) The effects of cross and self fertilisation in the vegetable kingdom. Digitally printed 2009, Cambridge University Press, Cambridge
50. Mendel G (1865) Experiments in plant hybridisation. English translation with introduction by R.A. Fisher, 1965, Oliver & Boyd, Edinburgh
51. Bateson W, Mendel G (1902) Mendel's principles of heredity. Digitally printed version 2009, Cambridge University Press, Cambridge
52. Salaman RN (1910) The inheritance of colour and other characters in the potato. J Genet 1:7–46
53. Whitehouse HLK (1973) Towards an understanding of the mechanism of heredity, 3rd edn. Edward Arnold, London
54. Lunden AP (1937) Arvelighetsundersøkelser i potet (inheritance studies in the potato), *Solanum tuberosum* L. Meld. fra Norges Landbrukshøiskole 1937, pp 1–156
55. Cadman CH (1942) Autotetraploid inheritance in the potato: some new evidence. J Genet 44:33–52
56. Toxopeus HJ (1953) On the significance of multiplex parental material in breeding for resistance to some diseases in the potato. Euphytica 2:139–146
57. Swaminathan MS (1954) Nature of polyploidy in some 48-chromosome species of the genus *Solanum*, section tuberarium. Genetics 39:59–76
58. Gottschalk W (1958) Über die anwendung cytologischer methoden ftir die bearbeimng phylogenetischer fragestellungen bei den Solanaceen. Eine erwiderung. Zeitschr ftir Pftanzenziicht 39:47–70
59. Lunden AP (1960) Some more evidence of autotetraploid inheritance in the potato (*Solanum tuberosum*). Euphytica 9:225–234
60. Allard RW (1960) Principles of plant breeding. Wiley International, New York
61. Bradshaw JE (2022) A brief history of the impact of potato genetics on the breeding of tetraploid potato cultivars for tuber propagation. Potato Res 65(3):461–501
62. Hougas RW, Peloquin SJ (1958) The potential of potato haploids in breeding and genetic research. Am Potato J 35:701–707
63. Hougas RW, Peloquin SJ, Ross RW (1958) Haploids of the common potato. J Hered 49:103–107
64. Hermundstad SA, Peloquin SJ (1987) Breeding at the 2*x* level and sexual polyploidization. In: Jellis GJ, Richardson DE (eds) The production of new potato varieties. Cambridge University Press, Cambridge, pp 197–210
65. Jansky SH, Yerk GL, Peloquin SJ (1990) The use of potato haploids to put 2*x* wild species germplasm into a usable form. Plant Breed 104:290–294
66. Ortiz R, Peloquin SJ (1994) Use of 24-chromosome potatoes (diploids and dihaploids) for geneti-

cal analysis. In: Bradshaw JE, Mackay GR (eds) Potato genetics. CAB International, Wallingford, pp 133–154
67. Bonierbale MW, Plaisted RL, Tanksley SD (1988) RFLP maps based on a common set of clones reveal modes of chromosomal evolution in potato and tomato. Genetics 120(4):1095–1103
68. Gebhardt C, Ritter E, Debener T, Schachtschabel U, Walkemeier B, Uhrig H, Salamini F (1989) RFLP analysis and linkage mapping in *Solanum tuberosum*. Theor Appl Genet 78(1):65–75
69. Bradshaw JE (2024) Potato genetics for crop improvement. In: Khurana SMP, Bradshaw JE, Bhardwaj V (eds) Approaches for potato crop improvement and stress management. Springer, Singapore, pp 1–27. https://doi.org/10.1007/978-981-97-1223-6_1
70. Yuan J, Cheng L, Wang Y, Zhang F (2024) Genome-wide association studies for key agronomic and quality traits in potato (*Solanum tuberosum* L.). Agronomy 14:2214. https://doi.org/10.3390/agronomy14102214
71. The Potato Genome Sequencing Consortium (2011) Genome sequence and analysis of the tuber crop potato. Nature 475:189–195
72. Marhadour S, Prodhomme C (2023) Recent trends in genetics studies and molecular breeding of potato. In: Caliskan ME, Bakhsh A, Jabran K (eds) Potato production worldwide. Academic, London, pp 273–301

10 DNA, Gene Editing and Genetic Transformation

Genetic transformation (since 1995) and gene editing (since 2015) have opened up new possibilities for potato improvement, beyond what can be achieved by conventional breeding. They should make valuable contributions to increasing potato production and the nutritional value of potatoes, and hence to the role of potatoes in providing food security during a period of climate change. Genetic transformation introduces one or more desired genes into the potato genome by combining naturally occurring processes with the genetic engineering of bacterial plasmids for use as vectors of the genes. The result is a genetically modified (GM) crop, also sometimes referred to as a biotech crop. In many countries, adoption of such crops has been delayed because politicians, pressure groups and the general public still need to be convinced that GM crops are not only safe to grow and eat but are also of real benefit to both farmers and consumers. Debates over regulation centre on whether the product or process should be regulated, whether or not the precautionary principle should apply, and the source(s) of the introduced gene(s). Gene editing should prove less controversial as it targets specific locations in the potato genome to make desired alterations to specific genes.

10.1 Beyond Conventional Breeding

The potato cultivars being grown today are mainly the result of deliberate hybridizations and selections by breeders from within the gene pool of cultivated potatoes and their wild relatives. In this chapter we shall see how genetic transformation and gene editing take us beyond this conventional breeding and open up new possibilities for potato improvement. First, however, we need a brief résumé of the molecular biology of the gene and gene expression.

10.2 DNA

On Saturday 28 February 1953, two scientists from Cambridge University's Cavendish Laboratory went for lunch in the nearby Eagle Pub. One of them, Francis Crick, bragged to those present that he and James Watson had discovered "the secret of life". They had worked out the molecular structure of DNA (Deoxyribose Nucleic Acid or Deoxyribonucleic Acid), the chemical carrier of hereditary information from

J. E. Bradshaw, *Can Potatoes Feed the World?*, Sustainable Development Goals Series,
https://doi.org/10.1007/978-3-031-92890-1_10

one generation to the next [1, 2]. DNA has a double helix structure in which the backbone comprises repeats of deoxyribose (a sugar) linked to a phosphate group. One of four bases, adenine (A), thymine (T), guanine (G) or cytosine (C), is attached to each sugar and faces inwards in such a way that adenine is opposite thymine and guanine is opposite cytosine (base pairs). The order of bases in each of the complementary strands is how the genetic information is coded (Fig. 10.1).

Today every genetically modified organism (GMO) that is produced by genetic transformation is proof that DNA carries the genetic information. The first proof, however, had been as late as 1944 from studying the transformation of types of pneumococcal bacteria [3]. If we now rapidly move forward to the 15 and 16 February 2001, we find that the order of bases in our human DNA had been worked out: the human genome had been sequenced, or more correctly, two draft sequences had been released, one publicly funded and the other from private entrepreneurship [4, 5]. Today we are happy to use the phrase "it's in our DNA, it's who we are".

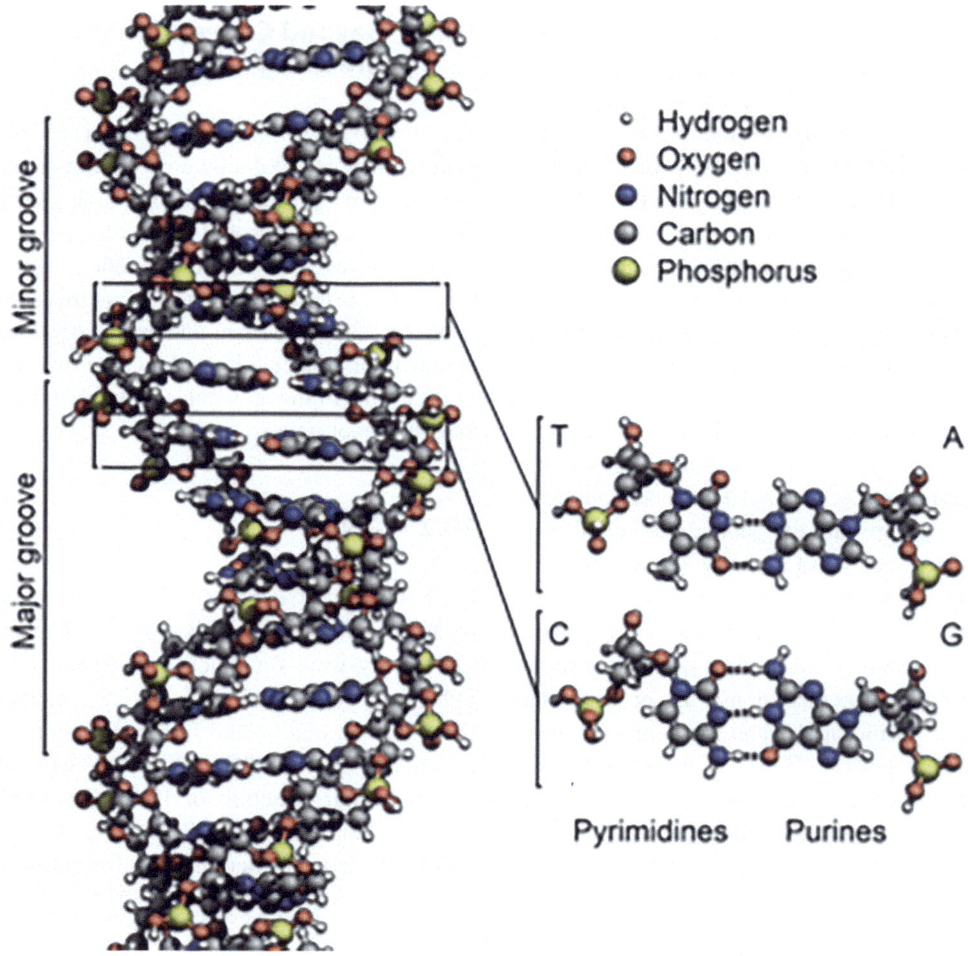

Fig. 10.1 The structure of the DNA double helix: T (thymine), A (adenine), C (cytosine) and G (guanine). (Source: Wikipedia) "DNA Structure+Key+Labelled.pn NoBB" by Zephyris Own work. Licensed under CC BY-SA 3.0 via Commons. https://commons.wikimedia.org/wiki/File:DNA_Structure%2BKey%2BLabelled.pn_NoBB.png#/media/File:DNA_Structure%2BKey%2BLabelled.pn_NoBB.png

Drafts of the base sequences of a potato (*S. tuberosum* Group Phureja DM1-3 516 R44) and a tomato (cultivar Heinz 1706) and one of its wild relatives (*S. pimpinellifolium*) were published in *Nature* on 14 July 2011 and 30 May 2012, respectively [6, 7]. In potato, 86% of the 844-megabase (Mb, mega = million) genome was sequenced (i.e., 726 Mb) with a predicted 39,031 genes (strictly speaking, protein-coding genes). In tomato, 84% of the 900-Mb genome was sequenced with a predicted 34,727 genes. For both species these genes are only a small proportion of the total DNA.

Group Phureja DM1-3 516 R44 was a diploid potato in which the two sets of 12 chromosomes were identical because it was what we call a doubled monoploid. This made the sequencing easier. It has proved an important resource for gene discovery and the development of new breeding methods. However, the draft sequence needed improving. Hence, in 2019, Solynta and Plant Breeding, Wageningen University and Research, sequenced Solyntus, a selection from the Solynta diploid hybrid breeding programme whose two sets of 12 chromosomes are almost identical as a result of a number of generations of self-pollination. The sequence covered 716 Mb of DNA (https://www.plantbreeding.wur.nl/Solyntus/).

The DNA in plant cells has three physical locations. Most occurs in the linear chromosomes in the cell nucleus. Much smaller amounts are found outside the nucleus in the cell cytoplasm, in the chloroplasts (plastids) where photosynthesis occurs, and in the mitochondria, where respiration takes place. The chloroplast genome (cpDNA) of potato is a circular DNA molecule of 155,296 base pairs (bp) encoding 132 genes, of which 87 are protein coding [8]. In contrast, the mitochondrial genome (mtDNA) comprises five circular DNA molecules of lengths from 49,171 to 297,014 bp and a total length of 756,058 bp, encoding 80 genes, of which 34 are protein coding [9]. We saw in Chaps. 3 and 4 that the presence or absence of a 24-bp deletion in the plastid DNA of cultivated potatoes provides valuable information about their history.

10.3 Protein-Coding Genes

How DNA determines the sequence of amino acids in proteins (polypeptides) is often called "the central dogma", as explained by Watson in his book *Molecular Biology of the Gene*, which was published in 1965 [10]. The information in the double-stranded helix of DNA (base pairs) is first transcribed (transcription) into single-stranded messenger RNA (mRNA) in the cell nucleus. The structure of RNA (ribonucleic acid) is similar to DNA but differs in that the sugar is D-ribose instead of 2-deoxy-D-ribose, and one of the pyrimidines is uracil (U) in place of thymine (T) (5-methyluracil); hence the letters (bases) of the RNA alphabet are adenine (A), uracil (U), guanine (G) and cytosine (C). The mRNA moves through the pores in the nuclear membrane to the site of protein synthesis in the cytoplasm, namely the ribosomes. Each of the 20 available amino acids has its own adaptor molecule (a transfer RNA), which associates it with a particular triplet of bases in the mRNA (the codon). The direction of translation of the nucleotide sequence in mRNA into the amino-acid sequence in a polypeptide is the same as the direction of transcription from DNA to RNA (Fig. 10.2). The relationship between the triplets of bases and the 20 amino acids is known as the genetic code. There are also start and stop codons. What is remarkable is that the same code applies in vivo to virtually all organisms; it is universal, providing a fundamental unity to all life. Equally remarkable was the realization that the linear sequence of amino acids in a protein determines its three-dimensional structure and hence its biological activity and function, simply as a result of the laws of physics and chemistry.

As the enzymes, which catalyse all of the chemical reactions of cellular metabolism, are proteins, which genes are transcribed at any particular time and in any particular type of cell has a major impact on this metabolism. Each protein-coding gene (structural gene) has its own regulatory regions, the most important of which is its promoter, which occurs about 25 base pairs

In nucleus

DNA encoding (sense) strand 5' ATG...............ACCGGC..............TAA 3'
DNA template (antisense) strand 3' TAC...............TGGCCG...............ATT 5'

RNA synthesis controlled by promoter and transcription factors (transcription)

mRNA 5' AUG..............ACCGGC..............UAA 3'
Moves to cytoplasm codon

In cytoplasm anticodon

tRNAs UAC UGG CCG

Polypeptide synthesis on ribosomes (translation). methione.....threonine glycine.........STOP
START

Fig. 10.2 Transcription and translation (DNA → mRNA → polypeptide). (From Bradshaw JE (2016) Plant Breeding: Past, Present and Future. Springer, Cham with permission)

before (upstream) the start codon (Fig. 10.2) and is the binding site for the transcription enzyme RNA polymerase II. The promoter can determine if a gene is transcribed (expressed) throughout the plant (constitutively) or in specific parts, such as a potato tuber (tuber-specific). In addition to the structural genes with their promoters, there are regulatory genes, which control their functioning. They encode proteins known as transcription factors, which can bind to DNA and affect transcription (turn genes on and off). For example, whether Mendel's peas had white or violet-red flowers was determined by a transcription factor, and likewise whether potato tubers have white or coloured skin (red or purple) [11, 12]. In both examples the transcription factor controls genes in the anthocyanin biosynthesis pathway. When mutated, the transcription factor becomes inactive, anthocyanin is not produced, and the result is white rather than colour. More information on the control of gene expression in plants can be found in the second (2008) edition of *Plant Biotechnology, The Genetic Manipulation of Plants* by Slater et al. and the fifth (2021) edition of *Plant Biochemistry* by Heldt and Piechulla [13, 14]. A short and readable book on modern biology for the non-specialist is Paul Nurse's *What Is Life*, which was published in 2020 [15].

10.4 Gene Editing

Gene editing is an alternative to sexual hybridization for the genetic improvement of potatoes, particularly already successful cultivars. In conventional potato breeding, two successful cultivars that complement each other for both desirable and undesirable characteristics are hybridized, and offspring are selected that combine as many of their desirable characteristics as possible, whilst also having as few as possible of the undesirable ones. In other words, sexual reproduction is used to generate new combinations of genetic variants on which to practise selection, the variants being the products of past mutations. Molecular markers can be used to locate (map) these genes on the 12 chromosomes of potato. Today the segregation of single base differences (the A, T, G, C letters of DNA), known as single nucleotide polymorphisms (SNPs), are the marker of choice for mapping because of their abundance. Furthermore, without going into details, the majority of SNPs can be assigned a unique genomic, as well as chromosomal, location in the published potato genome sequence. This establishes a link between the marker map and the genome sequence and hence a route from the postulated candidate gene to the actual causal gene. Once a causal gene and its base sequence have been found, the gene can be edited.

Gene editing uses enzymes called engineered nucleases to catalyse a double-strand break in the DNA at a specific location in the genome so that desired changes can be made at the break site. Zinc finger nucleases (ZFNs), transcription activator-like effector nucleases (TALENs) and mega nucleases (MNs) are examples of site-directed nucleases, but the most popular editing system is known as CRISPR/Cas9. It was discovered in *Streptococcus pyogenes*, where it protects the bacterium from invading nucleic acids, such as viruses, by cleaving their DNA in a sequence-dependent manner. It consists of an endonuclease protein (Cas9) that cleaves double-stranded DNA and a guide RNA (gRNA) with a critical 20-nucleotide sequence that binds to the target DNA sequence. It was first demonstrated in plants in 2015 [16]. Functional gRNA and Cas9 protein must be delivered into the plant cells targeted for editing, but without integration of their DNA into the plant DNA, which would make the edited plant a genetically modified organism (GMO). This can be done by techniques such as transient expression through PEG-mediated protoplast transfection or *Agrobacterium*-mediated leaf infiltration, a protoplast being a plant cell without its cell wall; but I am not going to give the technical details, which can be found elsewhere [17]. Guide RNAs with different sequences can be used to simultaneously modify a number of genes (traits). Careful selection of specific gRNA sequences should minimize the risk of unwanted genome modifications, thus making off-target effects rare and the changes more precise than those achieved with mutation breeding, which is exempt from the regulations governing genetically modified crops. In 2012, the European Food Safety Authority (EFSA) reached the same conclusion about off-target changes specifically for the third method of gene editing described in the next paragraph [18].

CRISPR/Cas9 can be designed to target all four alleles of a gene in a tetraploid potato and both alleles of a gene in a diploid potato. Three types of gene editing can be achieved. In simplistic terms, they are like adding or deleting a few letters in a word so that it becomes nonsense, making a small change to a word so that it has a different meaning (accent to ascent), and adding a word to a sentence. The first is a gene knockout where the double-strand break is repaired by a natural process called non-homologous end joining (NHEJ) in which short deletions or insertions of DNA occur at the break site, resulting in loss of gene function. The second is the replacement of an allele with a different one by a process called homology-directed repair (HDR), in which the new allele is provided by a DNA repair template with end sequences identical (homologous) to those flanking the break site. The donor template must be delivered to the cell at the same time as the nuclease mediates cleavage. The desired outcome currently has a low frequency. The third is gene insertion, in which a completely new gene (transgene) is inserted between the homologous sequences of the DNA repair template. In other words, it is site-directed genetic transformation. Hence, this third type of gene editing is still regarded as genetic transformation leading to a genetically modified (GM) crop whose environmental release and marketing in the United Kingdom, and elsewhere, are governed by GM regulations. This is because the genetic changes could not have occurred naturally or through conventional breeding. In contrast, in England, but not the other nations of the United Kingdom, the first two types of gene editing (GE) are exempt from the GM regulations as a result of "The Genetic Technology (Precision Breeding) Act", which became law on Thursday, 23 March 2023. The Act defines a Precision Bred Organism (PBO) as "a plant (or animal) whose DNA has been modified using a genetic technology but with a type of genetic change that could have occurred naturally and been selected in a traditional breeding programme". Secondary legislation is expected early in 2025 to unlock the benefits of the Precision Breeding Act.

10.5 Examples of Gene Editing

All three types of gene editing have been demonstrated in maize [19]. However, only the first type is now common in potatoes. Hence, my typical potato example is a "knockout" of the *GBSS* gene, which encodes the granular-bound starch synthase (GBSS) enzyme responsible for the synthesis of amylose. This was achieved through the transient expression in protoplasts (plant cells without their cell walls) of CRISPR/Cas9 constructs targeted at the allelic variants present in tetraploid cultivar Kuras [20]. Full knockout of GBSS enzyme activity required mutations in all four alleles, which occurred in 2% of regenerated lines. Lack of enzyme activity resulted in amylopectin starch, which stained red-brown with iodine. In contrast, one remaining wild-type allele was sufficient for enough GBSS activity to produce significant amounts of amylose. Subsequently, another group of researchers improved the efficacies of the CRISPR/Cas9 editing of *GBSS* in protoplasts [21].

A recent (2023) review of gene editing in potatoes revealed a lot of research with promising results on improving the methods used for CRISPR/Cas9 editing [22]. The review also reported encouraging results for the targets of such gene editing: resistance to late blight and to PVY, reductions in enzymic browning, cold-induced sweetening and steroidal glycoalkaloids, and altered starch composition. There was one example of the second type of gene editing involving an HDR template: a precise single nucleotide correction of the *StCCoAOMT* gene in cultivar Russet Burbank enhanced its resistance to late blight [23]. However, it is difficult to judge how soon we can expect to see commercial results from gene editing. Hence, I was drawn to a recent (22 March 2024) headline in the English farming press: "Scientists plan to use gene editing tech to transform potato sector" [24]. It was announcing a new research project, "Tubergene", funded as part of the UKRI's (UK Research and Innovation) National Engineering Biology Programme and led by B-hive Innovations, a research and development firm. It involves industry and research partners and will focus on two key goals: reducing bruising-related discolouration and making potatoes quicker to cook. The aim is to "enhance potato quality, cut down on food waste and meet the evolving needs of consumers". This new project comes a year after the "Genetic Technology (Precision Breeding) Act" became law in England. I think that we can expect to see an increasing number of such projects in countries where gene editing is permitted. In the meantime, there are a substantial number of demonstrations of potato improvements from non-site-directed genetic transformation and gene "knockouts" from gene silencing, but few examples of genetically modified (GM) potatoes being grown and marketed. In many countries, politicians, pressure groups and the general public still need to be convinced that GM crops are not only safe to grow and eat but are also of real benefit to both farmers and consumers.

10.6 Genetic Transformation

Genetic transformation is the name of the process used to insert one or more desired genes into the potato genome. The aim is to make targeted improvements of successful and widely grown cultivars. The resulting crops are referred to as genetically modified (GM), biotech or transgenic. Genetic transformation of potatoes combines naturally occurring processes with the genetic engineering of bacterial plasmids. Bacterial cells do not have a nucleus and typically contain only one circular chromosome but may possess smaller, circular DNA molecules called plasmids, which can replicate autonomously.

Genetic transformation in potatoes is done using *Agrobacterium tumefaciens* (now *Rhizobium radiobacter*), a motile soil-borne bacterium that causes crown gall disease when it infects wounded plants of crops such as apples and grapes. The crown gall formation results from the transfer of some genes (T-DNA, $\approx$ 20 kb) from the bacterium's tumour-inducing (Ti) plasmid ($\approx$ 200 kb) into the plant cell, and then into its nucleus, where it becomes integrated into the nuclear genome during repair of naturally occurring (random) chromosome breaks. Researchers in the mid-1980s worked out how to convert the Ti plasmid into a vector for plant transformation that does not result in crown gall disease [25–28]. The details need not concern us here but can be found in modern student textbooks on plant biochemistry [29] and the principles of plant genetics and breeding [30].

A prerequisite for transformation is finding a cellular source of the desired gene, transferring (cloning) it into plasmids, and maintaining the plasmids in colonies of a bacterium such as *Escherichia coli* until their final transfer into *A. tumefaciens*. In other words, the genetic engineering involved is complicated and involves a lot of work. The transformation takes place when *A. tumefaciens* is co-cultured with freshly cut tuber discs or leaf or internode explants of the potato. Regeneration of plants then takes place in tissue culture but can be a problem with some cultivars. This is why the easily transformed and regenerated cultivar Desiree is frequently used for research purposes [31]. In the early days of genetic transformation, the bacterial plasmid would contain a selectable marker such as resistance to an antibiotic or herbicide. Regeneration would take place in the presence of the selectable agent to ensure selection of transformants. However, public concerns arose over the use of selectable markers, particularly the risk to human and animal health of using antibiotic resistance. This is because genes for antibiotic resistance could, in theory, be transferred to bacteria in the gastro-intestinal tract, a process known as horizontal gene transfer. Resistant bacteria could then spread and transfer the resistance to pathogenic bacteria, thus making antibiotic treatments ineffective [32]. Despite the very low risk of this happening, marker-free transformation vectors are now commonly used, and transformants are detected by PCR (polymerase chain reaction) analysis for the presence of the transferred gene(s) [18, 33]. Transformants with undesirable characteristics can arise from the transformation per se or during the in vitro regeneration, when they are known as somaclonal variants, for which there are a number of possible causes [34]. As these undesirable transformants need to be eliminated, at least several hundred independently transformed clonal lines are usually produced and the best selected under field conditions, as normally done by breeders [31]. Each of Monsanto's NewLeaf™ Plus and NewLeaf™ Y programmes started with 3000 original lines, from which six were selected for commercialization some 7 years later [31].

Initially genetic transformation involved single genes, but now two or more are being stacked (combined) in new GM cultivars. Three methods can be used to stack genes: (1) two or more *A. tumefaciens* clones with different genes can be mixed and delivered to the potato tissue at the same time so that some co-transformation is possible; (2) two or more genes are linked together in the same vector (multigene cassette), with four to six considered the current maximum; and (3) re-transformation of a potato already containing an introduced gene(s). Cultivars containing different genes can also be hybridized and the offspring selected for ones containing all of the genes, but the outcome will be a completely new cultivar, which may or may not be the objective. Currently up to eight genes are being stacked (combined) in new GM cultivars, as can be seen in a project started in 2015 at the Sainsbury Laboratory in Norwich in the United Kingdom (http://www.tsl.ac.uk/news/new-potato-at-thesainsbury-laboratory/). The aim was to improve the disease resistance and processing characteristics of Maris Piper, the most widely grown cultivar in the United Kingdom. The latest update (4 April 2024) from their website (http://www.tsl.ac.uk/our-work/scientific-groups/jonathan-jones-group/potatoplus) reports success in combining resistance to late blight and PVY with resistance to cold-induced sweetening and tuber bruising in cultivar Hermes as well as Maris Piper. However, marketing is being held up by the GM regulations, which we will consider shortly.

10.7 Cisgenic, Intragenic and Transgenic Potatoes

The source of the genes used in genetic transformations has proved an important topic in discussions about the safety and regulation of GM crops. As a result, the European Food Safety Authority (EFSA) has come up with definitions of cisgenes, intragenes and transgenes [33, and online www.efsa.europa.eu/efsajournal]. My simplified explanations for potatoes (*Solanum tuberosum*) are as follows. A cisgene comes unaltered from the same or a sexually compatible species (including via bridging species) and is inserted in its normal sense (same direction) orientation [the full definition implies marker-free transformation (no foreign DNA) and also indicates whether or not T-DNA borders remain]. An intragene comprises different gene fragments from the same or a sexually compatible species arranged in a sense or antisense (reverse direction) orientation (e.g., inverted DNA repeats for RNA interference). Finally, a transgene can come from organisms (e.g., viruses, bacteria, fungi and plants) that are not sexually compatible with cultivated potatoes. The distinctions between cisgenic, intragenic and transgenic potatoes therefore lie in the genes and not the transformation process. Potentially new DNA sequence combinations, and hence novel proteins, could arise in transgenic and intragenic plants, which would not normally arise in conventionally bred and cisgenic plants. Novel proteins do raise health concerns, but these relate to novel foods in general, not just GM crops, namely, the possible toxicity or allergenicity of substances produced in the plant [32]. Taking all of these considerations into account, some researchers, including myself, think that cisgenesis should be excluded from the definition of genetic modification [35]. I covered cisgenesis in Chap. 5, as one of the best-known examples to date is the stacking of three *R*-genes for resistance to late blight from potato wild relatives. There is one other very important point to make about the genes used in transformations. Genes, which encode the amino acid sequences of proteins, have regulatory regions that control their expression, in particular their promoters. Sometimes the successful outcome of a genetic transformation depends on tuber-specific expression of the inserted gene, rather than constitutive expression throughout the plant, so that the correct choice of promoter is an essential part of the genetic engineering.

10.8 Gene Silencing

Transgene-mediated gene silencing through RNA interference (RNAi) is a way to inactivate a transcribed gene before protein synthesis can occur, and hence an alternative to a gene "knock-out" through gene editing. It operates by the degradation of targeted messenger RNAs (Fig. 10.2) and hence is sometimes referred to as post-transcriptional gene silencing (PTGS). In the simplest RNAi method, a transgene (usually an intragene) is inserted into the plant so that the sequence to be expressed is in the reverse orientation to the promoter. The promoter directs the production of antisense RNA, which is complementary to the endogenous mRNA, and hence they form double-stranded RNA, which is degraded by endonucleases. However, a more effective transgene sequence is a strong promoter (to drive expression) followed by a sense sequence, then a spacer sequence, and finally an antisense sequence, known as an inverted repeat. On transcription, the two complementary regions form a double-stranded hairpin RNA, which is very efficient in RNAi-induced destruction of any homologous mRNA via small (21–26 nucleotides) double-stranded interfering RNA molecules (siRNA). The mechanism is not obvious, but details can be found in books on the genetic manipulation of plants [13].

RNAi has been used successfully to achieve some desirable traits in potatoes: multiple virus resistance (PVY-O, PVA and PLRV), broad-spectrum resistance to root-knot nematodes (*Meloidogyne* spp.), prevention of enzymic browning (impact-induced black spot bruise), reduction in cold-induced sweetening (and hence acrylamide production in fried products as well as the desired golden-brown fry colours), reduction of acrylamide in fried products through lower levels of asparagine or reducing sugars in tubers, reduction in tuber glycoalkaloids, high amylopectin starch and high amylose starch [36–47]. More details for traits, which were not covered in Chap. 8 on nutritional value, can be found in my book on potato breeding [48].

10.9 Transgenes from Viruses, Bacteria, Microalga, Fungi and Non-Solanaceous Plants

Examples of desirable traits that have been achieved through transgenes from organisms that are not sexually compatible with cultivated potatoes are as follows. Resistance to virus PVY through expression of the PVY coat protein gene [31, 49]; resistance to the Colorado potato beetle (CPB, *Leptinotarsa decemlineata*) and resistance to the potato tuber moth (PTM, *Phthorimaea operculella*), through the introduction of *cry* genes from the soil bacterium *Bacillus thuringiensis* [31, 49–53]; production of beta-carotene (provitamin A) through the tuber-specific expression of three genes from the bacterium *Erwinia herbicola* [54], and production of astaxanthin, a feed additive in aquaculture, from beta-carotene through the tuber-specific expression of a gene from the green microalga *Haematococcus pluvialis* [55]; increased dietary fibre in the form of inulin through two genes from globe artichoke (*Cynara scolymus*) [56]; improved protein content and amino acid balance through tuber-specific expression of a gene encoding a seed protein, Amaranth Albumin 1 from *Amaranthus hypochondriacus* [57]; and partial resistance to cyst nematodes (*Globodera pallida* and *G. rostochiensis*) through expression of a plant cystatin [58]. Here, I have chosen three examples to make points about health and environmental issues.

B. thuringiensis was discovered by Shigetane Ishiwatari in 1901 in diseased silkworms. The *cry* genes are carried on plasmids, and the various strains of *B. thuringiensis* produce a wide range of different insecticidal crystal (Cry) proteins (δ-endotoxins), which are active against the larvae of particular classes of insects. The specificity of Cry proteins allowed the targeting of specific pests by particular transgenes. The toxicity of the crystals to other animals, including humans, is extremely low, and preparations of *B. thuringiensis* have been used as "organic" pesticides for over half a century. When the first GM cultivars with resistance to the Colorado potato beetle (CPB) were marketed in 1995 in North America, there was concern about the evolution of resistance to the *Bt-cry3A* gene by the beetle. Growers were therefore advised to plant 20% of their field area with a non-*Bt*-containing potato as a refuge for susceptible beetles and to rotate their fields in which transgenic crops were grown [49]. The refuges provide sources of large numbers of susceptible individuals that readily mix with any resistant individuals appearing in the transgenic crop before mating has occurred. Hence, the resistant individuals will mate with susceptible ones and produce susceptible offspring because resistance is a recessive trait. The durability of resistance could also be extended by the pyramiding of different *cry* genes.

The work on the seed protein, Amaranth Albumin 1 (AmA1), was done in India, where field trials showed a 25–60% increase in total protein content in the transgenic tubers and an increase in several of the essential amino acids, including methionine and cysteine (Fig. 10.3). The Indian researchers did in vitro and in vivo studies on experimental animals that demonstrated the safety of the transgenic tubers for human consumption. The public should also be reassured by the fact that AmA1 is a non-allergenic protein that originated from an edible crop.

The work on cyst nematodes showed that similar levels of partial resistance could be achieved with cystatins (cysteine proteinase inhibitors) from sunflower, rice and chicken egg white; the cystatins impair digestion of dietary protein by the nematodes. Subsequent work demonstrated that resistant plants did not harm above-ground organisms or soil microarthropods in potato fields [59]. The researchers found that high levels of resistance could be achieved by inserting the transgene into cultivars with partial resistance rather than into the susceptible cultivar Désirée. Finally, they found that they could also achieve high resistance to *G. pallida* (95% control) by combining a cystatin with another transgene that

Fig. 10.3 Transgenic Indian potato cultivars with improved nutritive value by tuber-specific expression of a gene (*AmA1*) encoding a seed protein, Amaranth Albumin 1 (AmA1), from *Amaranthus hypochondriacus*: Modipuram centre of the ICAR-Central Potato Research Institute (13 December 2008). (From Bradshaw JE (2021) Potato Breeding: Theory and Practice. Springer, Cham with permission)

suppressed root invasion by the nematode. The root-tip specific expression of the transgene resulted in secretion of a peptide that disrupts the chemoreception required by cyst nematodes to locate their host plants. Expression in the root-tips and not the tubers also means that there shouldn't be any human health issues. The overall conclusion from this work was that potato cyst nematode control by transgenic potatoes can replace pesticide use without impacting soil quality.

In future, genetic modification may deliver potatoes with increased yield and photosynthetic efficiency and improved nutrient and water use efficiency.

10.10 Regulations

The regulations governing GM crops are stricter than those for conventionally bred cultivars, both for their research and development in the laboratory, glasshouse and field and for their commercialization. The latter requires demonstration that GM crops and products are safe to grow and to eat or feed to animals. The regulations vary with the country, may address other public concerns, and may allow farmer and consumer choice over whether or not to grow GM crops and eat their products, for example, through labelling and traceability. There are two much-debated issues over the regulations. The first is whether the focus should be on the product or the process. An example is a gene for late blight resistance from a potato wild relative that can be introduced into a new cultivar by conventional breeding or genetic transformation. If the regulations are based on the product, you would say that there is no difference between the two outcomes, and hence the same regulations should apply. However, if the regulations are based on the process, you would say that genetic transformation is different from conventional breeding and that different regulations should apply. Thus, organic farmers would reject the transgenic route to a resistant cultivar even though they could not detect a difference in the product. The second issue concerns whether or not the "precautionary principle" should apply to GM crops. Some people have argued that the complexity and novelty of GM technology mean that the only concern should be appropriate measures to anticipate and prevent harm. Other people think that the consequences of not growing a GM crop should also be considered; for example, depriving a farmer of a GM crop with improved disease resistance or depriving consumers of a GM product with improved nutritional value. Another more technical issue concerns the stacking of transgenes; namely, should regulatory approval be required for the stack if the individual transgenes have already been approved? In other words, are there additional risks that could arise from the combined genes? The answer is currently yes in the EU and Japan and no in the United States and Canada. I am now going to briefly mention two contrasting sets of regulations, the very stringent ones for the European Union (EU) and the less stringent ones for the United States, bearing in mind that changes in regulations are continuously being debated. For example, now that the United Kingdom has left the EU, in a policy briefing in October 2023, The Royal Society argued for regulation based on "hypothesis-driven risk assessment of the specific properties conferred by each introduced trait, the intended use and the receiving environment" [60].

The European Commission (EC) is the executive body of the European Union responsible for proposing legislation, implementing decisions, upholding the EU treaties and managing the day-to-day business of the EU. There are three EU directives and regulations for GM crops, food and feed.

1. EU Directive 2001/18/EC of the European Parliament and of the Council of Ministers of 12 March 2001 concerns the deliberate release into the environment of genetically modified organisms. It sets out procedures for considering applications to release GMOs (genetically modified organisms) into the environment for research or commercial purposes.
2. Regulation (EC) No 1829/2003 of the European Parliament and of the Council of Ministers of 22 September 2003 concerns genetically modified food and feed. It creates a specific harmonized procedure for the scientific assessment and authorization of GM food and feed products. It also requires labelling of all GM food and feed, which contains or consists of GMOs, is produced from GMOs (like glucose syrup from GM maize), or contains ingredients produced from GMOs (e.g., tomato paste and lecithin from GM soya, which is used as an emulsifier in chocolate bars).
3. Regulation (EC) No 1830/2003 of the European Parliament and of the Council of Ministers of 22 September 2003 concerns the traceability and labelling of genetically modified organisms and the traceability of food and feed products produced from genetically modified organisms and amends Directive 2001/18/EC. It sets out EU requirements for a document audit trail to account for and identify approved GM products throughout the marketing chain, with the aim of facilitating accurate labelling.

The EU directives and regulations are implemented in member states through their own appropriate legislation. On 13 January 2015 members of the European Parliament passed a new law to allow EU member states from spring 2015 to restrict or ban the cultivation of GM crops in their own territory, even if this is allowed at the EU level, thus allowing greater flexibility within the EU. The European Food Safety Authority (EFSA) does scientific evaluations of potential GM crops and products for the European Commission and has produced useful guidance on the EU regulations [61, 62].

In the United States, regulations are implemented by three federal agencies (although some US states require additional regulation): the US Department of Agriculture (USDA), the US Environmental Protection Agency (EPA) and the US Food and Drug Administration (FDA). The nature of the GM modification and its intended use determine which agency or agencies are involved in its regulation. The Animal and Plant

Health Inspection Service (APHIS) of USDA, through the Federal Plant Pest Act, determines whether a transgenic crop is likely to be an agricultural or environmental pest. Field tests can often be conducted by simply notifying APHIS, provided safety criteria and standards to ensure biological confinement are met. If not, the process to obtain a permit is more detailed and involved. However, GM crops for commercialization can be granted non-regulated status by APHIS after presentation of extensive data on the nature of the genetic modification and its effects on the crop and the wider ecosystem. The GM cultivar and any offspring can then be released in the United States without APHIS review, but APHIS can prevent the continued use of a GM crop if it is considered a risk of becoming a pest.

The EPA, through the Federal Insecticide, Fungicide and Rodenticide Act, is responsible for ensuring the safety of pesticides, including those produced biologically, such as plants genetically engineered to be pest-resistant. The EPA regulates the testing and commercialization of pest-resistant GM crops and sets tolerances for herbicide residues on herbicide-resistant crops.

The FDA, through the Federal Food, Drug and Cosmetic Act, regulates foods and animal feed derived from GM crops. Foods from GM crops have to meet the same safety standards as all other foods but do not need FDA pre-market approval if their GM products are substantially equivalent to those in other foods.

There is also an international treaty that is relevant to GM crops. The Cartagena Protocol on Biosafety to the Convention on Biological Diversity is an international treaty governing the movements of living modified organisms (LMOs = GMOs) resulting from modern biotechnology from one country to another. It was adopted on 29 January 2000 in Montreal as a supplementary agreement to the Convention on Biological Diversity and entered into force on 11 September 2003, the Convention on Biological Diversity having previously come into force on 29 December 1993 (https://bch.cbd.int/protocol). By 28 April 2024, there were 173 out of 198 countries that were parties to the protocol, but it is not entirely clear to me if this means that they have all ratified and signed the protocol (the United Kingdom, but not the United States, is one of the parties). The protocol aimed to ensure that GMOs were safely handled, transported and used in terms of human health, the environment and genetic resources (potential risks were unintended harm to other organisms, unintended transfer of genes across organisms, the evolution of superweeds, allergens in foods and antibiotic resistance). The emphasis was on the precautionary principle. The Nagoya-Kuala Supplementary Protocol on Liability and Redress was adopted on 15 October 2010. It "requires that response measures are taken in the event of damage resulting from living modified organisms, or where there is sufficient likelihood that damage will result if timely response measures are not taken". In practice, the Cartagena Protocol has meant additional controls (regulations) for the import (and consumption) of GMOs to countries that are parties to the protocol. Furthermore, governments must also make their decisions regarding LMOs publicly available through the Biodiversity Clearing House.

The cultivation of GM potatoes in their centre of origin and diversity is strongly opposed. There are concerns that transgene flow into landraces would alter the natural gene pool and diminish biodiversity, as well as cultural values being at play. We saw in Chap. 3 that gene flow has occurred in South America between cultivated potatoes and their wild relatives. However, scientists from CIP were able to use molecular markers to confirm the integrity of native landraces and the fertile cultivar Yungay in three regions of the Peruvian Andes where they had coexisted for 15–25 years [63]. Nevertheless, these results are unlikely to allay the fears of those opposed to GM crops, although they provide some evidence for the view that Andean farmers could grow transgenic potatoes without destroying the integrity of their landraces.

10.11 Commercialization and Consumer Acceptance

The period from 1995 to 2013 saw a failure to successfully commercialize the first genetically modified (GM) potatoes in North America and

Europe. Monsanto attempted to commercialize GM processing cultivars with resistance to Colorado potato beetle and to viruses PLRV and PVY in North America from 1995, but leading fast-food outlet companies were reluctant to purchase them, and marketing stopped in 2001 [31]. In the United Kingdom, there was a public backlash against GM food in 1998 as a result of the controversial work of Dr Arpad Pusztai at the Rowett Research Institute [64]. He found that feeding transgenic potatoes carrying the snowdrop lectin GNA (*Galanthus nivalis agglutinin*) gene to rats resulted in unexpected (the natural lectin was not harmful to rats) and undesirable effects such as cell proliferation of the gastric mucosa (i.e., stomach damage). A number of critiques found the study inadequate, but the damage had been done to the public perception of GM crops [13]. There was resistance to GM potatoes throughout the European Union. BASF Plant Science failed to commercialize Amflora, a pure amylopectin starch potato, and Fortuna, a potato with two genes for resistance to late blight from the wild tuber-bearing Mexican species *Solanum bulbocastanum*. On 16 January 2012, BASF announced that it was stopping all of its GM crop development targeted at the European market due to lack of acceptance of GM technology in Europe, and in 2013 the company relocated from Germany to the United States. Whatever the arguments for and against GM crops and technology, in Europe there was suspicion over much of the work being done by multinational companies and the processes and products protected with patents. Furthermore, fears, whether real or imagined, were raised by past litigations involving Monsanto [65]. These arose as a result of pollution from the manufacture of polychlorinated biphenyls (PCBs), used as coolants and lubricants, and from the contamination of hormonal herbicides (2,4-D and 2,4,5-T) with dioxins (e.g., tetrachloro-p-dibenzodioxin or TCDD for short).

By 2015, GM crops had been grown outside Europe for 20 years and now occupied 180 million hectares of cropland (12% of the global total), albeit dominated by four crops [soybean (51%), maize (30%), cotton (13%) and canola (5%)] and two traits [herbicide tolerance (53%), *Bt* insect resistance (14%) and these combined (33%)] in five countries [United States (39%), Brazil (25%), Argentina (14%), India (6%) and Canada (6%)] [66]. This allowed scientific reviews of experiences with GM (= GE, genetically engineered) crops, such as the report in 2016 by the US National Academies of Sciences, Engineering and Medicine [67]. Interestingly the report pointed out that "the technologies, traits, and contexts of deployment of specific GE crop varieties are so diverse that generalizations about GE crops as a single defined entity are not possible". As a consequence, the report contained detailed analyses of all of the available information but no sweeping statements or generalizations. Nevertheless, I think it is fair to conclude that the report found little evidence to connect GE crops and their associated technologies with adverse agronomic or environmental problems, nor did it find a higher risk to human health from eating GE foods than from eating their non-GE counterparts, nor did it find persuasive evidence of adverse health effects directly attributable to the consumption of GE foods. The report also concluded that modern plant breeding and genetic engineering are complementary methods for improving crop yield, production efficiency and composition. Another 2016 publication was the book by Matin Qaim, *Genetically Modified Crops and Agricultural Development* [32]. He argued that despite public perceptions (in Europe) to the contrary, there was "strong evidence that GM crops are beneficial for farmers, consumers, and the environment, and that they are as safe as their conventionally bred counterparts". Environmentalist Mark Lynas was persuaded by the science. In 2013, he apologized for destroying GM crops, having been one of the original GM field wreckers in the 1990s. It was rigorous peer-reviewed science that first convinced him of the importance of climate change and then led him to change his mind about GM crops. He explains all of this in his 2018 book *Seeds of Science: Why We Got It So Wrong on GMOs* [68].

So where does this leave us with GM potatoes? Once again, they are being commercialized in the United States, this time by the J.R. Simplot Company with their Innate™ potatoes, which we met in Chap. 8. The programme took 14 years of scientific development, safety assessments and extensive field tests. It used marker-free transformants and intragenic approaches to silence genes by RNA interference that are key to the expression of black spot bruise and the production of asparagine and reducing sugars in tubers. No foreign genes, no antibiotic resistance markers and no vector backbone sequences were incorporated into the plant genome. The first Innate™ potatoes were deregulated by the US Department of Agriculture (USDA) on 7 November 2014 and voluntarily reviewed by the FDA (US Food and Drug Administration), who concluded on 20 March 2015 that they were as safe and nutritious as their conventional counterparts. By August 2017, the second generation of Innate™ potatoes had received approval in Canada as well as the United States. They combine resistance to late blight with processing quality: reduced bruising and blackspot, reduced asparagine, which contributes to reduced acrylamide, and lower reducing sugars for cold storage, as well as reduced acrylamide (www.simplot.com). In January 2024, Professor Dave Douches, director of the Michigan State University Potato Breeding and Genetics Program, received notice that their Kal91.3 potato did not pose an increased plant pest risk relative to its conventionally bred counterpart, Kalkaska, thus making it exempt from the USDA APHIS biotechnology regulations on genetically modified organisms/products. The gene that produces vacuolar acid invertase was silenced, thus preventing cold-induced sweetening from the breakdown of sucrose to the reducing sugars fructose and glucose and allowing storage at 4 °C for a number of months. The researchers can now proceed to marketing their new potato in collaboration with their state potato industry [69]. It remains to be seen how widely these new potatoes will be grown, as well as how many of the traits considered earlier will be incorporated into commercially successful GM potato cultivars. Genetic engineering will be required when a desired trait is not available through sexual hybridization, whereas when both genetic engineering and conventional breeding can be used, we will need to decide on the fastest and most efficient option, which may depend on the number of genes involved.

In summary, genetic transformation and gene editing open up new possibilities for potato improvement, beyond what can be achieved by conventional breeding. They should be able to make valuable contributions to increasing potato production and the nutritional value of potatoes, and hence to the role of potatoes in providing food security during a period of climate change. However, what will happen in practice is still difficult to predict because the science and its applications have outstripped public acceptance and regulations for crop improvement by these methods, but this is beginning to change.

References

1. Watson JD, Crick FHC (1953a) A structure for deoxyribose nucleic acid. Nature 171:737–738
2. Watson JD, Crick FHC (1953b) Genetical implications of the structure of deoxyribose nucleic acid. Nature 171:964–967
3. Avery OT, Macleod CM, McCarty M (1944) Studies on the chemical nature of the substance inducing transformation of pneumococcal types. Induction of transformation by a desoxyribonucleic acid fraction isolated from pneumococcus Type III. J Exp Med 79:137–158
4. The International Human Genome Sequencing Consortium (2001) Initial sequencing and analysis of the human genome. Nature 409:860–921
5. Venter et al (2001) The sequence of the human genome. Science 291:1304–1351
6. The Potato Genome Sequencing Consortium (2011) Genome sequence and analysis of the tuber crop potato. Nature 475:189–195
7. Tomato Genome Consortium (2012) The tomato genome sequence provides insights into fleshy fruit evolution. Nature 485:635–641
8. Zhang JY, Zhang JY, Zhao YF, Li S, Chen SS, Wang YP, Mou B, Ma HR, Han ZJ, Lu Y, Han YZ (2021) Characterization of the complete chloroplast genome of the *Solanum tuberosum* L. cv. Atlantic (Solanaceae). Mitochondrial DNA B Resour 6(1):73–75. https://doi.org/10.1080/23802359.2020.1845998
9. Cho KS, Cho JH, Im JS, Choi JG, Park YE, Hong SY, Kwon M, Kang JH, Park TH (2017) The complete mitochondrial genome sequences of potato (*Solanum tuberosum* L., Solanaceae). Mitochondrial DNA B Resour 2(2):781–782. https://doi.org/10.1080/23802359.2017.1398607
10. Watson JD (1965) Molecular biology of the gene. W.A. Benjamin, Inc., New York
11. Hellens RP, Moreau C, Lin-Wang K, Schwinn KE, Thomson SJ, Fiers MWEJ, Frew TJ, Murray SR, Hofer JMI, Jacobs JME, Davies KM, Allan AC, Bendahmane A, Coyne CJ, Timmerman-Vaughan

GM, Ellis THN (2010) Identification of Mendel's white flower character. PLoS One 5(10):e13230. https://doi.org/10.1371/journal.pone.0013230
12. Jung CS, Griffiths HM, De Jong DM, Cheng S, Bodis M, Kim TS, De Jong WS (2009) The potato *developer* (*D*) locus encodes an R2R3 MYB transcription factor that regulates expression of multiple anthocyanin structural genes in tuber skin. Theor Appl Genet 120:45–57
13. Slater A, Scott NW, Fowler MR (2008) Plant biotechnology: the genetic manipulation of plants, 2nd edn. Oxford University Press, Oxford
14. Heldt H-W, Piechulla B (2021) Plant biochemistry, 5th edn. Academic Press (Elsevier), London
15. Nurse P (2020) What is life? David Fickling Books, Oxford
16. Bortesi L, Fischer R (2015) The CRISPR/Cas9 system for plant genome editing and beyond. Biotechnol Adv 33:41–52
17. Ghislain M, Douches DS (2020) The genes and genomes of the potato. In: Campos H, Ortiz O (eds) The potato crop: its agricultural, nutritional and social contribution to humankind. Springer, ebook, pp 139–162
18. EFSA (2012) EFSA Panel on Genetically Modified Organisms (GMO). Scientific opinion. Addressing the safety assessment of plants developed using Zinc Finger Nuclease 3 and other Site-Directed Nucleases with similar function. EFSA J 10(10):2943
19. Svitashev S, Young JK, Schwartz C, Gao H, Falco SC, Cigan AM (2015) Targeted mutagenesis, precise gene editing, and site-specific gene insertion in maize using Cas9 and guide RNA. Plant Physiol 169:931–945
20. Andersson M, Turesson H, Nicolia A, Fält A-S, Samuelsson M, Hofvander P (2017) Efficient targeted multiallelic mutagenesis in tetraploid potato (*Solanum tuberosum*) by transient CRISPR-Cas9 expression in protoplasts. Plant Cell Rep 36:117–128
21. Johansen IE, Liu Y, Jørgensen B, Bennett EP, Andreasson E, Nielsen KL, Blennow A, Petersen BL (2019) High efficacy full allelic CRISPR/Cas9 gene editing in tetraploid potato. Sci Rep 9:17715. https://doi.org/10.1038/s41598-019-54126-w
22. Chincinska IA, Miklaszewska M, Sołtys-Kalina D (2023) Recent advances and challenges in potato improvement using CRISPR/Cas genome editing. Planta 257:25. https://doi.org/10.1007/s00425-022-04054-3
23. Hegde N, Joshi S, Soni N, Kushalappa AC (2021) The cafeoyl-CoA O-methyltransferase gene SNP replacement in Russet Burbank potato variety enhances late blight resistance through cell wall reinforcement. Plant Cell Rep 40:237–254. https://doi.org/10.1007/S00299-020-02629-6
24. Gillbard E (2024) Scientists plan to use gene-editing tech to transform potato sector. Farmers Weekly 22 March 2024
25. De Block M, Herrera-Estrella L, Van Montagu M, Schell J, Zambryski P (1984) Expression of foreign genes in regenerated plants and in their progeny. EMBO J 3:1681–1689
26. Horsch RB, Fraley RT, Rogers SG, Sanders PR, Lloyd A, Hoffmann N (1984) Inheritance of functional foreign genes in plants. Science 223:496–498
27. Ooms G, Bossen ME, Burrell MM, Carp A (1986) Genetic manipulation in potato with *Agrobacterium rhizogenes*. Potato Res 29:367–379
28. Stiekema WJ, Heidekamp F, Louwerse JD, Verhoeven HA, Dijkhuis P (1988) Introduction of foreign genes into potato cultivars Bintje and Desiree using an *Agrobacterium tumefaciens* binary vector. Plant Cell Rep 7:47–50
29. Heldt HW, Piechulla B (2021) Plant biochemistry, 5th edn. Academic, London
30. Acquaah G (2021) Principles of plant genetics and breeding, 3rd edn. Wiley Blackwell, Chichester
31. Davies HV (2002) Commercial developments with transgenic potato. In: Valpuesta V (ed) Fruit and vegetable biotechnology. Woodhead Publishing Limited, Cambridge, pp 222–249
32. Qaim M (2016) Genetically modified crops and agricultural development. Palgrave Macmillan, Basingstoke
33. EFSA (2012) EFSA Panel on Genetically Modified Organisms (GMO). Scientific opinion. Addressing the safety assessment of plants developed through cisgenesis and intragenesis. EFSA J 10(2):2561
34. Bairu MW, Aremu AO, Van Staden J (2011) Somaclonal variation in plants: causes and detection methods. Plant Growth Regul 63:147–173
35. Jacobsen E (2007) The Canon of Potato Science. 6. Genetic modification and cis- and transgenesis. Potato Res 50:227–230
36. Chung BN, Yoon J-Y, Palukaitis P (2013) Engineered resistance in potato against potato leafroll virus, potato virus A and potato virus Y. Virus Genes 47:86–92. https://doi.org/10.1007/s11262-013-0904-4
37. Dinh PTY, Zhang L, Mojtahedi H, Brown CR, Elling AA (2015) Broad *Meloidogyne* resistance in potato based on RNA interference of effector gene *16D10*. J Nematol 47(1):71–78
38. Rommens CM, Yan H, Swords K, Richael C, Ye J (2008) Low-acrylamide French fries and potato chips. Plant Biotechnol J 6:843–853
39. Rommens CM, Ye J, Richael C, Swords K (2006) Improving potato storage and processing characteristics through all-native DNA transformation. J Agric Food Chem 54:9882–9887
40. Chawla R, Shakya R, Rommens CM (2012) Tuber-specific silencing of *asparagine synthetase-1* reduces the acrylamide-forming potential of potatoes grown in the field without affecting tuber shape and yield. Plant Biotechnol J 10:913–924
41. Zhu X, Gong H, He Q, Zeng Z, Busse JS, Jin W, Bethke PC, Jiang J (2016) Silencing of vacuolar invertase and asparagine synthetase genes and its impact on acrylamide formation of fried potato products. Plant Biotechnol J 14(2):709–718
42. Itkin M, Heinig U, Tzfadia O, Bhide AJ, Shinde B, Cardenas PD, Bocobza SE, Unger T, Malitsky S, Finkers R, Tikunov Y (2013) Biosynthesis of antinu-

tritional alkaloids in solanaceous crops is mediated by clustered genes. Science 341:175–179
43. Sawai S, Ohyama K, Yasumoto S, Seki H, Sakuma T, Yamamoto T, Takebayashi Y, Kojima M, Sakakibara H, Aoki T, Muranaka T, Saito K, Umemoto N (2014) Sterol side chain reductase 2 is a key enzyme in the biosynthesis of cholesterol, the common precursor of toxic steroidal glycoalkaloids in potato. Plant Cell 26:3763–3774
44. Shepherd LVT, Hackett CA, Alexander CJ, McNicol JW, Sungurtas JA, Stewart D, McCue KF, Belknap WR, Davies HV (2015) Modifying glycoalkaloid content in transgenic potato–metabolome impacts. Food Chem 187:437–443
45. Umemoto N, Nakayasu M, Ohyama K, Yotsu-Yamashita M, Mizutani M, Seki H, Saito K, Muranaka T (2016) Two cytochrome P450 mono-oxygenases catalyse early hydroxylation steps in the potato steroid glycoalkaloid biosynthetic pathway. Plant Physiol 171:2458–2467
46. Visser RGF, Somhorst I, Kuipers GJ, Ruys NJ, Feenstra WJ, Jacobsen E (1991) Inhibition of the expression of the gene for granule-bound starch synthase in potato by antisense constructs. Mol Gen Genet 225:289–296
47. Schwall GP, Safford R, Westcott RJ, Jeffcoat R, Tayal A, Shi Y-C, Gidley MJ, Jobling SA (2000) Production of very high amylose potato starch by inhibition of SBE A and B. Nat Biotech 18:551–554
48. Bradshaw JE (2021) Potato breeding: theory and practice. Springer, Cham
49. Duncan DR, Hammond D, Zalewski DJ, Cudnohufsky J, Kaniewski W, Thornton M, Bookout JT, Lavrik PB, Rogan GJ, Feldman-Riebe J (2002) Field performance of transgenic potato, with resistance to Colorado potato beetle and viruses. HortScience 37:275–276
50. Mohammed A, Douches DS, Pett W, Grafius E, Coombs J, Liswidowati J, Li WS, Madkour M (2000) Evaluation of potato tuber moth (Lepidoptera: Gelechiidae) resistance in tubers of Bt-cry5 transgenic potato lines. J Econ Entomol 93:472–476
51. Davidson MM, Butler RC, Wratten SD, Conner AJ (2004) Resistance of potatoes transgenic for a *cry*1Ac9 gene, to *Phthorimaea operculella* (Lepidoptera: Gelechiidae) over field seasons and between plant organs. Ann Appl Biol 145:271–277
52. Douches DS, Pett W, Santos F, Coombs J, Grafius E, Li W, Metry EA, Nasr El-Din T, Madkour M (2004) Field and storage testing *Bt* potatoes for resistance to potato tuber worm (Lepidoptera: Gelichiidae). J Econ Entomol 97(4):1425–1431
53. Douches DS, Grafius EJ (2005) Transformation for insect resistance. In: Razdan MK, Mattoo AK (eds) Genetic improvement of solanaceous crops, Volume I: Potato. Science Publishers Inc, Enfield, pp 235–266
54. Diretto G, Al-Babili S, Tavazza R, Papacchioli V, Beyer P, Giuliano G (2007) Metabolic engineering of potato carotenoid content through tuber-specific over-expression of a bacterial mini-pathway. PLoS One 2(4):p.e350
55. Morris WL, Ducreux JM, Fraser PD, Millam S, Taylor MA (2006) Engineering ketocarotenoid biosynthesis in potato tubers. Metab Eng 8:253–263
56. Hellwege EM, Czapla S, Jahnke A, Willmitzer L, Heyer AG (2000) Transgenic potato (*Solanum tuberosum*) tubers synthesize the full spectrum of inulin molecules naturally occurring in globe artichoke (*Cynara scolymus*) roots. PNAS 97(15):8699–8704. https://doi.org/10.1073/pnas.150043797
57. Chakraborty S, Chakraborty N, Agrawal L, Ghosh S, Narula K, Shekhar S, Naik PS, Pande PC, Chakrborti SK, Datta A (2010) Next-generation protein-rich potato expressing the seed protein gene *AmA1* is a result of proteome rebalancing in transgenic tuber. Proc Natl Acad Sci USA 107:17533–17538. https://doi.org/10.1073/pnas.1006265107
58. Urwin PE, Green J, Atkinson HJ (2003) Expression of a plant cystatin confers partial resistance to *Globodera*, full resistance is achieved by pyramiding a cystatin with natural resistance. Mol Breed 12:263–269
59. Green J, Wang D, Lilley CJ, Urwin PE, Atkinson HJ (2012) Transgenic potatoes for potato cyst nematode control can replace pesticide use without impact on soil quality. PLoS One 7(2):e30973. https://doi.org/10.1371/journal.pone.0030973
60. The Royal Society (2023) Enabling genetic technologies for food security: policy briefing issued: October 2023 DES8542 ISBN: 978-1-78252-682-7 © The Royal Society
61. EFSA (2010) EFSA Panel on Genetically Modified Organisms (GMO). Scientific opinion. Guidance on the environmental risk assessment of genetically modified plants. EFSA J 8(11):1879
62. EFSA (2011) EFSA Panel on Genetically Modified Organisms (GMO). Scientific opinion. Guidance for risk assessment of food and feed from genetically modified plants. EFSA J 9(5):2150
63. Ghislain M, Montenegro JD, Juarez H, del Rosario HM (2015) Ex-post analysis of landraces sympatric to a commercial variety in the center of origin of the potato failed to detect gene flow. Transgenic Res 24(3):519–528. https://doi.org/10.1007/s11248-014-9854-4
64. Ewen SWB, Pusztai A (1999) Effect of diet containing genetically modified potatoes expressing *Galanthus nivalis* lectin on rat small intestine. Lancet 354:1353–1354
65. Robin MM (2010) The world according to Monsanto (English translation). The New Press, New York
66. ISAAA (2015) 20th Anniversary (1996 to 2015) of the global commercialization of biotech crops and biotech crop highlights in 2015. ISAAA Brief 51
67. National Academies of Sciences, Engineering, and Medicine (2016) Genetically engineered crops: experiences and prospects. The National Academies Press, Washington, DC. https://doi.org/10.17226/23395
68. Lynas M (2018) Seeds of science: why we got it so wrong on GMOs. Bloomsbury Sigma, London
69. Falinski J (2024) MSU potato breeder develops new genetically engineered potato. Michigan State University AgBioResearch 4 June 2024

Conclusions

"Can potatoes feed the world?" became a serious question in 2008 when a global food crisis coincided with the United Nations International Year of the Potato. There was a realization that poor countries could achieve greater food security by growing more potatoes to supplement or even replace imported grains, given that potatoes have a relatively low requirement for land. It remains a serious question today in its longer and more accurate version: "Can potatoes make a greater contribution to global food security during a period of world population growth, lack of new land for food production, and climate change?" It is a scientific and technological question set in a political, economic, societal and historical context. The answer requires an understanding of the history and biology of potato cultivation, as well as farming and food systems. Increased potato production and improved nutritional value would certainly be important contributions to achieving the United Nations "2030 Agenda for Sustainable Development", particularly goal two, which is to end hunger, achieve food security and improved nutrition, and promote sustainable agriculture. The Food and Agriculture Organization (FAO) of the United Nations (UN) observes the contribution potatoes can make to the 2030 Agenda, and beyond, on May 30 each year through an "International Day of Potato". To quote from their website: "We spotlight the potato, *Solanum tuberosum* (L.) – a crop regularly consumed by billions of people. Potatoes are a key crop across diverse farming systems globally, ranging from smallholders producing diverse heirloom varieties by hand in the Andes, to vast commercial, mechanized farms in different continents. Potatoes contribute to the food security and nutrition as well as the livelihoods and employment of people in rural and urban areas the world over".

The wild tuber-bearing *Solanum* species never intended to feed us. They evolved swollen subterranean stems to survive from season to season as dormant storage organs in a wide range of habitats in the Americas, with centres of diversity in Mexico and Peru. The indigenous peoples of the Andes, who first started to cultivate, select and hence domesticate potatoes, could not have imagined them becoming a global crop that one day would be considered food for space travel and life on Mars. There is no written history of potato domestication nor of the spread of potato cultivation around the world, but the latter was intimately connected with the course of human history and population growth. A number of key periods during this history are worth recalling. There is archaeological evidence for the early use of potatoes at Jiskairumoko, an early village site in the western Titicaca Basin of the high south-central Andes (3890 masl), dating to 5420–3620 years before present. Domestication was complete by the end of this period, with potatoes part of the small-scale farming around Lake Titicaca in the harsh conditions of the Andes mountains (Cordillera). More extensive farming presumably started with the arrival of more favourable climatic conditions. Historical accounts from the time of the Spanish conquest of the Incas (1532 CE onwards) tell us that

J. E. Bradshaw, *Can Potatoes Feed the World?*, Sustainable Development Goals Series,
https://doi.org/10.1007/978-3-031-92890-1

potatoes were being grown throughout the Andes, from Colombia to Peru, and also in southern Chile. Outside of South America, potato cultivation started in the Canary Islands around 1562 and in Spain, near Seville, around 1570, but we will never know who were the first sailors and ships to bring potatoes to Europe and who were the first farmers to grow them. Cultivation spread from Spain to Portugal and Italy. During the period 1600–1750, potatoes came to be cultivated in many European countries in the cottage gardens of villagers, peasants and labourers; but there is no historical account of the spread and just glimpses of potatoes being grown in fields. However, from 1750 onwards, potato cultivation on a larger scale was stimulated by a combination of food shortages, the need to feed the armies of Europe, and the cheap labour required for the Industrial Revolution, which began in the second half of the eighteenth century and gained momentum in the first half of the nineteenth century. Home-grown potatoes were seen by poor families as a way of staying alive when wheat was seized by invading armies. They also provided food security during both the First and Second World Wars. By 1925, the USSR, Germany, Poland and France accounted for 70% of world potato production, which stood at 188 million tonnes. As late as 1961, the potato was still primarily a European crop, with 82% of the 271 million tonnes of global production. This was despite the fact that during the period 1600–1800, British, Dutch, French, Portuguese and Spanish colonists and missionaries had taken potatoes with them to many parts of the world. Sixty years later, in 2021, the world of potato cultivation was very different, and potatoes could be considered a global crop. World production was up to 374 million tonnes, with 52.6% in Asia, 27.0% in Europe, 7.6% in Africa, 6.7% in North America, 5.6% in Latin America and 0.5% in Australia and New Zealand. It is too simplistic to think that production is divided between developing and industrialized countries with their contrasting rural and industrial farming systems. The possibility of increasing potato production needs to be viewed country by country in terms of the best ways of improving their farming and food systems to ensure food security. Production has increased in three ways: by bringing new land into cultivation, by using existing cultivated land for potatoes rather than other crops, and by increasing production per unit area of land. All three ways could be used in the future, and their importance will probably vary from country to country, particularly during a period of climate change. However, there is global political pressure for increasing production per unit area of land.

As potato cultivation spread around the world, a combination of human and natural selection resulted in potatoes adapted to different day lengths, climates and growing seasons. Adaptation was first to the short summer days of the highland tropics and subtropics, then to the long summer days of the lowland temperate regions, and finally to the short winter days of the lowland subtropics and tropics. Today potatoes are grown globally as a summer crop in the tropical highlands of Bolivia, Peru and Mexico and in the lowland temperate regions of North America, Europe, Northern China, Australia and New Zealand; all year round in South Africa, parts of China and Brazil, and the equatorial highlands of South America and East Africa (e.g., Ecuador and Colombia and Kenya and Uganda); as spring and autumn crops in the Mediterranean (e.g., North Africa and Southern Spain); and as a winter crop in the lowland subtropics (e.g., Northern India, Bangladesh and Southern China). Although most potato production is from seed tubers, selection for adaptation to different environments and end uses required genetic variation. This came from sexual reproduction in the form of plants derived from the botanical seeds in potato berries, either from natural insect pollination of flowers or more recently from deliberate artificial hybridizations. In future, new and desirable combinations of genetic variants will come from conventional breeding complemented by gene editing and genetic transformations. The latter two processes will need to be controlled by regulations that are fit for purpose in terms of making rapid progress whilst ensuring that crops are safe to grow and eat. The latter is particularly important for consumer acceptance given the failure from 1995 to 2013 to successfully commercialize the first genetically modified (GM) potatoes in North America and Europe. Gene editing should prove less controversial than genetic modification, as it targets specific locations in the potato genome to make desired alterations to specific genes, rather than introducing genes from other organisms that do not naturally hybridize with potatoes. The outcome should be new cultivars better adapted to evolving farming and food systems as well as to climate change.

There is no written account of the natural hybridizations and human and natural selection that went into transforming the potato into a global crop. Fortunately, there is a biological history written in the DNA of extant cultivars, landraces and wild relatives, as well as in herbarium specimens. Today we can make some important inferences from the DNA data. There are 107 wild tuber-bearing *Solanum* species that grow in a wide range of habitats and altitudes in 16 countries from the southwest of the United States (38°N), through Central America and the Andes, to Argentina and adjacent Chile (41°S), with centres of diversity in Mexico and Peru. However, potatoes were domesticated from just one diploid wild species, *S. candolleanum*, in the high Andes of southern Peru, where the summers are cool and wet and day lengths are 12–13 h. Although the first cultigen was a diploid potato, *S. tuberosum* Group Stenotomum, it was the cultivation of tetraploid Group Andigena potatoes that spread throughout the Andes and the genetically distinct tetraploid Group Tuberosum potatoes that became established in the longer days of southern Chile. There were natural hybridizations between cultivated potatoes and their wild relatives in South America which could not subsequently occur elsewhere in the world as there were no wild relatives. Further hybridizations with wild relatives had to await the advent of collecting expeditions and modern plant breeding in the twentieth century. DNA extracted from herbarium specimens revealed that since around 1800, new potato introductions came from coastal Chile, rather than the Andes, and today their offspring predominate in global potato cultivation. The most important conclusion from all of these inferences is that the wild tuber-bearing *Solanum* species and the cultivated landraces of the Andes are an underutilized plant genetic resource for breeding new potato cultivars better able to provide global food security during a period of climate change. They need to be conserved, both in situ and ex situ in genebanks, studied, and better utilized in breeding programs through knowledge of their taxonomy, reproductive biology and adaptive traits. This needs to be done in compliance with the International Treaty on Plant Genetic Resources for Food and Agriculture (ITPGRFA), which came into force on 29 June 2004, and by 16 July 2024 had 152 contracting parties (out of 194 countries). The treaty recognized the contribution of farmers to the diversity of crops that feed the world, established a global system to provide farmers, plant breeders and scientists with access to plant genetic materials, and ensured that recipients shared the derived benefits with the countries of origin. The conserved germplasm should also be thought of as an insurance policy for the future: a reservoir of genetic variants that may prove useful in combating increasing abiotic stresses as well as new disease and pest problems.

In 2008, there was a caveat to "How the humble potato could feed the world". Reliance on a single crop for food security is dangerous if widespread crop failures can occur and result in famine, as happened in Ireland with the late blight epidemics of 1845, 1846, 1848 and 1849. Potatoes have now been protected from late blight with fungicides for 140 years, but insensitivity to some modern ones has arisen, necessitating the use of more than one chemistry in spraying regimes. Furthermore, there is political pressure to reduce the use of fungicides. New sources of genes for blight resistance are therefore an attractive proposition but will require more thoughtful deployment than in the past to deal with the likely evolution of new strains of blight. When blight outbreaks do occur, they will need to be analyzed for knowledge of the prevailing pathogen populations. Good farm hygiene will also be required because diseased tubers left lying around and volunteer plants from diseased tubers left in the ground are sources of inoculum. *Phytophthora infestans* (Mont.) de Bary, the cause of late blight and destroyer of plants, retains the potential for exponential production of pathogenic spores and hence to cause devastating epidemics when the weather and other conditions are right. Soil-borne pests such as cyst nematodes and diseases such as wart spread more slowly but can build up over time and cause real problems. Hence, short crop rotations should be avoided, and resistant cultivars used to halt and reverse any increases in pest and pathogen populations that do occur. Disease-free planting material is also essential if potato crops are going to deliver the yield and quality of tubers required of a food security crop, with virus diseases a particular problem despite the availability of effective resistance genes. Statutory seed certification schemes for the multiplication of seed tubers and propagation through true potato seed (TPS) are two important ways of achieving healthy planting material. TPS propagation could be an attractive proposition in Asian and African countries with a high proportion of farm-saved seed tubers, particularly now that genetically uniform diploid F_1 hybrid cultivars are available. However, it is clear that wherever potatoes are grown and tubers stored, they will be under attack from various pests and pathogens which will need to be controlled. More generally, diversity in crops for food security is highly desirable, however enthusiastic we may be for an increasing use of potatoes.

If potatoes are going to make a greater contribution to food security through increased yields, their basic needs need to be understood and met, whether grown by smallholders or on vast commercial, mechanized farms. Like all plants, potatoes require water (H_2O) from the soil and carbon dioxide (CO_2) from the air for photosynthesis, in which the sun's light energy is converted into chemical energy through the synthesis of sugars and other carbohydrates made of carbon (C), oxygen (O) and hydrogen (H). Potatoes also require 14 mineral elements in the form of ions in aqueous solution, which they need to extract from the soil unless being grown hydroponically, with growth limited by the essential mineral that is in the shortest supply. The main driving force used to move water and minerals from the soil to all parts of the plant is the transpiration of water from the plant's leaves. The water retained in the plant maintains the turgor of its living cells, which is necessary for their functioning, and prevents wilting of tissues and organs. It is also the medium in which cellular metabolism (chemical reactions) takes place and comprises about 80% by weight of a potato plant with tubers. The rate and efficiency of photosynthesis are determined by temperature and carbon dioxide concentration as well as being dependent on the light intensity. Plant physiologists are able to make predictions about the effects of a likely increase in carbon dioxide concentrations in the air from 380 ppm in 2000 to 550 ppm in 2050. Potential potato yields could, on average, increase by around 28%. In practice, higher yields will depend on adequate supplies of fresh or desalinated water; absence or tolerance of abiotic stresses such as nutrient-poor soil, heat and salinity; and control of pests and diseases, which includes healthy planting material. Furthermore, as the average global temperature in 2050 is likely to be 2 °C above pre-industrial times and not the hoped-for 1.5 °C, yields are likely to decrease in some climatic regions of the world unless changes in planting and harvest dates can accommodate new seasonal periods which are suitable for potato growth. Plant physiologists are also able to explore the reasons for the gaps between actual and predicted yields for growing seasons in different climatic and agro-ecological regions. These gaps (yield losses) vary from country to country for a variety of reasons, but globally there is scope to double yields from the 2016 to 2020 average of 21.41 t/ha. Although modern industrial farming has given relatively large short-term gains in yields to over 40 t/ha in a number of countries, its sustainability is being questioned together with its impact on the environment, biodiversity and climate change. There is increasing debate about the importance of regenerative (restorative) agriculture, with its emphasis on crop rotations to maintain soil health, including the more restrictive requirements of organic farming, but the outcomes in yields per hectare are not clear. It is difficult and costly, but necessary, to do research to compare farming systems that provides important answers about potato production that are both statistically significant and biologically meaningful. Longer-term research may also result in new potato cultivars with increased resource use efficiency through the better capture and utilization of light, carbon dioxide, water and nutrients, and precision agriculture and smart farming may help them achieve their potential.

Lastly, there is the issue of the nutritional value of potatoes. They are seen primarily as a source of energy in the form of starch and hence an alternative to wheat and rice in meals nutritionally balanced with other food. Many countries are seeing the increasing importance of processed potato products as convenience foods and snacks, but ultra-processed foods are raising health concerns. A baked potato in the jacket remains one of the healthiest types of cooked potato and the form in which they were first eaten in the Andes, having been baked in the embers of a fire or cooked in an earth oven on hot stones. We need to recognize that people globally want to enjoy eating potatoes and hence want them to have a nice flavour and texture, as well as being cheap and nutritious. There are potential problems that arise from starch as the energy source, but they do have solutions, as can be seen in two examples. Firstly, acrylamide, a Group 2A carcinogen, can be reduced to acceptable levels in roast potatoes and fried potato products by selecting or engineering tubers to have lower contents of asparagine and reducing sugars. Secondly, selection of potatoes with higher contents of resistant starch can lower their glycaemic index and help prevent obesity, glucose intolerance, type-2 diabetes and cardiovascular disease. The other important contributions of potatoes to a nutritionally balanced meal are fibre; proteins with a good amino acid balance; minerals, particularly potassium; vitamins, particularly B vitamins and vitamin C; and antioxidants. However, when considering ways to make potatoes even more nutritious, it is important to determine the effect of storage, cooking and processing on their chemical composition, as well as the bioavailability of ingested nutrients for utilization by the human body. The latter needs to be determined in vivo using human subjects, although there are some in vitro proxies. Only then can one quantify the contributions of potatoes to a healthy diet and set targets for further improvements. Currently, there is interest in increasing the amounts of micronutrients in potatoes, as well as grains, to combat the hidden hunger that occurs in areas where they make a large contribution to meals lacking complementary components. The emphasis has been on increasing iron and zinc content by conventional breeding (genetic biofortification). Improved cultivars from CIP are currently in their advanced stages of trialling in Peru, Bolivia, Nepal, Bhutan, Rwanda, Kenya, Ethiopia, Uganda, Malawi, China and India. This is CIP's contribution to the HarvestPlus Biofortification Challenge, which started in 2004 as part of the work of the International Food Policy Research Institute (IFPRI). Potatoes can also be genetically engineered to produce provitamin A (β-carotene) and hence address vitamin A deficiency as a major cause of poor eyesight and blindness in children in Asia, Africa and Latin America. However, such "Golden Potatoes" may not be grown for a long time, given the continuing arguments over the release of "Golden Rice", which was first conceived of in the late 1980s as a solution to a serious health problem.

My overall conclusion is that potatoes can help to feed the world in 2050 through increased production of tubers with improved nutritional values. This will mainly be through higher yields per hectare rather than more land in potato production, but who will make it happen? Will it be national leaders worki ng together towards common goals through the international organizations of the United Nations, or multinational companies seeing global opportunities in farming and farm products, or national governments wanting to achieve food security for their nations, or non-governmental organizations concerned with food production, or simply the result of millions of farmers worldwide wanting to feed their families and wider communities and to generate income from their farms? The answer will probably be complex interactions between all of the players in our evolving farm and food systems during a period of continuing human-made climate change. But there does need to be a sense of urgency because nobody should go to bed hungry and malnourished.

Index

J. E. Bradshaw, *Can Potatoes Feed the World?*, Sustainable Development Goals Series, https://doi.org/10.1007/978-3-031-92890-1

C

D

The manufacturer's authorised representative in the EU is Springer Nature Customer Service Centre GmbH, Europaplatz 3, 69115 Heidelberg, Germany. If you have any concerns regarding our products, please contact ProductSafety@springernature.com

Printed and bound by CPI Group (UK) Ltd, Croydon, CR0 4YY
30/09/2026
02225562-0019